I0053783

Bee and Wasp Venoms: Biological Characteristics and Therapeutic Application

Special Issue Editor
Sok Cheon Pak

Special Issue Editor
Sok Cheon Pak
School of Biomedical Sciences, Charles Sturt University
Australia

Editorial Office
MDPI AG
St. Alban-Anlage 66
Basel, Switzerland

This edition is a reprint of the Special Issue published online in the open access journal *Toxins* (ISSN 2072-6651) from 2015–2016 (available at: http://www.mdpi.com/journal/toxins/special_issues/bee-wasp).

For citation purposes, cite each article independently as indicated on the article page online and as indicated below:

Author 1; Author 2; Author 3 etc. Article title. *Journalname*. **Year**. Article number/page range.

ISBN 978-3-03842-340-9 (Pbk)
ISBN 978-3-03842-341-6 (PDF)

Articles in this volume are Open Access and distributed under the Creative Commons Attribution license (CC BY), which allows users to download, copy and build upon published articles even for commercial purposes, as long as the author and publisher are properly credited, which ensures maximum dissemination and a wider impact of our publications. The book taken as a whole is © 2017 MDPI, Basel, Switzerland, distributed under the terms and conditions of the Creative Commons by Attribution (CC BY-NC-ND) license (http://creativecommons.org/licenses/by-nc-nd/4.0/).

Table of Contents

About the Guest Editor

Sok Cheon Pak began his academic career in 1995 at Chosun University Medical School (Gwangju, South Korea) as an invited lecturer. Then, he was employed as a full-time lecturer at Dongshin Univeristy Oriental Medical School (Naju, South Korea). Dr Pak's new teaching career began at New Zealand College of Oriental Medicine (Hamilton, New Zealand) in 2002. Since starting another new teaching career at Charles Sturt University (Bathurst, Australia) from 2007, he has completed many teaching programs as continuing professional development. Currently, he is teaching Evidence Based Complementary Medicine. His area of research expertise specifically relates to introducing evidence based practice to Complementary Medicine research and practice; that has been based on laboratory experiments which incorporate modern medical technologies to identify and evidence the underlying mechanisms of treatment of diseases.

Preface to "Bee and Wasp Venoms: Biological Characteristics and Therapeutic Application"

Given the rapid expansion of the therapeutic application of bee and wasp venoms in recent years, it is an opportune time to present this Special Issue on 'bee and wasp venoms: biological characteristics and therapeutic application'. Modern approaches of venomics have allowed the discovery of venom constituents which were proven to be of pharmacological significance and have opened the way to optimization of therapeutic strategies through the use of active compounds such as melittin, apamin and mastoparan. Recent studies using these compounds have demonstrated diverse mechanisms on a range of conditions. However, further identification of intricate mechanisms, justification of the route of application and formulation of those constituents are essential in the future. Understanding the signalling pathways associated with the compound-mediated in vivo dynamics and further communication between cells at the molecular level will facilitate the development of new therapeutics. It is my hope that this Special Issue will be a source for many years to those interested in the agricultural and pharmacological relevance of bee and wasp venoms.

Sok Cheon Pak
Guest Editor

Article

Bee Venom Protects against Rotenone-Induced Cell Death in NSC34 Motor Neuron Cells

So Young Jung [1], Kang-Woo Lee [1], Sun-Mi Choi [2] and Eun Jin Yang [3,*]

[1] Department of Medical Research, Korea Institute of Oriental Medicine, 483 Expo-ro, Yuseong-gu, Daejeon 305-811, Korea; syzzim84@gmail.com (S.Y.J.); neurokangwoo@kaist.ac.kr (K.-W.L.)
[2] Executive Director of R&D, Korea Institute of Oriental Medicine, 483 Expo-ro, Yuseong-gu, Daejeon 305-811, Korea; smchoi@kiom.re.kr
[3] Department of Clinical Research, Korea Institute of Oriental Medicine, 483 Expo-ro, Yuseong-gu, Daejeon 305-811, Korea
* Correspondence: yangej@kiom.re.kr; Tel.: +82-42-863-9497; Fax: +82-42-863-9464.

Academic Editor: Ren Lai
Received: 8 July 2015; Accepted: 31 August 2015; Published: 21 September 2015

Abstract: Rotenone, an inhibitor of mitochondrial complex I of the mitochondrial respiratory chain, is known to elevate mitochondrial reactive oxygen species and induce apoptosis via activation of the caspase-3 pathway. Bee venom (BV) extracted from honey bees has been widely used in oriental medicine and contains melittin, apamin, adolapin, mast cell-degranulating peptide, and phospholipase A_2. In this study, we tested the effects of BV on neuronal cell death by examining rotenone-induced mitochondrial dysfunction. NSC34 motor neuron cells were pretreated with 2.5 µg/mL BV and stimulated with 10 µM rotenone to induce cell toxicity. We assessed cell death by Western blotting using specific antibodies, such as phospho-ERK1/2, phospho-JNK, and cleaved capase-3 and performed an MTT assay for evaluation of cell death and mitochondria staining. Pretreatment with 2.5 µg/mL BV had a neuroprotective effect against 10 µM rotenone-induced cell death in NSC34 motor neuron cells. Pre-treatment with BV significantly enhanced cell viability and ameliorated mitochondrial impairment in rotenone-treated cellular model. Moreover, BV treatment inhibited the activation of JNK signaling and cleaved caspase-3 related to cell death and increased ERK phosphorylation involved in cell survival in rotenone-treated NSC34 motor neuron cells. Taken together, we suggest that BV treatment can be useful for protection of neurons against oxidative stress or neurotoxin-induced cell death.

Keywords: bee venom (BV); NSC34 motor neuron cell; rotenone; phospho-JNK; cleaved caspase-3

---

1. Introduction

Rotenone is a naturally-occurring plant compound and a specific inhibitor of complex I of the mitochondrial respiration chain. It had been found to induce cell death in a variety of cells [1]. Features include mitochondrial impairment, microglial activation, oxidative damage, dopaminergic degeneration, and L-dopa-responsive motor deficit. Rotenone was thought to elevate mitochondrial reactive oxygen species production [1], decreasing cellular ATP levels [2] and mitochondrial membrane potential [3]. Other reports have indicated that rotenone induces apoptosis via an increase in mitochondria reactive oxygen species production and neurotoxicity associated with increased levels of caspase family gene expression [4] in PC12 cells.

Bee venom (BV) is known to be a very complex mixture of active peptides including melittin, phospholipase A_2, apamin, adolapin, and mast cell-degranulating peptide [5]. Apamin, phospholipase A_2 and melittin had a neuroprotective effect [6–8] and adolapin had an anti-inflammatory effect [9].

Many experimental studies on the biological activities of BV have been reported and its anti-inflammatory effects, as well as its activities in relieving pain of rheumatoid arthritis and in immune modulation, have been described [10–13]. In addition, BV has also been reported to enhance activation of the apoptotic signaling pathway in experimental osteosarcoma, breast cancer, and lung cancer cell lines [14–16]. Although previous studies have clearly demonstrated that BV possesses anti-proliferative and pro-apoptotic effects, those studies have primarily been performed in cancer cell lines. Recently, Doo *et al.* have shown that BV protected neuronal cells against MPP$^+$-induced apoptotic cell death via activation of PI3K/Akt-mediated signaling and inhibition of cell death signaling [17].

Therefore, in this study, we investigated the effects of BV on rotenone-induced cell toxicity in NSC34 motor neuron cells. The MAPK family is known to regulate neuronal survival and death [18–20]; ERK1/2 is activated by growth factors, whereas JNKs are activated by cell stress-induced signaling. We examined the effect of rotenone on the activation of JNK and ERK1/2 related to cell death and cell survival, respectively. In our previous study, we demonstrated that BV had a neuro-protective effect against glutamate-induced toxicity via inhibition of the expression of phospho-JNK and phopho-ERK in neuronal cells [21]. We report that pretreatment of BV significantly attenuated rotenone-mediated toxicity via inhibition of the activation of c-Jun *N*-terminal kinase (JNK) and an increase of extracellular signal-regulated kinase (ERK) signaling pathway. In addition, pretreatment with BV significantly inhibited mitochondria impairment and the expression of cleaved caspase-3 in NSC34 neuronal cells. The present results may have clinical implications and suggest that BV may be a potential treatment for the prevention of oxidative stress-induced cell death in neurodegenerative diseases.

2. Materials and Methods

2.1. Cell Culture

The motor neuron cell line NSC34 was obtained from Cellutions Biosystems Inc. (Toronto, ON, Canada). Cells were cultured in Dulbecco's modified Eagle's medium (DMEM) supplemented with 10% FBS, 100 U/mL penicillin, and 100 μg/mL streptomycin at 37 °C with 5% CO_2. Cells were subcultured in a fresh culture dish when growth reached 70%–90% confluence (*i.e.*, every 2–3 days) as recommended by Cellutions Biosystems Inc. In all experiments, cells were incubated in the presence or absence of 2.5 μg/mL of BV before the addition of 10 μM rotenone to the culture media.

2.2. Cell Viability Assay

Cell viability was determined by a modified MTT (3-[4,5-dimethyl-thiazol-2-yl]-2,5-diphenyltetrazolium bromide) reduction assay. This assay is based on the ability of active mitochondrial dehydrogenase to convert dissolved MTT into water-insoluble purple formazan crystals. NSC34 motor neuron cells were plated in 96-well plates (2×10^4 cells/well). After 24 h, the cells were treated with the indicated concentration of BV for 24 h prior to 10 μM rotenone treatment for 24 h. Briefly, MTT was added to each well at a final concentration of 0.5 mg/mL, and the plates were incubated for 1 h at 37 °C. After removing the culture medium, DMSO was added, and the plates were shaken for 10 min to solubilize the formazan reaction product. The absorbance at 570 nm was measured using a microplate reader (Bio-rad, Hercules, CA, USA).

2.3. Preparation of Primary Cortical Neuronal Culture

Mixed primary cortical neuronal cells were prepared from embryonic day 15 (E15) ICR mouse embryos. Briefly, the cortical region of mouse brain was dissected and cleaned of meningeal tissue, minced, and dissociated mechanically by flamed polished Pasteur pipettes in minimal essential medium (MEM). Dissociated cortical cells were then plated in Neurobasal medium with B-27 supplement, 5% FBS (Gibco, Grand Island, NY, USA), 5% horse serum, and 2 mM glutamine onto laminin- and poly-D-lysine-coated 12-well plates. Primary cortical cultures at 14 days *in vitro* (DIV) were used.

2.4. Western Blot

Cells were washed twice with ice-cold phosphate-buffered saline and harvested into 1.5 mL tube. Cells were lysed with lysis buffer containing 50 mM Tris HCl, pH 7.4, 1% NP-40, 0.1% SDS, 150 mM NaCl, and the Complete Mini Protease Inhibitor Cocktail (Roche, Basel, Switzerland). The protein concentration was measured with a BCA Protein Assay Kit (Pierce, Rockford, IL, USA). Extracted samples (20 μg total protein per lane) were separated using SDS-polyacrylamide gel electrophoresis (SDS-PAGE) and then transferred to nitrocellulose membranes (Whatman, Lawrence, KS, USA). The membranes were blocked with 5% skim milk to prevent nonspecific protein binding and incubated with primary antibodies against p-ERK (1:1000, cell signaling), p-JNK (1:1000, cell signaling), total ERK (1:1000, cell signaling), total JNK (1:1000, cell signaling), α-tubulin (1:5000, Abcam, Cambridge, MA, USA), and cleaved caspase-3 (1:1000, cell signaling) in 5% skim milk overnight. After washing three times with TBS-T (pH7.5, 1 M Tris-HCl, 1.5 M NaCl, 0.5% tween-20), the membranes were hybridized with horseradish peroxidase-conjugated secondary antibodies for 1 h. Following five washes with TBS-T, specific protein bands were detected using the SuperSignal West Femto Chemiluminescent Substrate (Pierce, IL, USA) and enhanced chemiluminescence reagents (Amersham Pharmacia, Piscataway, NJ, USA). α-tubulin was used as an internal control to normalize protein loading. Protein bands were detected and analyzed using the FusionSL4-imaging system. Quantification of the blotting bands was performed using Bioprofil (Bio-1D version 15.01, Eberhardzell, Germany).

2.5. Mitochondria Staining

MitoTracker Red CMXRos (Molecular Probes, Eugene, OR, USA) is a red fluorescent dye that stains mitochondria in live cells. NSC34 motor neuron cells were plated in 12-well plates (5×10^4 cells/well). After 24 h of cell seeding, the cells were treated with the indicated concentration of BV for 24 h prior to 10 μM rotenone treatment for 24 h. Briefly, cells were stained with MitoTracker at a final concentration of 100 nM according to the standard protocol. The cells were fixed with 4% paraformaldehyde at room temperature, and then washed with phosphate-buffered saline twice. The cover slip was fixed using Fluoromount to the slide glass.

2.6. Data Analysis

Data are expressed as means ±S.E.M. Comparisons were evaluated by one-way analysis of variance (ANOVA) and followed by Newman-keuls post-hoc test for multiple comparisons using GraphPad Prism 5.0 (GraphPad software, La Jolla, CA, USA). Probability values less than 0.05 were considered statistically significant.

3. Results

3.1. Effect of Bee Venom on Rotenone-Induced Cell Death in NSC34 Cells

To examine the cytotoxicity of NSC34 cells after rotenone treatment, we incubated NSC34 motor neuron cells with 10 μM rotenone for various time periods (0, 3, 6, 12, and 24 h). Our results showed that rotenone decreased cell survival in a time-dependent manner (Figure 1A). Compared to untreated NSC34 cells, 10 μM rotenone treatment for 24 h resulted in a 37% decrease in cell viability. To investigate whether BV attenuates rotenone-induced cytotoxicity, we performed a cell viability test on the cells treated with 10 μM rotenone for 24 h with or without 2.5 μg/mL BV pretreatment. BV pretreatment prevented cell death induced by 10 μM rotenone, with about 29% protection for a dose of 2.5 μg/mL BV compared to rotenone-treated cells (Figure 1B). These results indicate that BV pretreatment had a protective effect against rotenone-induced cytotoxicity.

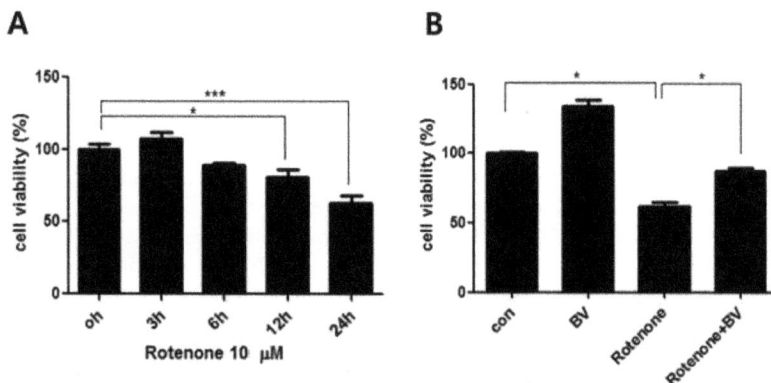

Figure 1. BV pretreatment prevents rotenone-induced cytotoxicity in NSC34 neuronal cells. (**A**) NSC34 cells were treated with 10 μM rotenone for 0, 3, 6, 12, and 24 h and cell viability was determined using MTT assay. 10 μM rotenone treatment induced time-dependent cytotoxicity in NSC34 cells; (**B**) NSC34 cells were treated with 2.5 μg/mL BV prior to stimulation with 10 μM rotenone for 24 h. BV pretreatment prevented 10 μM rotenone-induced cytotoxicity in NSC34 cells. The values shown are the means ±S.E.M. of data obtained from three independent experiments. * $p < 0.05$, *** $p < 0.001$.

3.2. Bee Venom Affects ERK and JNK Phosphorylation

For this study, NSC34 cells were treated with 10 μM rotenone for various time periods (0, 0.25, 6, and 24 h). JNK phosphorylation, indicating JNK activation, increased after 1 h of rotenone treatment in NSC34 cells. However, the expression of ERK1/2 phosphorylation, which promoted cell survival, was significantly reduced at 15 min by rotenone treatment compared to untreated cells (Figure 2A). Both significant increase in JNK phosphorylation and decrease in ERK phosphorylation were observed after 1 h of rotenone treatment in NSC34 cells. To examine the BV effect on cell death signaling, we pretreated with 2.5 μg/mL BV for 24 h and then stimulated with 10 μM rotenone for 1 h in the presence or absence of BV. As shown in Figure 2B,C, pretreatment with BV attenuated by 1.2 fold JNK phosphorylation compared to rotenone-treated cells. In addition, BV pretreatment recovered by 1.4 fold ERK under-phosphorylation compared rotenone-treated NSC34 cells (Figure 2B,C). These results suggest that BV pretreatment could prevent oxidative stress induced-neuronal cell death by inhibiting cell death-related MAPK signaling.

3.3. Bee Venom Attenuates Rotenone- Induced Caspase-3 Activation in NSC34 Neuronal Cells

To investigate the effects of BV on oxidative stress-induced mitochondria dysfunction, capase-3 activation was assessed after exposure to rotenone in NSC34 neuronal cells. The expression level of cleaved caspase-3 was significantly increased after 6 h of 10 μM rotenone treatment compared to untreated NSC34 neuronal cells (Figure 3A). However, pretreatment with 2.5 μg/mL BV suppressed by two-fold the increase of caspase-3 activation induced by rotenone treatment compared with rotenone-treated NSC34 cells (Figure 3B,C). To confirm BV effects against rotenone, we performed the same experiment with primary cortical neuronal cells. Consistent with the result in NSC34 cell line, 10 μM rotenone treatment induced the activation of caspase-3 but 2.5 μg/mL BV treatment reduced rotenone-induced cleaved caspase-3 expression. Those results suggest that BV pretreatment could prevent oxidative stress-induced mitochondria dysfunction.

A

Rotenone (10 μM)

0 15' 1h 6h 24h

p-ERK

ERK

p-JNK

JNK

B

Rotenone - - + +

BV - + - +

pJNK

JNK

pERK

ERK

C

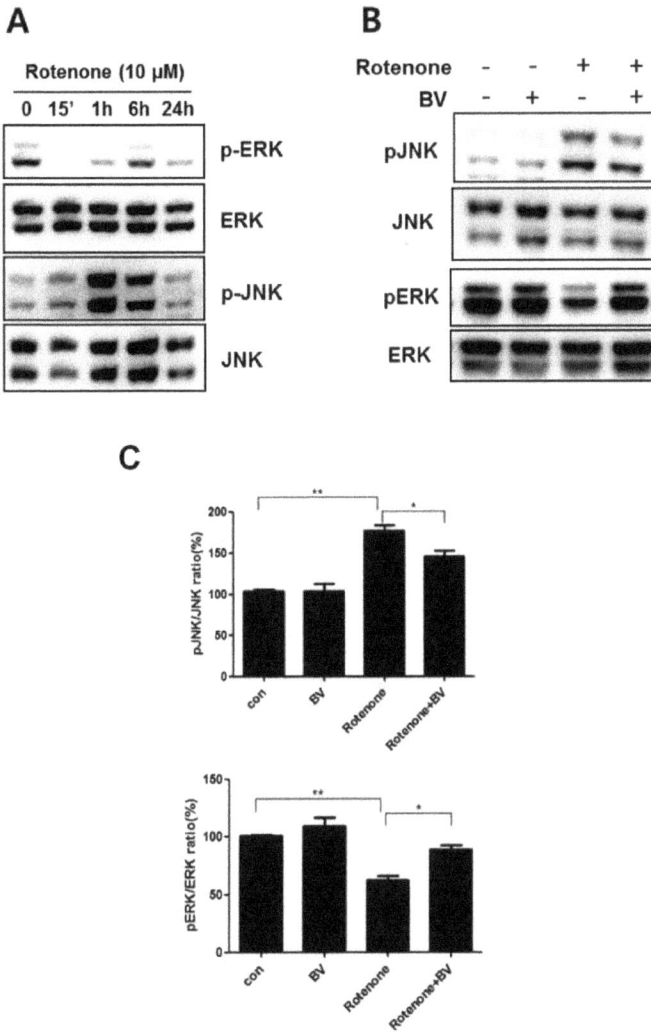

Figure 2. BV pretreatment regulates the activation of rotenone-mediated signaling in NSC34 neuronal cells. Effect of rotenone on the phosphorylation of the MAPK proteins ERK and JNK in NSC34 cells. (**A**) NSC34 cells were treated with 10 μM rotenone for the indicated time. Total cell lysates were separated with SDS-PAGE and Western blots were performed using anti-phospho JNK, anti-phospho ERK1/2, JNK, and ERK antibodies; (**B**) NSC34 cells were pretreated with 2.5 μg/mL BV for 24 h and then stimulated with 10 μM rotenone for 1 h in the presence or absence of BV. Western blots were performed with specific antibodies, including those for the phosphorylated forms of ERK and JNK. Total ERK and JNK were used as loading controls for the cell lysates; (**C**) Immune blots were quantified with the relative phospho-/nonphospho ratio. The values shown are the means ±S.E.M. of data obtained from three independent experiments. * $p < 0.05$, ** $p < 0.01$.

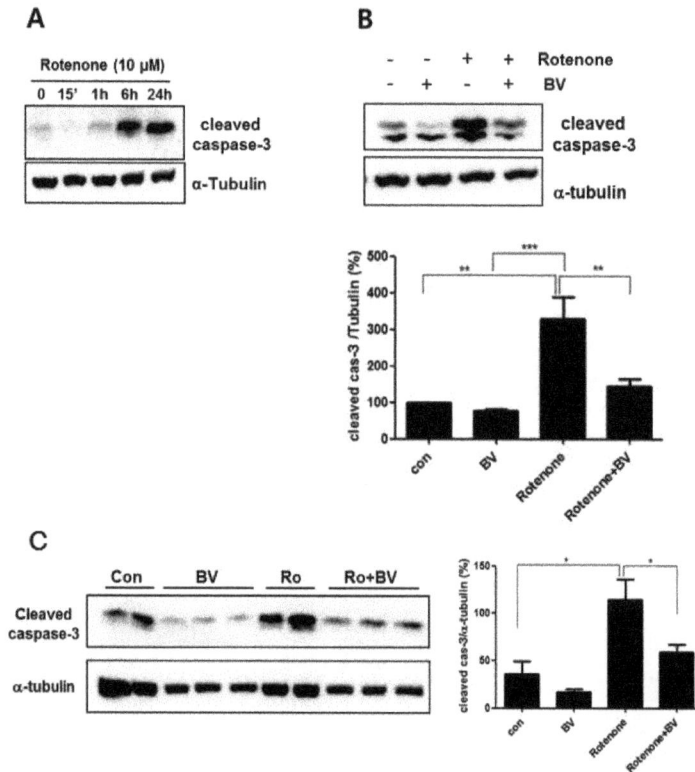

Figure 3. BV inhibits the expression of cleaved caspase-3 in NSC34 cells. (**A**) Expression of cleaved caspase-3 proteins was detected in rotenone-induced NSC34 cells. NSC34 cells incubated with 10 μM rotenone for the indicated time periods. The loading control for the cell lysates was determined by re-probing the membranes with α-tubulin antibody; (**B**) BV attenuates the expression level of cleaved caspase-3 protein in NSC34 neuronal cells. NSC34 cells were pretreated with 2.5 μg/mL BV for 24 h and then stimulated with 10 μM rotenone for 24 h in the presence or absence of BV; (**C**) BV inhibits the expression of cleaved caspase-3 protein in rotenone-treated in primary cortical neuronal cells system. BV pretreated with 2.5 μg/mL for 24 h and stimulated with 10 μM rotenone for 24 h in the presence or absence of BV in primary cortical neuronal cells. The loading control for the cell lysates was determined by re-probing the membranes with α-tubulin antibody. The values shown are the means ±S.E.M. of data obtained from three independent experiments. * $p < 0.05$, ** $p < 0.01$, *** compared to * and ** and indicates $p < 0.001$.

3.4. Bee Venom Suppresses Rotenone-Mediated Mitochondrial Impairment

To evaluate the effect of BV pretreatment on rotenone-induced mitochondria alteration, we stained neuronal cells with Mitotracker® Red probes that passively diffuse across the plasma membrane and accumulate mitochondria of live cells and observed mitochondria using confocal microscopy. As shown in Figure 4, mitochondria showed broad cytoplasmic distribution in both control and BV-treated NSC34 neuronal cells. However, 10 μM rotenone treatment for 24 h induced aggregated mitochondria in neuronal cells (Figure 4C). Interestingly, pretreatment with 2.5 μg/mL BV inhibited mitochondrial aggregation resulting from rotenone treatment (Figure 4D,E). These findings suggest that BV pretreatment can block interference with the electron transport chain in mitochondria induced by rotenone.

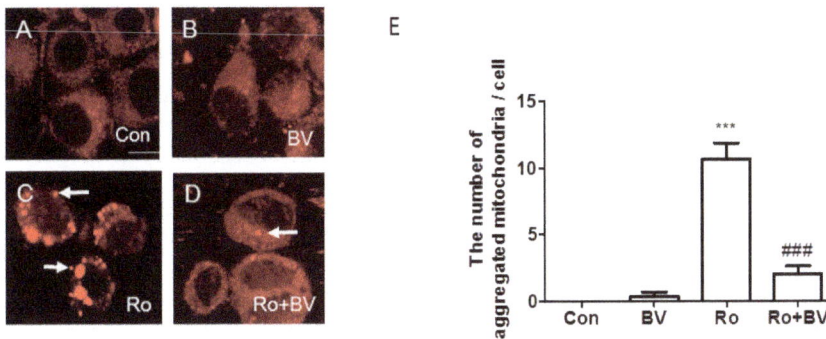

Figure 4. Confocal microscopy images of mitochondria using Mitotracker® Red. Mitochondria stained with Mitotracker® Red in NSC 34 cells untreated (**A**), treated with 2.5 μg/mL BV for 24 h (**B**) or 10 μM rotenone for 24 h (**C**), pretreated with 2.5 μg/mL BV for 24 h and 10 μM rotenone for 24 h (**D**). Arrows indicate aggregated mitochondria. BV pretreatment reduces mictochondria impairment induced by rotenone treatment in NSC34 cells. The bar indicates 200 μm. (**E**) Quantification of the number of aggregated mitochondria in a cell of 10 microscopic visual fields randomly selected. The values shown are the means ±S.E.M. of data obtained from three independent experiments. *** and ### indicate $p < 0.001$. *** compared to Con and ### compared to Ro. Con: control, BV: BV-treated cell, Ro: rotenone-treated cell, and Ro + BV: BV pretreated-cells prior to rotenone treatment.

4. Discussion

The present study demonstrates that BV reduces rotenone-induced cell death in NSC34 cells by blocking the JNK and ERK1/2 signaling pathways. The MAPK family, including ERK1/2, stress-activated protein kinase/c-jun NH4-terminal kinase (SAPK/JNK), and p38 MAPKs [22,23], play central roles in the signaling pathways involved in cell proliferation, survival, and apoptosis. ERK1/2 are activated by mitogens and growth factors leading to cell growth and survival, whereas JNK and p38 MAPK are preferentially activated by pro-inflammatory cytokines and oxidative stress resulting in cell differentiation and apoptosis [24]. Here, we show that rotenone treatment causes decreased phosphorylation of ERK1/2 and increased phosphorylation of JNK. However, BV pretreatment induces the reduction of JNK and an increase of ERK1/2 activation in rotenone-treated NSC34 cells. After rotenone treatment, cell death was induced by the up-regulation of capase-3 activation and mitochondria aggregation leading to mitochondria dysfunction. However, BV treatment significantly suppressed the rotenone-induced activation of caspase-3 and mitochondria impairment in NSC34 cells. Taken those data together, we suggest that BV treatment could be useful for the inhibition of oxidative stress induced cell death in neurodegenerative diseases.

BV extracted from honeybees has been used in traditional oriental medicine. BV is traditionally used for the treatment of chronic inflammatory diseases such as rheumatoid arthritis and for relief of pain in oriental medicine [25]. In this study, BV attenuated rotenone-induced cell toxicity through inhibition of the JNK and activation of the ERK in neuronal cells. The optimized concentration of 2.5 μg/mL BV increased the expression of phospho-AKT [26]. The activation of MAPKs, including ERK, p38, and JNK [27,28], activated by interactions with a small GTPase and/or phosphorylation by protein kinases downstream from cell surface receptors. Previous reports have demonstrated that ERK is involved in growth-associated responses and cell differentiation. However, JNK and p38 are phosphorylated by environmental stresses, including the cytokine-like lipopolysaccharide (LPS) and oxidative stress [28].

Rotenone is used worldwide as an inhibitor of mitochondrial respiratory chain complex I, and induces PD-like symptoms in neurons through disrupting ATP supply [29]. Since it also increases ROS release and cause cell apoptosis, exposure of animals of rotenone induces dopaminergic cell death production [30,31].

Wagdy *et al.* have shown that BV acupuncture (BVA) has a neuroprotective effect against rotenone-induced oxidative stress, neuroinflammation, and cell death in PD-like animal model [32]. BV treatment inhibited rotenone-induced activation of caspase-3 and mitochondria aggregation inducing mitochondrial dysfunction. Those results provide that it is possible that multiple signaling pathways are involved in protective effects of BV, which deserves to be further investigated. Mitochondrial structure alternation by rotenone plays principal roles in cell death [33,34], which are critical pathological factors in neurodegenerative diseases including Alzheimer's disease (AD), and amyotrophic lateral sclerosis (ALS) [35,36]. In further study, it is necessary to evaluate the effect of BV on genetically-engineered neurodegenerative disease animal models.

Acknowledgments: This work was supported by grants (K13012) from the Korea Institute of Oriental Medicine (KIOM), South Korea.

Author Contributions: E.J.Y. designed the experiments and analyzed the data as well as wrote the manuscript; S.Y.J. executed MTT assay, Western blot analysis, and Mitotracker staining and K.W.L. performed primary neuronal cell culture experiment. S.M.C. discussed with the manuscript. All authors have read and approved the final manuscript.

Conflicts of Interest: The authors declare no conflicts of interest.

References

1. Li, N.; Ragheb, K.; Lawler, G.; Sturgis, J.; Rajwa, B.; Melendez, J.A. Robinson Mitochondrial complex I inhibitor rotenone induces apoptosis through enhancing mitochondrial reactive oxygen species production. *J. Biol. Chem.* **2003**, *278*, 8516–8525. [CrossRef] [PubMed]
2. Im, A.R.; Kim, Y.H.; Uddin, M.R.; Chae, S.; Lee, H.W.; Kim, Y.H.; Kim, Y.S.; Lee, M.Y. Betaine protects against rotenone-induced neurotoxicity in PC12 cells. *Cell. Mol. Neurobiol.* **2013**, *33*, 625–635. [CrossRef] [PubMed]
3. Tamilselvam, K.; Braidy, N.; Manivasagam, T.; Essa, M.M.; Prasad, N.R.; Karthikeyan, S.; Thenmozhi, A.J.; Selvaraju, S.; Guillemin, G.J. Neuroprotective effects of hesperidin, a plant flavanone, on rotenone-induced oxidative stress and apoptosis in a cellular model for Parkinson's disease. *Oxid. Med. Cell. Longev.* **2013**, *2013*, 102741. [CrossRef] [PubMed]
4. Samantaray, S.; Knaryan, V.H.; Guyton, M.K.; Matzelle, D.D.; Ray, S.K.; Banik, N.L. The parkinsonian neurotoxin rotenone activates calpain and caspase-3 leading to motoneuron degeneration in spinal cord of Lewis rats. *Neuroscience* **2007**, *146*, 741–755. [CrossRef] [PubMed]
5. Eiseman, J.L.; von Bredow, J.; Alvares, A.P. Effect of honeybee (*Apis mellifera*) venom on the course of adjuvant-induced arthritis and depression of drug metabolism in the rat. *Biochem. Pharmacol.* **1982**, *31*, 1139–1146. [CrossRef]
6. Alvarez-Fischer, D.; Noelker, C.; Vulinovic, F.; Grunewald, A.; Chevarin, C.; Klein, C.; Oertel, W.H.; Hirsch, E.C.; Michel, P.P.; Hartmann, A. Bee Venom and its component apamin as neuroprotective agents in a Parkinson disease mouse model. *PLoS ONE* **2013**, *8*, e61700. [CrossRef] [PubMed]
7. Armugam, A.; Cher, C.D.; Lim, K.; Koh, D.C.; Howells, D.W.; Jeyaseelan, K. A secretory phospholipase A$_2$-mediated neuroprotection and anti-apoptosis. *BMC Neurosci.* **2009**, *10*, 120. [CrossRef] [PubMed]
8. Yang, E.J.; Kim, S.H.; Yang, S.C.; Lee, S.M.; Choi, S.M. Melittin restores proteasome function in an animal model of ALS. *J. Neuroinflammation* **2011**, *8*, 69. [CrossRef] [PubMed]
9. Koburova, K.L.; Michailova, S.G.; Shkenderov, S.V. Further investigation on the antiinflammatory properties of adolapin—Bee venom polypeptide. *Acta Physiol. Pharmacol. Bulg.* **1985**, *11*, 50–55. [PubMed]
10. Kwon, Y.B.; Lee, H.J.; Han, H.J.; Mar, W.C.; Kang, S.K.; Yoon, O.B.; Beitz, A.J.; Lee, J.H. The water-soluble fraction of bee venom produces antinociceptive and anti-inflammatory effects on rheumatoid arthritis in rats. *Life Sci.* **2002**, *71*, 191–204. [CrossRef]
11. Yang, E.J.; Jiang, J.H.; Lee, S.M.; Yang, S.C.; Hwang, H.S.; Lee, M.S.; Choi, S.M. Bee venom attenuates neuroinflammatory events and extends survival in amyotrophic lateral sclerosis models. *J. Neuroinflammation* **2010**, *7*, 69. [CrossRef] [PubMed]

12. Putz, T.; Ramoner, R.; Gander, H.; Rahm, A.; Bartsch, G.; Thurnher, M. Antitumor action and immune activation through cooperation of bee venom secretory phospholipase A_2 and phosphatidylinositol-(3,4)-bisphosphate. *Cancer Immunol. Immunother.* **2006**, *55*, 1374–1383. [CrossRef] [PubMed]

13. Chung, E.S.; Kim, H.; Lee, G.; Park, S.; Kim, H.; Bae, H. Neuro-protective effects of bee venom by suppression of neuroinflammatory responses in a mouse model of Parkinson's disease: Role of regulatory T cells. *Brain Behav. Immun.* **2012**, *26*, 1322–1330. [CrossRef] [PubMed]

14. Chu, S.T.; Cheng, H.H.; Huang, C.J.; Chang, H.C.; Chi, C.C.; Su, H.H.; Hsu, S.S.; Wang, J.L.; Chen, I.S.; Liu, S.I.; *et al.* Phospholipase A_2-independent Ca^{2+} entry and subsequent apoptosis induced by melittin in human MG63 osteosarcoma cells. *Life Sci.* **2007**, *80*, 364–369. [CrossRef] [PubMed]

15. Ip, S.W.; Liao, S.S.; Lin, S.Y.; Lin, J.P.; Yang, J.S.; Lin, M.L.; Chen, G.W.; Lu, H.F.; Lin, M.W.; Han, S.M.; *et al.* The role of mitochondria in bee venom-induced apoptosis in human breast cancer MCF7 cells. *In Vivo* **2008**, *22*, 237–245. [PubMed]

16. Jang, M.H.; Shin, M.C.; Lim, S.; Han, S.M.; Park, H.J.; Shin, I.; Lee, J.S.; Kim, K.A.; Kim, E.H.; Kim, C.J. Bee venom induces apoptosis and inhibits expression of cyclooxygenase-2 mRNA in human lung cancer cell line NCI-H1299. *J. Pharmacol. Sci.* **2003**, *91*, 95–104. [CrossRef] [PubMed]

17. Doo, A.R.; Kim, S.N.; Kim, S.T.; Park, J.Y.; Chung, S.H.; Choe, B.Y.; Chae, Y.; Lee, H.; Yin, C.S.; Park, H.J. Bee venom protects SH-SY5Y human neuroblastoma cells from 1-methyl-4-phenylpyridinium-induced apoptotic cell death. *Brain Res.* **2012**, *1429*, 106–115. [CrossRef] [PubMed]

18. Chang, L.M. Karin Mammalian MAP kinase signalling cascades. *Nature* **2001**, *410*, 37–40. [CrossRef] [PubMed]

19. Junn, E.M.M. Mouradian Apoptotic signaling in dopamine-induced cell death: The role of oxidative stress, p38 mitogen-activated protein kinase, cytochrome c and caspases. *J. Neurochem.* **2001**, *78*, 374–383. [CrossRef] [PubMed]

20. Davis, R.J. Signal transduction by the JNK group of MAP kinases. *Cell* **2000**, *103*, 239–252. [CrossRef]

21. Lee, S.M.; Yang, E.J.; Choi, S.M.; Kim, S.H.; Baek, M.G.; Jiang, J.H. Effects of bee venom on glutamate-induced toxicity in neuronal and glial cells. *Evid. Based Complement. Altern. Med.* **2012**, *2012*, 368196. [CrossRef] [PubMed]

22. Johnson, G.L.R. Lapadat mitogen-activated protein kinase pathways mediated by ERK, JNK, and p38 protein kinases. *Science* **2002**, *298*, 1911–1912. [CrossRef] [PubMed]

23. Fan, Y.; Chen, H.; Qiao, B.; Luo, L.; Ma, H.; Li, H.; Jiang, J.; Niu, D.; Yin, Z. Opposing effects of ERK and p38 MAP kinases on Hela cell apoptosis induced by dipyrithione. *Mol. Cells* **2007**, *23*, 30–38. [PubMed]

24. Minden, A.M. Karin Regulation and function of the JNK subgroup of MAP kinases. *Biochim. Biophys. Acta* **1997**, *1333*, F85–F104. [PubMed]

25. Kim, J.I.; Yang, E.J.; Lee, M.S.; Kim, Y.S.; Huh, Y.; Cho, I.H.; Kang, S.; Koh, H.K. Bee venom reduces neuroinflammation in the MPTP-induced model of Parkinson's disease. *Int. J. Neurosci.* **2011**, *121*, 209–217. [CrossRef] [PubMed]

26. Almeida, R.D.; Manadas, B.J.; Melo, C.V.; Gomes, J.R.; Mendes, C.S.; Graos, M.M.; Carvalho, R.F.; Carvalho, A.P.; Duarte, C.B. Neuroprotection by BDNF against glutamate-induced apoptotic cell death is mediated by ERK and PI3-kinase pathways. *Cell Death Differ.* **2005**, *12*, 1329–1343. [CrossRef] [PubMed]

27. Nadra, I.; Mason, J.C.; Philippidis, P.; Florey, O.; Smythe, C.D.; McCarthy, G.M.; Landis, R.C.; Haskard, D.O. Proinflammatory activation of macrophages by basic calcium phosphate crystals via protein kinase C and MAP kinase pathways: A vicious cycle of inflammation and arterial calcification? *Circ. Res.* **2005**, *96*, 1248–1256. [CrossRef] [PubMed]

28. Clerk, A.P.H. Sugden Untangling the Web: Specific signaling from PKC isoforms to MAPK cascades. *Circ. Res.* **2001**, *89*, 847–849. [PubMed]

29. Sherer, T.B.; Betarbet, R.; Testa, C.M.; Seo, B.B.; Richardson, J.R.; Kim, J.H.; Miller, G.W.; Yagi, T.; Matsuno-Yagi, A.; Greenamyre, J.T. Mechanism of toxicity in rotenone models of Parkinson's disease. *J. Neurosci.* **2003**, *23*, 10756–10764. [PubMed]

30. Martins, J.B.; Bastos Mde, L.; Carvalho, F.; Capela, J.P. Differential Effects of Methyl-4-Phenylpyridinium Ion, Rotenone, and Paraquat on Differentiated SH-SY5Y Cells. *J. Toxicol.* **2013**, *2013*, 347312. [CrossRef] [PubMed]

31. Tanner, C.M.; Kamel, F.; Ross, G.W.; Hoppin, J.A.; Goldman, S.M.; Korell, M.; Marras, C.; Bhudhikanok, G.S.; Kasten, M.; Chade, A.R.; *et al.* Langston Rotenone, paraquat, and Parkinson's disease. *Environ. Health Perspect.* **2011**, *119*, 866–872. [CrossRef] [PubMed]

32. Khalil, W.K.; Assaf, N.; ElShebiney, S.A.; Salem, N.A. Neuroprotective effects of bee venom acupuncture therapy against rotenone-induced oxidative stress and apoptosis. *Neurochem. Int.* **2015**, *80*, 79–86. [CrossRef] [PubMed]
33. Borutaite, V. Mitochondria as decision-makers in cell death. *Environ. Mol. Mutagen.* **2010**, *51*, 406–416. [CrossRef] [PubMed]
34. Tsujimoto, Y.S. Shimizu Role of the mitochondrial membrane permeability transition in cell death. *Apoptosis* **2007**, *12*, 835–840. [CrossRef] [PubMed]
35. Mena, N.P.; Urrutia, P.J.; Lourido, F.; Carrasco, C.M.; Nunez, M.T. Mitochondrial iron homeostasis and its dysfunctions in neurodegenerative disorders. *Mitochondrion* **2015**, *21*, 92–105. [CrossRef] [PubMed]
36. Cozzolino, M.; Rossi, S.; Mirra, A.; Carri, M.T. Mitochondrial dynamism and the pathogenesis of Amyotrophic Lateral Sclerosis. *Front. Cell Neurosci.* **2015**, *9*, 31. [CrossRef] [PubMed]

© 2015 by the authors; licensee MDPI, Basel, Switzerland. This article is an open access article distributed under the terms and conditions of the Creative Commons Attribution (CC-BY) license (http://creativecommons.org/licenses/by/4.0/).

toxins

MDPI

Article

Effects of Melittin Treatment in Cholangitis and Biliary Fibrosis in a Model of Xenobiotic-Induced Cholestasis in Mice

Kyung-Hyun Kim [1], Hyun-Jung Sung [1], Woo-Ram Lee [1], Hyun-Jin An [1], Jung-Yeon Kim [1], Sok Cheon Pak [2], Sang-Mi Han [3] and Kwan-Kyu Park [1,*]

[1] Department of Pathology, College of Medicine, Catholic University of Daegu, 3056-6, Daemyung-4-Dong, Nam-gu, Daegu 705-718, Korea; khkim1@cu.ac.kr (K.-H.K.); ewingsar@gmail.com (H.-J.S.); woolamee@cu.ac.kr (W.-R.L.); ahj119@cu.ac.kr (H.-J.A.); kjy1118@cu.ac.kr (J.-Y.K.)

[2] School of Biomedical Sciences, Charles Sturt University, Panorama Avenue, Bathurst, NSW 2795, Australia; spak@csu.edu.au

[3] Department of Agricultural Biology, National Academy of Agricultural Science, RDA, 300, Nongsaengmyeong-ro, Wansan-gu, Jeonju-si, Jeollabuk-do 55365, Korea; sangmih@korea.kr

* Correspondence: kkpark@cu.ac.kr; Tel.: +82-53-650-4149; Fax: +82-53-650-4843

Academic Editor: Ren Lai

Received: 23 June 2015; Accepted: 20 August 2015; Published: 25 August 2015

Abstract: Cholangiopathy is a chronic immune-mediated disease of the liver, which is characterized by cholangitis, ductular reaction and biliary-type hepatic fibrosis. There is no proven medical therapy that changes the course of the disease. In previous studies, melittin was known for attenuation of hepatic injury, inflammation and hepatic fibrosis. This study investigated whether melittin provides inhibition on cholangitis and biliary fibrosis *in vivo*. Feeding 3,5-diethoxycarbonyl-1,4-dihydrocollidine (DDC) to mice is a well-established animal model to study cholangitis and biliary fibrosis. To investigate the effects of melittin on cholangiopathy, mice were fed with a 0.1% DDC-containing diet with or without melittin treatment for four weeks. Liver morphology, serum markers of liver injury, cholestasis markers for inflammation of liver, the degree of ductular reaction and the degree of liver fibrosis were compared between with or without melittin treatment DDC-fed mice. DDC feeding led to increased serum markers of hepatic injury, ductular reaction, induction of pro-inflammatory cytokines and biliary fibrosis. Interestingly, melittin treatment attenuated hepatic function markers, ductular reaction, the reactive phenotype of cholangiocytes and cholangitis and biliary fibrosis. Our data suggest that melittin treatment can be protective against chronic cholestatic disease in DDC-fed mice. Further studies on the anti-inflammatory capacity of melittin are warranted for targeted therapy in cholangiopathy.

Keywords: cholangiopathy; melittin; DDC-fed mice

1. Introduction

Liver fibrosis refers to a classical outcome of many chronic liver diseases irrespective of the etiology of injury. It is characterized by changes in the composition and quantity of extracellular matrix (ECM) deposits distorting the normal structure by forming fibrotic scars. Failure to degrade the accumulated ECM is a major reason why fibrosis progresses to cirrhosis in liver. Various insults, such as viral infection, drugs or metabolic disorders, contribute to the progression of liver fibrosis. Among them, improper regulation of bile flow (*i.e.*, cholestasis), which causes hepatic inflammation and subsequent tissue injury, is one of the main insults for cholangiopathies [1].

As the main cause of liver-related death, cholangiopathies are also the leading cause of liver transplantations in paediatric patients (50%) and the third leading cause in adults (20%) [1,2].

Cholangiopathies, such as primary sclerosing cholangitis (PSC), primary biliary cirrhosis (PBC) and drug-induced bile duct damage, may result in cholestasis, which is characterized by the loss of cholangiocytes through necrosis by apoptosis [3]. Cholangiocyte proliferation can occur during cholangiopathies, resulting in the formation of new side branches to ducts in an effort to regain function and portal/periportal inflammation [4,5].

Cholangiocyte proliferation is described as the expanded population of epithelial cells at the interface of the biliary tree, which refers to the proliferation of pre-existing ductules, progenitor cell activation and the appearance of intermediate hepatocytes [6,7]. The ability of cholangiocytes to proliferate is important in many different human pathological conditions, such as the regaining of proper liver function and the remodelling of biliary cirrhosis in chronic cholestatic conditions.

Lately, 3,5-diethoxycarbonyl-1,4-dihydrocollidine (DDC)-fed mice have been suggested to be used as a novel model for sclerosing cholangitis and biliary fibrosis. Chronic feeding of DDC in mice causes cholangitis with a pronounced ductular reaction, onionskin-type periductal fibrosis and, finally, biliary fibrosis, reflecting several specific pathological hallmarks of human PSC [8]. This model is therefore especially useful to investigate the mechanisms of chronic cholangiopathies and their consequences, including biliary fibrosis, and to test novel therapeutic approaches, such as melittin, for these diseases [8,9].

Melittin is a cationic, haemolytic peptide that is the major bioactive component in honey bee (*Apis mellifera*) venom, and it has been shown to play a role in attenuating fibrosis in various animal models [10,11]. Previous studies have shown that melittin treatment reduced the expression of inflammatory proteins in inflammatory diseases [12]. These studies are informative, but they are not enough to demonstrate that melittin can prevent the development of the inflammatory molecular mechanisms of sclerosing cholangitis. Therefore, the aim of this work was to determine how melittin could become a profibrogenic control and to characterize the underlying mechanism for this effect. The biological properties of melittin were examined in chronic liver injury using DDC-fed model.

2. Results

2.1. Effects of Melittin on DDC-Fed Mice

After four weeks of DDC feeding, liver showed ductules and small bile ducts, which frequently contained pigment plugs (Figure 1). Additionally, pronounced hepatic inflammatory response was observed near bile ducts with predominating neutrophil granulocytes. Furthermore, spontaneous DDC feeding resulted in the deposition of collagen fibres near fibrous septae and expanded bile ducts (Figure 2). These changes were improved by melittin treatment. The DDC + melittin (Mel) group showed the reduction of collagen deposition. DDC feeding also resulted in increased serum AST and ALT levels as an indicator of hepatocyte injury followed by significant elevations of cholestasis parameters of AP and bilirubin. Serum AST and ALT revealed no significant differences between DDC mice and DDC + Mel mice. However, DDC + Mel mice showed significantly lower serum AP and bilirubin levels (Figure 3). These results suggest that melittin treatment effects a decreased susceptibility of cholestasis in DDC-fed mice.

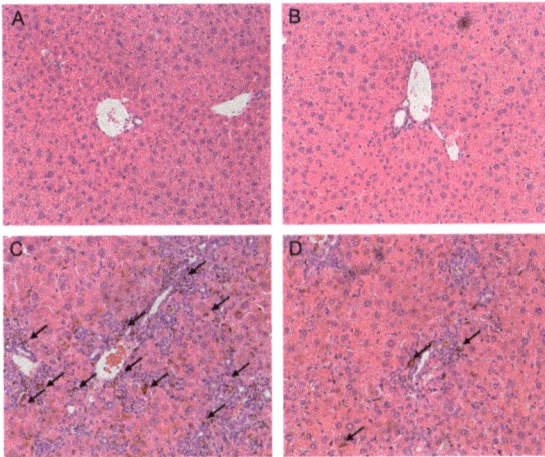

Figure 1. Effect of melittin in DDC-induced liver fibrosis. Hematoxylin and eosin (H&E) staining results show that melittin effectively suppresses inflammation and ductular reaction (arrowheads in (**C,D**)) in response of DDC feeding. Representative H&E images from each study group (five mice per group): (**A**) NC, normal control group; (**B**) Mel, melittin (0.1 mg/kg)-treated group with normal diet; (**C**) DDC, 0.1% DDC-supplemented diet group; (**D**) DDC + Mel, melittin (0.1 mg/kg)-treated group with 0.1% DDC-supplemented diet. Magnification ×200.

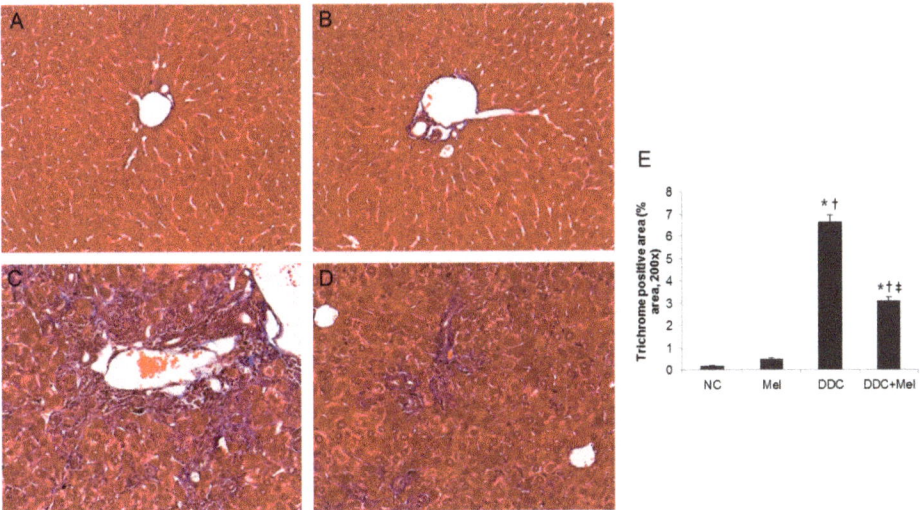

Figure 2. Effect of melittin in DDC-induced liver fibrosis. Masson's trichrome staining results show that melittin effectively attenuates ECM remodelling following chronic DDC feeding in mice. Representative trichrome staining images from each study group (five mice per group): (**A**) NC, normal control group; (**B**) Mel, melittin (0.1 mg/kg)-treated group with normal diet; (**C**) DDC, 0.1% DDC-supplemented diet group; (**D**) DDC + Mel, melittin (0.1 mg/kg)-treated group with 0.1% DDC-supplemented diet; magnification ×200; (**E**) morphometric assessment of the trichrome staining positive areas. Results are expressed as the mean ± SE of three independent determinations. * $p < 0.05$ compared to the NC group. † $p < 0.05$ compared to the Mel group. ‡ $p < 0.05$ compared to the DDC group.

2.2. Melittin Treatment Attenuates DDC-Induced Inflammatory Changes

DDC-fed mice demonstrated pronounced hepatic inflammatory response characterized by an increase of infiltrating inflammatory cells near bile ducts (Figure 1C). In contrast, treatment with melittin changed the inflammatory response in portal tracts (Figure 1D). To investigate whether melittin could influence inflammatory changes in DDC-fed mice, pro-inflammatory cytokines were examined in the livers of experimental mice. Pro-inflammatory cytokines, such as TNF-α and IL-6, are key players in eliciting an inflammatory reaction during liver fibrogenesis [13,14]. The expressions of TNF-α and IL-6 were significantly increased in the DDC-fed mice, whereas melittin treatment markedly abrogated this activation (Figure 4). During liver injury, IL-6 activates STAT3 in liver parenchymal and non-parenchymal cells [15,16]. While chronic DDC feeding activated p-STAT3 in the DDC group, the expression level of p-STAT3 was significantly reduced by treatment with melittin. Moreover, the expression level of MCP-1, which promotes liver fibrosis by recruitment of macrophages, was determined by immunohistochemical staining. As shown in Figure 5A,B, MCP-1-positive cells were barely detected in liver sections from the NC and Mel groups. However, MCP-1-positive areas in the DDC group were significantly increased in liver sections, especially near portal tracts (Figure 5C). Compared to the DDC group, treatment with melittin inhibited MCP-1 expression in DDC + Mel liver (Figure 5D). These results indicate that melittin markedly attenuates the levels of pro-inflammatory cytokines during chronic DDC feeding, which may result in the suppression of DDC-induced cholangitis.

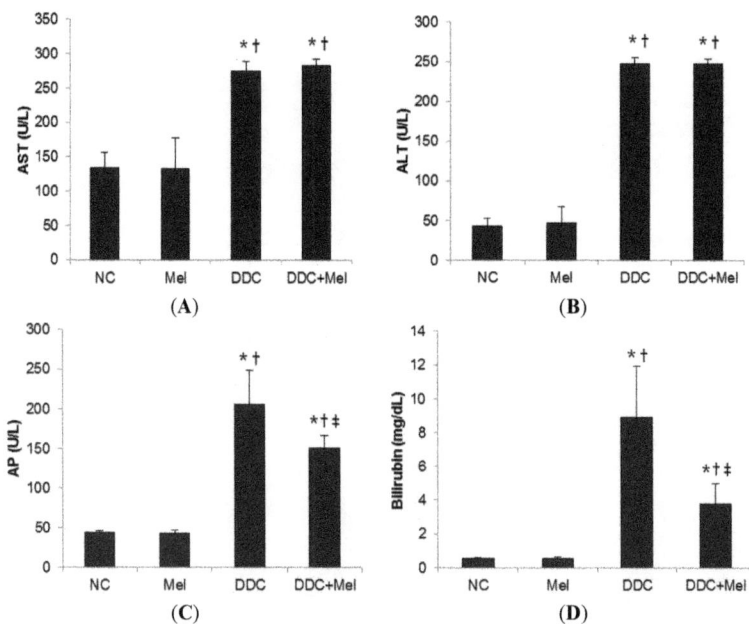

Figure 3. Serum biochemistry of DDC-fed mice. Melittin treatment effectively suppresses serum alkaline phosphatase (AP) and bilirubin, but not aspartate aminotransferase (AST) and alanine transaminase (ALT). (**A**) Serum AST; (**B**) serum ALT; (**C**) serum AP; (**D**) serum bilirubin. NC, normal control group; Mel, melittin (0.1 mg/kg)-treated group with normal diet; DDC, 0.1% DDC-supplemented diet group; DDC + Mel, melittin (0.1 mg/kg)-treated group with 0.1% DDC-supplemented diet. Results are expressed as the mean ± SE of three independent determinations. * $p < 0.05$ compared to the NC group. † $p < 0.05$ compared to the Mel group. ‡ $p < 0.05$ compared to the DDC group.

(A)

(B)

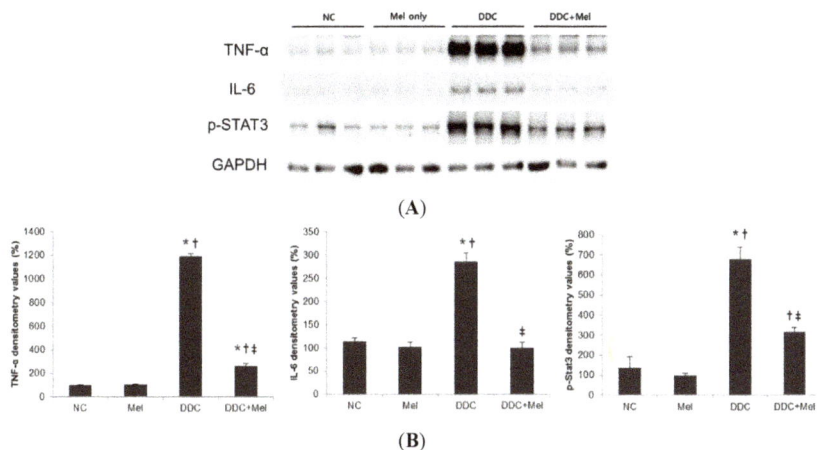

Figure 4. Melittin inhibits pro-inflammatory cytokine expression in DDC-fed mice. (**A**) Western blotting results demonstrated that melittin effectively suppresses the expressions of TNF-α, IL-6 and p-STAT3; (**B**) graphical presentation of the ratio of TNF-α, IL-6 and p-STAT3 to GAPDH in various groups. NC, normal control group; Mel, melittin (0.1 mg/kg)-treated group with normal diet; DDC, 0.1% DDC-supplemented diet group; DDC + Mel, melittin (0.1 mg/kg)-treated group with 0.1% DDC-supplemented diet. Results are expressed as the mean ± SE of three independent determinations. * $p < 0.05$ compared to the NC group. † $p < 0.05$ compared to the Mel group. ‡ $p < 0.05$ compared to the DDC group.

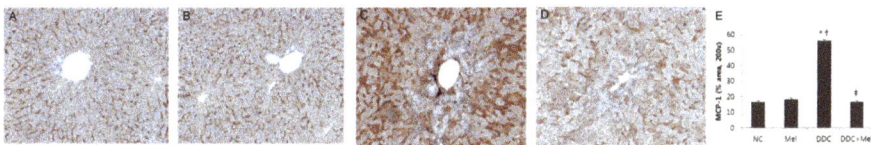

Figure 5. Melittin inhibits inflammatory changes in DDC-fed mice. Immunohistochemical staining results demonstrated that melittin effectively suppresses the expression of MCP-1. Representative immunohistochemical images from each study group (five mice per group) (**A**) NC, normal control group; (**B**) Mel, melittin (0.1 mg/kg)-treated group with normal diet; (**C**) DDC, 0.1% DDC-supplemented diet group; (**D**) DDC + Mel, melittin (0.1 mg/kg)-treated group with 0.1% DDC-supplemented diet; magnification ×200; (**E**) morphometric assessment of the trichrome staining positive areas. Results are expressed as the mean ± SE of three independent determinations. * $p < 0.05$ compared to the NC group. † $p < 0.05$ compared to the Mel group. ‡ $p < 0.05$ compared to the DDC group.

2.3. Melittin Suppresses Liver Fibrosis in the Livers of DDC-Fed Mice

Chronic exposure to injuries in cholangitis becomes a progressive course of cholestatic liver with inflammation and fibrosis of bile ducts. Much attention has been focused on the central role of TGF-β1 upregulation as a prototypical fibrogenic cytokine in liver fibrosis [17]. Western blotting results showed that the expression of TGF-β1 was significantly increased in DDC mice, whereas melittin treatment markedly decreased the expression of TGF-β1 in DDC + Mel livers (Figure 6). The expressions of TGF-β1-regulated ECM protein, fibronectin and vimentin were increased in DDC-fed mice. Treatment with melittin effectively abrogated this increase in the DDC + Mel group. During tissue remodelling in liver fibrosis, FSP-1 is considered a marker of fibroblasts in fibrotic liver. DDC feeding significantly increased the number of cells positive for FSP-1 (Figure 7). However, melittin treatment resulted in a reduction in the FSP-1-positive cells in DDC + Mel livers. Along with the upregulation of TGF-β1 and ECM proteins, the expression level of p-Smad2 was increased by chronic DDC feeding in the DDC group (Figure 8). Melittin treatment attenuated the expression of p-Smad2 in the DDC + Mel group. Taken together, these results show that melittin might protect liver during DDC feeding by attenuating fibrotic gene expression.

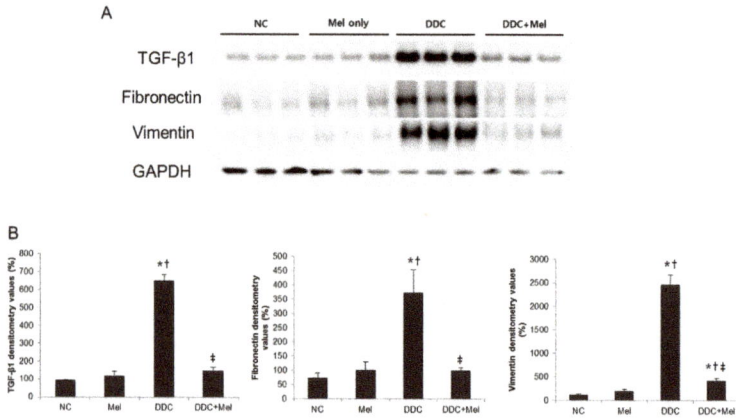

Figure 6. Melittin inhibits fibrosis-related gene expression in DDC-fed mice. (**A**) Western blotting results demonstrated that melittin effectively suppresses the expression of TGF-β1, fibronectin and vimentin; (**B**) graphical presentation of the ratio of TGF-β1, fibronectin and vimentin to GAPDH in various groups. NC, normal control group; Mel, melittin (0.1 mg/kg)-treated group with normal diet; DDC, 0.1% DDC-supplemented diet group; DDC + Mel, melittin (0.1 mg/kg)-treated group with 0.1% DDC-supplemented diet. Results are expressed as the mean ± SE of three independent determinations. * $p < 0.05$ compared to the NC group. † $p < 0.05$ compared to the Mel group. ‡ $p < 0.05$ compared to the DDC group.

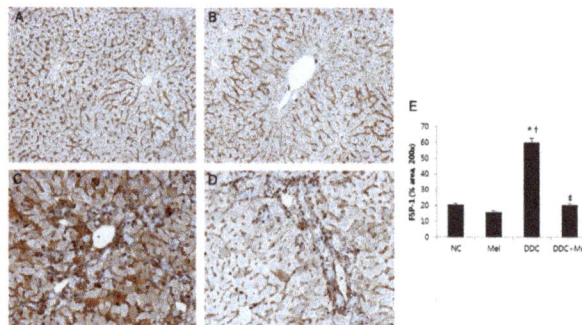

Figure 7. Melittin inhibits fibrotic changes in DDC-fed mice. Immunohistochemical staining findings demonstrated that melittin effectively suppresses the expression of FSP-1. Representative immunohistochemical images from each study group (five mice per group) (**A**) NC, normal control group; (**B**) Mel, melittin (0.1 mg/kg)-treated group with normal diet; (**C**) DDC, 0.1% DDC-supplemented diet group; (**D**) DDC + Mel, melittin (0.1 mg/kg)-treated group with 0.1% DDC-supplemented diet; magnification × 200; (**E**) morphometric assessment of the trichrome staining positive areas. Results are expressed as the mean ± SE of three independent determinations. * $p < 0.05$ compared to the NC group. † $p < 0.05$ compared to the Mel group. ‡ $p < 0.05$ compared to the DDC group.

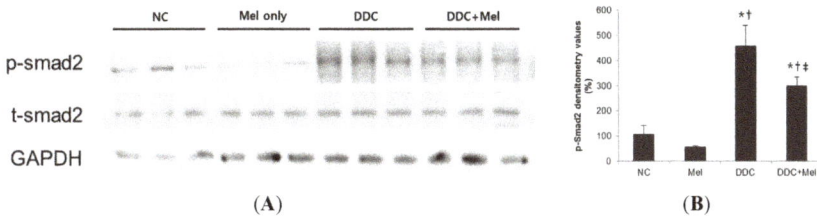

(A)   (B)

Figure 8. Melittin inhibits phosphorylation of Smad2 in DDC-fed mice. (**A**) Western blotting results demonstrated that melittin effectively suppresses the expression of p-Smad2; (**B**) graphical presentation of the ratio of p-Smad2 to GAPDH in various groups. NC, normal control group; Mel, melittin (0.1 mg/kg)-treated group with normal diet; DDC, 0.1% DDC-supplemented diet group; DDC + Mel, melittin (0.1 mg/kg)-treated group with 0.1% DDC-supplemented diet. Results are expressed as the mean $\pm$ SE of three independent determinations. * $p < 0.05$ compared to the NC group. † $p < 0.05$ compared to the Mel group. ‡ $p < 0.05$ compared to the DDC group.

2.4. Melittin Effectively Suppresses Proliferating Cholangiocytes in DDC-Fed Mice

Proliferating cholangiocytes are characteristic for DDC-fed mice and are a source of growth factors, chemokines, cytokines and other soluble factors. Cholangiocytes as specialized epithelial cells that line the biliary tree are considered as pace markers of biliary fibrosis [18]. Cholangiocyte-specific markers, including CK-7, are expressed in the epithelial cells of bile ductules in enlarged portal tracts. Chronic DDC feeding resulted in a comparable amount of ductular reaction and an increased number of CK-7- and PCNA-positive cells (Figure 9). In contrast, melittin treatment significantly suppressed the proliferation of cholangiocytes, which was shown as the expression of PCNA-positive cells in DDC + Mel livers. In summary, these data proved the ability of melittin to suppress cholangiocyte proliferation in response of DDC-induced liver injury and liver fibrosis.

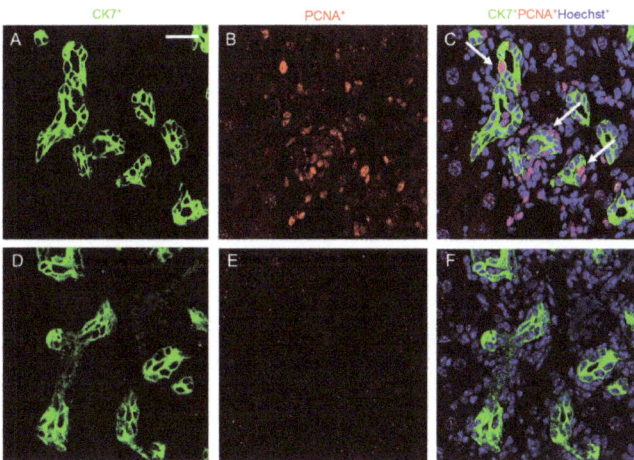

Figure 9. Melittin inhibits cholangiocyte proliferation in DDC-fed mice. Immunofluorescence staining shows co-localization of PCNA staining with CK-7 (arrow head) following DDC treatment. Immunofluorescence staining results demonstrated that melittin effectively suppresses the expression of PCNA. CK-7 and PCNA immune complexes were detected by anti-mouse FITC (green) and anti-rabbit Texas red (red). Nuclei were counterstained with Hoechst 33342 (blue). Representative immunofluorescence staining images from each study group ((**A–C**) DDC group; (**D–F**) DDC + Mel group). Scale bar = 20 μm.

3. Discussion

The bulk of the liver is occupied by parenchymal hepatocytes, but little is known about the physiology of cholangiocytes. Cholangiocytes are epithelial cells that line the biliary system and make up 3%–5% of the liver cell population. The major function of cholangiocytes is closely related to bile flow. The ability of cholangiocytes to proliferate is important in many different human pathological liver conditions that involve this cell type, and those conditions are referred to as cholangiopathies. Pro-inflammatory cytokines may be critically involved in the pathophysiology of several cholangiopathies [19,20]. Thus, this study aimed at investigating the specific effects of melittin on the pathogenesis of DDC-induced sclerosing cholangitis and biliary-type liver fibrosis.

Melittin is a major component of bee venom, which makes up 50%–60% of the dry weight of bee venom. It has been studied for its antibacterial, antiviral and anti-inflammatory properties in various cell types [21]. A low concentration of bee venom has been reported to protect TNF-α/actinomycin D-induced hepatocyte apoptosis [22]. Although melittin has lytic effects on biological and cell membranes when inserted into the phospholipid layer at a high concentration, a concentration of melittin lower than 2 µM does not disrupt the cell membrane [21,23,24]. During the progression of atherosclerosis and restenosis, melittin inhibited aortic vascular smooth muscle cell proliferation by suppressing NF-κB, Akt activation and the mitogen-activated protein kinase pathway [25]. Recently, melittin effectively suppressed skin inflammation in the *P. acnes*-induced *in vitro* and *in vivo* inflammatory models [26]. However, there have been no reports on the effects of *in vivo* cholangiopathy-associated molecular mechanisms of melittin. The current study confirmed the anti-inflammatory function of melittin as an effective inhibitor of inflammatory cytokines and biliary fibrosis. This study demonstrated that melittin reduced the ductular reaction in DDC-induced liver injury, suggesting a potential therapeutic use of this compound in the treatment of cholangitis-related liver fibrosis.

Xenobiotics, which are foreign to an organism, include chemical compounds, detergents and pharmaceuticals. Liver plays a major role in the detoxification and elimination of xenobiotics. DDC feeding is widely used to study xenobiotic-induced liver injury. DDC-induced liver injury is associated with chronic cholestatic liver diseases, which are further related to the induction of a reactive phenotype of bile duct epithelial cells with the development of bile duct injury, leading to portal-portal fibrosis and large duct disease [8,27].

To further evaluate the function of melittin in liver fibrosis, this study used a DDC-induced liver fibrosis model. Chronic DDC feeding elevated the hepatic inflammatory responses in portal fields and hepatocyte injuries and increased collagen deposition and periductal portal-portal fibrosis. Increased serum AST and ALT are known to be major risk factors related to the development of chronic liver disease. Especially, the elevations of serum AP and bilirubin are well-known parameters for cholestatic liver disease. The reductions of serum AP and bilirubin play a major role in mediating the repression of biliary disease [28,29]. Chronic DDC feeding in mice increased the serum levels of AST, ALT, AP and bilirubin. Of particular interest, this study showed that treatment with melittin appeared to decrease AP and bilirubin concentrations in the serum of DDC-fed mice. Elevations of serum AST, ALT, AP and bilirubin, from liver metabolic disorder, play important roles in the initiation of liver fibrosis, and liver metabolic disorder was affected by pro-inflammatory cytokines [30].

Pro-inflammatory cytokines and chemokines have an important role in the initiation and perpetuation of various types of liver fibrosis. Although a previous study has demonstrated that melittin has anti-inflammatory and anti-fibrotic activity in thioacetamide-induced liver fibrosis [31], the precise mechanisms of action of melittin in cholangiopathies remain to be elucidated. TNF-α is a key molecule in the hepatic inflammatory response, the execution of apoptosis and the regulation of liver regeneration. Moreover, the TNF superfamily may represent major players in the immunobiology of sclerosing cholangitis and associated biliary fibrosis [20,32]. A recent study showed that genetic loss of TNFR1 significantly affects the pathogenesis of DDC-induced sclerosing cholangitis and ductular reaction. Consistent with these results, DDC-induced injury led to increased production of TNF-α expression. Along with TNF-α upregulation, the expressions of IL-6, p-STAT3 and MCP-1 were also increased in DDC-fed mice. STAT3 is the major downstream signaling molecule of IL-6 in hepatocytes. The hepatoprotective function of STAT3 in the liver has been well documented in many murine models. Especially, conditionally-inactivated STAT3 in hepatocytes and cholangiocytes led to strongly aggravated hepatocellular damage and fibrosis in a sclerosing cholangitis animal model using mice lacking the multidrug resistance gene 2 (mdr2$^{-/-}$) [33]. Our present study showed that melittin effectively suppressed the expressions of TNF-α, IL-6, p-STAT3 and MCP-1 in DDC-fed mice. These results demonstrate that melittin mediates the anti-inflammatory effect during the resolution of biliary fibrosis in liver.

During liver injury, periportal hepatocytes are damaged, and their proliferation is impaired. Damaged liver parenchyma become a source of regenerating hepatocytes, biliary epithelial cells and draining ductules in order to restore the functional liver mass [34,35]. When liver parenchyma is damaged, the ductular reaction progresses as an alternative pathway for liver restoration. Ductular reaction has been regarded as a barometer of portal fibrosis, since proliferating biliary epithelial cells are a source of molecules that mobilize ECM deposition and secrete pro-inflammatory cytokines and chemokines, which further activate hepatic stellate cells and portal fibroblast [36,37]. This remodelling process of liver milieu demonstrates a strong capacity to increase myofibroblast proliferation and ECM deposition, thus contributing to the fibrogenic response to liver injury [38]. Our current study investigated the question of whether melittin could affect ductular reaction during chronic liver injuries. Chronic DDC feeding led to increased ductular reaction at the portal tract interface, including small biliary ductules with bile plugs and inflammatory cells. In addition, CK-7, which is a cholangiocyte-specific epithelial cell marker, was increased in ductular reaction near the portal tract in DDC-fed mice. Furthermore, proliferating cholangiocytes were also increased by chronic DDC feeding. However, melittin treatment withdrew ductular reactions and cholangiocyte proliferation in DDC-fed mice.

Several studies have demonstrated that proliferating cholangiocytes secrete profibrogenic factors. During biliary fibrosis, proliferating bile duct epithelial cells are the predominant source of the profibrogenic gene, such as connective tissue growth factor (CTGF) [39]. Moreover, epithelial cells of newly-formed bile ducts express mRNA for α1 (IV) procollagen, suggesting that proliferating cholangiocytes are a source of hepatic collagen during fibrosis [40]. The expression of TGF-β1, which is a key upstream signaling molecule of CTGF and the major fibrogenic cytokine in liver fibrosis, was increased in chronic DDC-induced liver injury. Following TGF-β1 upregulation, the expressions of fibronectin, vimentin, FSP-1 (a key marker of early fibroblast lineage) and p-Smad2 were also increased in DDC-fed mice. In contrast, melittin markedly reduced the responses induced by DDC in mice. In fact, melittin treatment downregulated the expression of matrix components during biliary fibrosis.

Overall, this study demonstrated the protective effects of melittin on DDC-induced biliary fibrosis *in vivo*. Treatment of melittin markedly decreased the expression of inflammatory cytokines in DDC-fed mice. Melittin also suppressed biliary fibrosis by attenuating the expression of fibrogenic cytokines and ECM proteins. Moreover, melittin modulated ductular reaction and subsequent fibrosis. These results collectively suggest that melittin may regulate the remodelling process, which involves crosstalk between mesenchymal cells and cholangiocytes in biliary fibrosis. In summary, these findings expand the role for melittin in regulating the expression of ECM and begin to provide insight into its regulatory involvement in liver fibrosis-related pathologies.

4. Experimental Section

4.1. Animal Model

For the induction of liver injury, 6–8-week-old C57BL/6 male mice (20–25 g; Samtako, Kyungki do, Korea) were used. Mice were fed with a 0.1% DDC-supplemented diets for 4 weeks, housed with a 12-h light/dark cycle and permitted *ad libitum* consumption of water. All animal protocols were approved by the Institutional Animal Care and Use Committee of Catholic University of Daegu (Daegu, Korea). Mice were randomly divided into four groups as follows: (1) untreated group (normal control, NC); (2) melittin-treated group with normal diet (Mel); (3) 0.1% DDC-supplemented diet group (DDC); (4) melittin-treated group with 0.1% DDC-supplemented diet (DDC + Mel). The Mel only group and the DDC + Mel group received intraperitoneal injection of melittin (0.1 mg/kg; Sigma-Aldrich, St. Louis, MO, USA) dissolved in saline twice a week. Mice were sacrificed after 4 weeks of treatment, and the livers were removed.

4.2. Histopathology and Immunohistochemistry

Small pieces of liver from each lobe were kept in 10% formalin solution. Paraffin-embedded liver tissues were sectioned and stained with H&E and Masson's trichrome according to the standard protocol. For immunohistochemistry, sections were incubated with anti-monocyte chemoattractant protein (MCP)-1 and anti-fibroblast specific protein (FSP)-1. After three serial washes with phosphate-buffered saline (PBS), the sections were processed by an indirect immunoperoxidase technique using a commercial kit (LSAB kit; Dako, Glostrup, Denmark). The slides were examined with an Eclipse 80i microscope (Nikon, Tokyo, Japan) and analysed with iSolution DT software (IMT i-Solution, Coquitlam, BC, Canada).

4.3. Serum Biochemical Analysis

Serum samples were stored at $-70\ ^{\circ}$C until analysed using a QuantiChrom™ kit (BioAssay Systems, Atlanta, GA, USA) for alanine transaminase (ALT), aspartate aminotransferase (AST), alkaline phosphatase (AP) and bilirubin.

4.4. Western Blot Analysis

Liver tissues were homogenized in radioimmunoprecipitation assay (RIPA) buffer (Cell Signaling Technology, Danvers, MA, USA) for 15 min on ice and centrifuged at 12,000 rpm for 15 min at 4 $^{\circ}$C. The supernatant was collected, and the residual protein concentration was measured by the Bradford protein assay (Bio-Rad Laboratories, Berkeley, CA, USA). Sodium dodecyl sulphate polyacrylamide gel electrophoresis was performed with 8%–12% polyacrylamide gels at 100 V for three hours. The resolved proteins were transferred from the gel onto a polyvinylidene fluoride (PVDF) membrane (Millipore Corporation, Bedford, MA, USA) and probed with anti-TNF-α, anti-fibronectin, anti-IL-6 (Abcam, Cambridge, UK), anti-t-Smad 2 (Santa Cruz Biotechnology, Dallas, TX, USA), anti-TGF-β1 (R&D Systems, Minneapolis, MN, USA), anti-fibronectin, anti-vimentin (BD Biosciences, San Jose, CA, USA), anti-p-Smad2 (Novus Biologicals, Littleton, CO, USA), anti-p-signal transducer and activator of transcription (STAT)3 and anti-glyceraldehyde-3-phosphate dehydrogenase (GAPDH) (Cell Signaling, Danvers, MA, USA), followed by secondary antibody conjugated to horseradish peroxidase (1:2000) and detected with enhanced chemiluminescence reagents (Amersham Bioscience, Piscataway, NJ, USA). Signal intensity was quantified by an image analyser (Las3000; Fuji, Tokyo, Japan).

4.5. Immunofluorescence Staining and Confocal Microscopy

Paraffin-embedded mouse liver sections (3-μm thickness) were prepared by a routine procedure. After blocking with 10% donkey serum for 30 min, the slides were immunostained with primary antibodies against proliferating cell nuclear antigen (PCNA, Santa Cruz Biotechnology, Dallas, TX, USA) and cytokeratin-(CK)-7 (Millipore, Darmstadt, Germany). To visualize the primary antibodies, sections were stained with secondary antibodies conjugated with FITC or Texas red (Invitrogen, Carlsbad, CA, USA). Sections were then counterstained with Hoechst 33342. Stained slides were viewed under a Nikon A1 microscope equipped with a digital camera (Nikon, Tokyo, Japan).

4.6. Statistical Analyses

Data are presented as the mean ± SE. Student's *t*-test was used to assess the significance of independent experiments. The criterion $p < 0.05$ was used to determine statistical significance.

5. Conclusions

These data expand the role for melittin in regulating the expression of ECM and begin to provide insight into its regulation by liver fibrosis and related pathologies, PSC and biliary fibrosis.

Acknowledgments: This work was supported by the National Research Foundation of Korea Grant funded by the Korean Government (NRF-2014R1A1A2008955).

Author Contributions: Kyung Hyun Kim and Kwan Kyu Park designed the study and prepared the manuscript. Hyun-Jung Sung, Woo Ram Lee, Hyun Jin An, Jung Yeon Kim and Sang Mi Han performed the overall experiments. Sok Cheon Pak discussed the study. All authors have read and approved the final version of this manuscript.

Conflicts of Interest: The authors declare no conflict of interest.

References

1. Lazaridis, K.N.; Strazzabosco, M.; Larusso, N.F. The cholangiopathies: Disorders of biliary epithelia. *Gastroenterology* **2004**, *127*, 1565–1577. [CrossRef] [PubMed]
2. Patel, T.; Gores, G.J. Apoptosis and hepatobiliary disease. *Hepatology* **1995**, *21*, 1725–1741. [PubMed]
3. Peters, M.G. Pathogenesis of primary biliary cirrhosis, primary sclerosing cholangitis, and autoimmune cholangiopathy. *Clin. Liver Dis.* **1998**, *2*, 235–247, vii–viii. [CrossRef]
4. Glaser, S.S.; Gaudio, E.; Miller, T.; Alvaro, D.; Alpini, G. Cholangiocyte proliferation and liver fibrosis. *Expert Rev. Mol. Med.* **2009**, *11*, e7. [CrossRef] [PubMed]
5. LeSage, G.; Glaser, S.; Alpini, G. Regulation of cholangiocyte proliferation. *Liver* **2001**, *21*, 73–80. [CrossRef] [PubMed]
6. Alvaro, D.; Mancino, M.G.; Glaser, S.; Gaudio, E.; Marzioni, M.; Francis, H.; Alpini, G. Proliferating cholangiocytes: A neuroendocrine compartment in the diseased liver. *Gastroenterology* **2007**, *132*, 415–431. [CrossRef] [PubMed]
7. Sell, S. Comparison of liver progenitor cells in human atypical ductular reactions with those seen in experimental models of liver injury. *Hepatology* **1998**, *27*, 317–331. [CrossRef] [PubMed]
8. Fickert, P.; Stoger, U.; Fuchsbichler, A.; Moustafa, T.; Marschall, H.U.; Weiglein, A.H.; Tsybrovskyy, O.; Jaeschke, H.; Zatloukal, K.; Denk, H.; *et al.* A new xenobiotic-induced mouse model of sclerosing cholangitis and biliary fibrosis. *Am. J. Pathol.* **2007**, *171*, 525–536. [CrossRef] [PubMed]
9. Zatloukal, K.; Stumptner, C.; Fuchsbichler, A.; Fickert, P.; Lackner, C.; Trauner, M.; Denk, H. The keratin cytoskeleton in liver diseases. *J. Pathol.* **2004**, *204*, 367–376. [CrossRef] [PubMed]
10. Habermann, E. Bee and wasp venoms. *Science* **1972**, *177*, 314–322. [CrossRef] [PubMed]
11. Dempsey, C.E. The actions of melittin on membranes. *Biochim. Biophys. Acta* **1990**, *1031*, 143–161. [CrossRef]
12. Jang, H.S.; Kim, S.K.; Han, J.B.; Ahn, H.J.; Bae, H.; Min, B.I. Effects of bee venom on the pro-inflammatory responses in raw264.7 macrophage cell line. *J. Ethnopharmacol.* **2005**, *99*, 157–160. [CrossRef] [PubMed]
13. Yin, M.; Wheeler, M.D.; Kono, H.; Bradford, B.U.; Gallucci, R.M.; Luster, M.I.; Thurman, R.G. Essential role of tumor necrosis factor alpha in alcohol-induced liver injury in mice. *Gastroenterology* **1999**, *117*, 942–952. [CrossRef]
14. Burra, P.; Hubscher, S.G.; Shaw, J.; Elias, E.; Adams, D.H. Is the intercellular adhesion molecule-1/leukocyte function associated antigen 1 pathway of leukocyte adhesion involved in the tissue damage of alcoholic hepatitis? *Gut* **1992**, *33*, 268–271. [CrossRef] [PubMed]
15. Taub, R. Liver regeneration: From myth to mechanism. *Nat. Rev. Mol. Cell Biol.* **2004**, *5*, 836–847. [CrossRef] [PubMed]
16. Wang, H.; Lafdil, F.; Kong, X.; Gao, B. Signal transducer and activator of transcription 3 in liver diseases: A novel therapeutic target. *Int. J. Biol. Sci.* **2011**, *7*, 536–550. [CrossRef] [PubMed]
17. Meindl-Beinker, N.M.; Dooley, S. Transforming growth factor-beta and hepatocyte transdifferentiation in liver fibrogenesis. *J. Gastroenterol. Hepatol.* **2008**, *23*, S122–S127. [CrossRef] [PubMed]

18. Penz-Osterreicher, M.; Osterreicher, C.H.; Trauner, M. Fibrosis in autoimmune and cholestatic liver disease. *Best Pract. Res. Clin. Gastroenterol.* **2011**, *25*, 245–258. [CrossRef] [PubMed]

19. Aron, J.H.; Bowlus, C.L. The immunobiology of primary sclerosing cholangitis. *Semin. Immunopathol.* **2009**, *31*, 383–397. [CrossRef] [PubMed]

20. Aoki, C.A.; Bowlus, C.L.; Gershwin, M.E. The immunobiology of primary sclerosing cholangitis. *Autoimmun. Rev.* **2005**, *4*, 137–143. [CrossRef] [PubMed]

21. Raghuraman, H.; Chattopadhyay, A. Melittin: A membrane-active peptide with diverse functions. *Biosci. Rep.* **2007**, *27*, 189–223. [CrossRef] [PubMed]

22. Park, J.H.; Kim, K.H.; Kim, S.J.; Lee, W.R.; Lee, K.G.; Park, K.K. Bee venom protects hepatocytes from tumor necrosis factor-alpha and actinomycin d. *Arch. Pharm. Res.* **2010**, *33*, 215–223. [CrossRef] [PubMed]

23. Hider, R.C. Honeybee venom: A rich source of pharmacologically active peptides. *Endeavour* **1988**, *12*, 60–65. [CrossRef]

24. Pratt, J.P.; Ravnic, D.J.; Huss, H.T.; Jiang, X.; Orozco, B.S.; Mentzer, S.J. Melittin-induced membrane permeability: A nonosmotic mechanism of cell death. *Vitro Cell. Dev. Biol. Anim.* **2005**, *41*, 349–355. [CrossRef] [PubMed]

25. Son, D.J.; Ha, S.J.; Song, H.S.; Lim, Y.; Yun, Y.P.; Lee, J.W.; Moon, D.C.; Park, Y.H.; Park, B.S.; Song, M.J.; *et al.* Melittin inhibits vascular smooth muscle cell proliferation through induction of apoptosis via suppression of nuclear factor-kappab and akt activation and enhancement of apoptotic protein expression. *J. Pharmacol. Exp. Ther.* **2006**, *317*, 627–634. [CrossRef] [PubMed]

26. Lee, W.R.; Kim, K.H.; An, H.J.; Kim, J.Y.; Chang, Y.C.; Chung, H.; Park, Y.Y.; Lee, M.L.; Park, K.K. The protective effects of melittin on propionibacterium acnes-induced inflammatory responses in vitro and in vivo. *J. Investig. Dermatol.* **2014**, *134*, 1922–1930. [CrossRef] [PubMed]

27. Osterreicher, C.H.; Trauner, M. Xenobiotic-induced liver injury and fibrosis. *Expert Opin. Drug Metab. Toxicol.* **2012**, *8*, 571–580. [CrossRef] [PubMed]

28. Johnston, D.E. Special considerations in interpreting liver function tests. *Am. Family Phys.* **1999**, *59*, 2223–2230.

29. European Association for the Study of the Liver. Easl clinical practice guidelines: Management of cholestatic liver diseases. *J. Hepatol.* **2009**, *51*, 237–267. [CrossRef] [PubMed]

30. Marchesini, G.; Brizi, M.; Bianchi, G.; Tomassetti, S.; Bugianesi, E.; Lenzi, M.; McCullough, A.J.; Natale, S.; Forlani, G.; Melchionda, N. Nonalcoholic fatty liver disease: A feature of the metabolic syndrome. *Diabetes* **2001**, *50*, 1844–1850. [CrossRef] [PubMed]

31. Park, J.H.; Kum, Y.S.; Lee, T.I.; Kim, S.J.; Lee, W.R.; Kim, B.I.; Kim, H.S.; Kim, K.H.; Park, K.K. Melittin attenuates liver injury in thioacetamide-treated mice through modulating inflammation and fibrogenesis. *Exp. Biol. Med. (Maywood)* **2011**, *236*, 1306–1313. [CrossRef] [PubMed]

32. Adams, D.H.; Afford, S.C. The role of cholangiocytes in the development of chronic inflammatory liver disease. *Front. Biosci.* **2002**, *7*, e276–e285. [PubMed]

33. Mair, M.; Zollner, G.; Schneller, D.; Musteanu, M.; Fickert, P.; Gumhold, J.; Schuster, C.; Fuchsbichler, A.; Bilban, M.; Tauber, S.; *et al.* Signal transducer and activator of transcription 3 protects from liver injury and fibrosis in a mouse model of sclerosing cholangitis. *Gastroenterology* **2010**, *138*, 2499–2508. [CrossRef] [PubMed]

34. Richardson, M.M.; Jonsson, J.R.; Powell, E.E.; Brunt, E.M.; Neuschwander-Tetri, B.A.; Bhathal, P.S.; Dixon, J.B.; Weltman, M.D.; Tilg, H.; Moschen, A.R.; *et al.* Progressive fibrosis in nonalcoholic steatohepatitis: Association with altered regeneration and a ductular reaction. *Gastroenterology* **2007**, *133*, 80–90. [CrossRef] [PubMed]

35. Theise, N.D.; Badve, S.; Saxena, R.; Henegariu, O.; Sell, S.; Crawford, J.M.; Krause, D.S. Derivation of hepatocytes from bone marrow cells in mice after radiation-induced myeloablation. *Hepatology* **2000**, *31*, 235–240. [CrossRef] [PubMed]

36. Alvaro, D.; Invernizzi, P.; Onori, P.; Franchitto, A.; De Santis, A.; Crosignani, A.; Sferra, R.; Ginanni-Corradini, S.; Mancino, M.G.; Maggioni, M.; *et al.* Estrogen receptors in cholangiocytes and the progression of primary biliary cirrhosis. *J. Hepatol.* **2004**, *41*, 905–912. [CrossRef] [PubMed]

37. Hsieh, C.S.; Huang, C.C.; Wu, J.J.; Chaung, H.C.; Wu, C.L.; Chang, N.K.; Chang, Y.M.; Chou, M.H.; Chuang, J.H. Ascending cholangitis provokes il-8 and mcp-1 expression and promotes inflammatory cell infiltration in the cholestatic rat liver. *J. Pediatr. Surg.* **2001**, *36*, 1623–1628. [CrossRef] [PubMed]

38. Friedman, S.L. Mechanisms of hepatic fibrogenesis. *Gastroenterology* **2008**, *134*, 1655–1669. [CrossRef] [PubMed]
39. Sedlaczek, N.; Jia, J.D.; Bauer, M.; Herbst, H.; Ruehl, M.; Hahn, E.G.; Schuppan, D. Proliferating bile duct epithelial cells are a major source of connective tissue growth factor in rat biliary fibrosis. *Am. J. Pathol.* **2001**, *158*, 1239–1244. [CrossRef]
40. Milani, S.; Herbst, H.; Schuppan, D.; Kim, K.Y.; Riecken, E.O.; Stein, H. Procollagen expression by nonparenchymal rat liver cells in experimental biliary fibrosis. *Gastroenterology* **1990**, *98*, 175–184. [PubMed]

© 2015 by the authors; licensee MDPI, Basel, Switzerland. This article is an open access article distributed under the terms and conditions of the Creative Commons Attribution (CC-BY) license (http://creativecommons.org/licenses/by/4.0/).

Article

Molecular Cloning and Functional Studies of Two Kazal-Type Serine Protease Inhibitors Specifically Expressed by *Nasonia vitripennis* Venom Apparatus

Cen Qian [1,2], Qi Fang [2], Lei Wang [1,2] and Gong-Yin Ye [2,*]

[1]   College of Life Science, Anhui Agricultural University, Hefei 230036, China; qiancenqiancen@163.com (C.Q.); wanglei20041225@163.com (L.W.)
[2]   State Key Laboratory of Rice Biology, Key Laboratory of Agricultural Entomology, Institute of Insect Sciences, Zhejiang University, Hangzhou 310058, China; fangqi@zju.edu.cn
*   Author to whom correspondence should be addressed; chu@zju.edu.cn; Tel./Fax: +86-571-8898-2696.

Academic Editors: Sokcheon Pak and Ren Lai
Received: 13 May 2015; Accepted: 27 July 2015; Published: 4 August 2015

Abstract: Two cDNA sequences of Kazal-type serine protease inhibitors (KSPIs) in *Nasonia vitripennis*, *NvKSPI-1* and *NvKSPI-2*, were characterized and their open reading frames (ORFs) were 198 and 264 bp, respectively. Both *NvKSPI-1* and *NvKSPI-2* contained a typical Kazal-type domain. Real-time quantitative PCR (RT-qPCR) results revealed that *NvKSPI-1* and *NvKSPI-2* mRNAs were mostly detected specifically in the venom apparatus, while they were expressed at lower levels in the ovary and much lower levels in other tissues tested. In the venom apparatus, both *NvKSPI-1* and *NvKSPI-2* transcripts were highly expressed on the fourth day post eclosion and then declined gradually. The *NvKSPI-1* and *NvKSPI-2* genes were recombinantly expressed utilizing a pGEX-4T-2 vector, and the recombinant products fused with glutathione S-transferase were purified. Inhibition of recombinant GST-*NvKSPI-1* and GST-*NvKSPI-2* to three serine protease inhibitors (trypsin, chymotrypsin, and proteinase K) were tested and results showed that only *NvKSPI-1* could inhibit the activity of trypsin. Meanwhile, we evaluated the influence of the recombinant GST-*NvKSPI-1* and GST-*NvKSPI-2* on the phenoloxidase (PO) activity and prophenoloxidase (PPO) activation of hemolymph from a host pupa, *Musca domestica*. Results showed PPO activation in host hemolymph was inhibited by both recombinant proteins; however, there was no significant inhibition on the PO activity. Our results suggested that *NvKSPI-1* and *NvKSPI-2* could inhibit PPO activation in host hemolymph and trypsin activity *in vitro*.

Keywords: *Nasonia vitripennis*; Kazal-type; serine protease inhibitors; humoral immunity

1. Introduction

The Kazal-type serine protease inhibitors (KSPIs) comprise a large family of protease inhibitors. They are present widely in mammals, birds, crayfish, and insects and are named in reference to the work on the pancreatic secretory trypsin inhibitor first isolated by Kazal *et al.* [1]. During the 1950s–1980s, KSPIs were explosively studied in vertebrates, particularly mammals and birds [2]. Studies on KSPIs from invertebrates began in the 1990s when Friedrich *et al.* first reported a double-headed Kazal-type thrombin inhibitor, rhodniin, from *Rhodnius prolixus* (Hemiptera: Reduviidae) [3]. In 1994, a four-domain Kazal protease inhibitor from the blood cells of crayfish *Pacifastacus leniusculus* (Crustacea: Decapoda) [4] and a leech-derived tryptase inhibitor from the medicinal leech *Hirudo medicinalis* (Hirudinea: Hirudinidae) [5] were documented, respectively. After that, studies on invertebrate KSPIs have extended to other invertebrate species including shrimp, blood-sucking insects, silk moths, locusts, and so on [6–12].

KSPIs have the conserved structures of one or more Kazal domains (KDs). A typical KD is composed of 40–60 amino acids including six cysteine residues and the following conservative motif: C1-X(1-7)-C2-X(5)-PVC3-X(4)-TYXNXC4-X(2-6)-C5-X(9-16)-C6. These six cysteine residues formed three intra-domain disulfide bridges between cysteine numbers 1–5, 2–4, and 3–6, resulting in a characteristic three-dimensional structure [13]. So far, hundreds of KSPIs with various functions have been reported [14]. KSPIs are involved in many important physiological processes, such as embryogenesis, development, excessive autophagy, microbial invasion, inflammation, and immune responses [15–19]. Native KSPIs from blood-feeding arthropods can inhibit trypsin, thrombin, elastase, chymotrypsin, plasmin, subtilisin A, and factor XIIa [3,20–23].

Parasitic wasps are natural resources that play an important role in biological control. They lay eggs into hosts or on the surface of hosts along with maternal and embryonic factors such as venom, polydnavirus (PDV), virus-like particles (VLP), ovarian proteins, teratocytes, and proteins secreted from larvae to interfere with the host immune responses for successful parasitization [24–26]. Unlike ichneumonid and braconid parasitoids, *Nasonia vitripennis* (Hymenoptera: Pteromalidae) injects venom, but not PDVs, into its host after oviposition. *N. vitripennis* venom is responsible for multiple functions in regulating the physiological processes of its host including induction of pathological and ultrastructural changes in cultured cells, interfering with the cellular immunity of host hemocytes, causing cell death, stimulation of intracellular calcium release in cultured cells, and disruption of the pupariation and eclosion behavior of the host [27–29]. Using proteomics methods, De Graaf *et al.* [30] previously identified 79 venom proteins from *N. vitripennis*, two of which are KSPIs (NCBI accession numbers NM_001161523 and NM_001170879). Recently, our group determined the transcriptome and proteome of venom gland and other residual tissues in *N. vitripennis* by RNA-seq (RNA sequencing) and LC-MS/MS (liquid chromatography–tandem mass spectrometry) methods [31], and we also detected two KSPIs in venom, as described by de Graaf *et al.* [30].

N. vitripennis is not only an advantageous ectoparasitoid to flies but an ideal perfect model insect for genetic and developmental biology studies [32,33]. In addition, genome sequences and comparative analyses have been reported for *N. vitripennis* and two other closely related parasitoid wasps, *N. giraulti*, and *N. longicornis* (Hymenoptera: Pteromalidae) [34]. While these reports document progress in understanding the composition of *N. vitripennis* venom, information on the activity and functional molecular mechanisms of venom actions are still lacking.

Here, we molecularly characterized two KSPIs (*NvKSPI-1* and *NvKSPI-2*) in *N. vitripennis* and determined their tissue and developmental expression patterns. We also tested the inhibition of recombinant *NvKSPI-1* and *NvKSPI-2* on three serine protease inhibitors' *in vitro* and PO activity and PPO activation of host hemocytes. These results will provide further insight into the role of KSPIs in insects, especially in parasitoid wasps.

2. Results

2.1. Molecular Characterization of NvKSPI-1 and NvKSPI-2

Fragments of *NvKSPI-1* and *NvKSPI-2* containing 198 and 264 nucleotides were amplified and sequenced; they respectively encoded 65 and 87 amino acids with predicted secretory N-terminal signal peptides (Figure 1). BLASTn results showed that these two cDNA sequences were completely consistent with *N. vitripennis* KSPIs in the NCBI database (NM_001161523 and NM_001170879). A multiple sequence alignment of KDs from *NvKSPI-1*, *NvKSPI-2*, and other KSPIs showed that although the numbers of KDs in these species varied from one to seven, typical KD motifs were highly similar, including the six cysteine residues that formed disulfide bonds between cysteine numbers 1–5, 2–4, and 3–6 (Figure 2). Phylogenetic analyses of *NvKSPI-1* and *NvKSPI-2* with homologs in 12 other species using their mature peptide region sequences indicated that *NvKSPI-1* and *NvKSPI-2* were not classified into the same cluster and showed close genetic distances with *Danaus plexippus* (Lepidoptera: Danaidae) and *Panstrongylus megistus* (Hemiptera: Reduviidae), respectively (Figure 3). In Diptera insects, *Lutzomyia longipalpis* (Diptera: Psychodidae) and *Phlebotomus papatasi* (Diptera: Psychodidae) are classified into one cluster, while *Glossina morsitans* (Diptera, Glossinidae) and *Stomoxys calcitrans* (Diptera: Muscidae) are classified into another cluster. In general, *NvKSPI-1* and *NvKSPI-2* were genetically closer to insects, and farther from crustaceans and vertebrates.

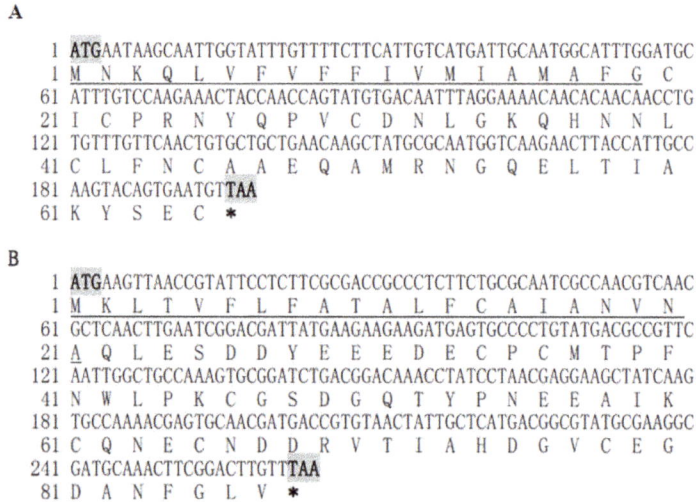

A

```
  1 ATGAATAAGCAATTGGTATTTGTTTTCTTCATTGTCATGATTGCAATGGCATTTGGATGC
  1 M   N   K   Q   L   V   F   V   F   F   I   V   M   I   A   M   A   F   G   C
 61 ATTTGTCCAAGAAACTACCAACCAGTATGTGACAATTTAGGAAAACAACACAACAACCTG
 21 I   C   P   R   N   Y   Q   P   V   C   D   N   L   G   K   Q   H   N   N   L
121 TGTTTGTTCAACTGTGCTGCTGAACAAGCTATGCGCAATGGTCAAGAACTTACCATTGCC
 41 C   L   F   N   C   A   A   E   Q   A   M   R   N   G   Q   E   L   T   I   A
181 AAGTACAGTGAATGTTAA
 61 K   Y   S   E   C   *
```

B

```
  1 ATGAAGTTAACCGTATTCCTCTTCGCGACCGCCCTCTTCTGCGCAATCGCCAACGTCAAC
  1 M   K   L   T   V   F   L   F   A   T   A   L   F   C   A   I   A   N   V   N
 61 GCTCAACTTGAATCGGACGATTATGAAGAAGAAGATGAGTGCCCCTGTATGACGCCGTTC
 21 A   Q   L   E   S   D   D   Y   E   E   E   D   E   C   P   C   M   T   P   F
121 AATTGGCTGCCAAAGTGCGGATCTGACGGACAAACCTATCCTAACGAGGAAGCTATCAAG
 41 N   W   L   P   K   C   G   S   D   G   Q   T   Y   P   N   E   E   A   I   K
181 TGCCAAAACGAGTGCAACGATGACCGTGTAACTATTGCTCATGACGGCGTATGCGAAGGC
 61 C   Q   N   E   C   N   D   D   R   V   T   I   A   H   D   G   V   C   E   G
241 GATGCAAACTTCGGACTTGTTTAA
 81 D   A   N   F   G   L   V   *
```

Figure 1. Nucleotide and deduced amino sequences of *NvKSPI-1* (**A**) and *NvKSPI-2* (**B**) cDNAs from *N. vitripennis*. Signal peptides are underlined; the initiator codon "ATG" and terminator codon "TAA" are bolded and highlighted.

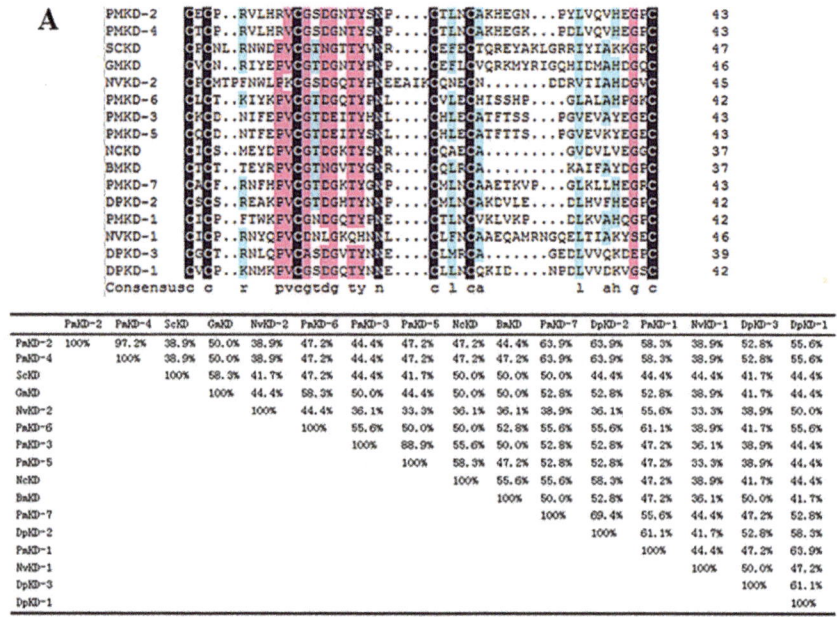

A

PMKD-2	GEGP..RVLHRVGGSDGNTYSNP....GTLNGAKHEGN...PYLVCVHEGFC	43
PMKD-4	GTCP..RVLHRVCGSDGNTYSNP....GTLNGAKHEGK...PDLVCVHEGFC	43
SCKD	GFCNL.RNWDFVCGTNGTTYVNR....CEFDGTQREYAKLGRRIYIAKKGFC	47
GMKD	GVCN..RIYEFVCGTCGNTYFNP....GEFLGVCRKMYRIGGHICMAHDGQC	46
NVKD-2	GFCMTPFNWLPKCGSDGQTYFNEEAIKCQNECN.......CDRVTIAHDGVC	45
PMKD-6	GLCT..KIYKPVCGTDGQTYFNL...CVLECHISSHP....GLALAHPGKC	42
PMKD-3	GKCD..NIFEFVCGTDEITYHNL...CHLECATFTSS...FGVEVAYEGEC	43
PMKD-5	GCCD..NTFEFVCGTDEITYSNL...CHLECATFTTS...FGVEVKYEGEC	43
NCKD	GICS..MEYDFVCGTCGKTYSNR...CQAECA.........GVCVLVEGGC	37
BMKD	GTCT..TEYRPVCGTNGVTYGNR...CQLFCA.........KAIFAYDGFC	37
PMKD-7	GACF..RNFHPVCGTDGKTYGNP...CMLNGAATKVP...GLKLLHEGRC	43
DPKD-2	GSCS..REAKPVCGTDGHTYNNP...CMLNGAKDVLE...CLHVFHEGFC	42
PMKD-1	GICP..FTWKFVCGNDGQTYFNE...CTLNGVKLVKP....DLKVAHQGFC	42
NVKD-1	GICP..RNYQFVCDNLGKQHNNL....CLFNCAAEQAMRNGQELTIAKYSEC	46
DPKD-3	GGCT..RNLQFVCASDGVTYNNE...CLMFCA........GEDLVVCKDEFC	39
DPKD-1	GVCP..KNMKFVCGSDGQTYNNE...CLLNGQKID....NPDLVVDKVGSC	42
Consensus	sc c r pvcgtdg ty n c l ca l ah g c	

	PaKD-2	PaKD-4	ScKD	GaKD	NvKD-2	PaKD-6	PaKD-3	PaKD-5	NcKD	BaKD	PaKD-7	DpKD-2	PaKD-1	NvKD-1	DpKD-3	DpKD-1
PaKD-2	100%	97.2%	38.9%	50.0%	38.9%	47.2%	44.4%	47.2%	47.2%	44.4%	63.9%	63.9%	58.3%	38.9%	52.8%	55.6%
PaKD-4		100%	38.9%	50.0%	38.9%	47.2%	44.4%	47.2%	47.2%	47.2%	63.9%	63.9%	58.3%	38.9%	52.8%	55.6%
ScKD			100%	58.3%	41.7%	47.2%	44.4%	41.7%	50.0%	50.0%	50.0%	44.4%	44.4%	44.4%	41.7%	44.4%
GaKD				100%	44.4%	58.3%	50.0%	44.4%	50.0%	50.0%	52.8%	52.8%	52.8%	38.9%	41.7%	44.4%
NvKD-2					100%	44.4%	36.1%	33.3%	36.1%	36.1%	38.9%	36.1%	55.6%	33.3%	38.9%	50.0%
PaKD-6						100%	55.6%	50.0%	50.0%	52.8%	55.6%	55.6%	61.1%	38.9%	41.7%	55.6%
PaKD-3							100%	88.9%	55.6%	50.0%	52.8%	52.8%	47.2%	36.1%	38.9%	44.4%
PaKD-5								100%	58.3%	47.2%	52.8%	52.8%	47.2%	33.3%	38.9%	44.4%
NcKD									100%	55.6%	55.6%	58.3%	47.2%	38.9%	41.7%	44.4%
BaKD										100%	50.0%	52.8%	47.2%	36.1%	50.0%	41.7%
PaKD-7											100%	69.4%	55.6%	44.4%	47.2%	52.8%
DpKD-2												100%	61.1%	41.7%	52.8%	58.3%
PaKD-1													100%	44.4%	47.2%	63.9%
NvKD-1														100%	50.0%	47.2%
DpKD-3															100%	61.1%
DpKD-1																100%

Figure 2. *Cont.*

B

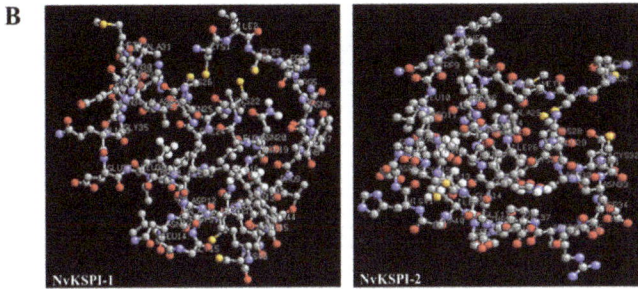

Figure 2. Sequence analysis of *NvKSPI-1* and *NvKSPI-2*. (**A**) Multiple sequences alignment and amino acid identity analysis of Kazal domains (KDs) between *NvKSPI-1*, *NvKSPI-2*, and other KSPIs. (**B**) Predicted tertiary structure of *NvKSPI-1* and *NvKSPI-2* KDs by phyre^2 on line. Yellow spheres represent cysteines forming three intra-domain disulfide bridges between cysteine numbers 1–5, 2–4, and 3–6. The corresponding species of abbreviations and their GenBank accession numbers are as follows: PMKD-1,2,3,4,5,6,7: *Panstrongylus megistuss* (ADF97836); SCKD: *Stomoxys calcitrans* (AAY98015); GMKD: *Glossina morsitans* (AFG28187); NCKD: *Neospora caninum* (AAM29188); BMKD: *Bombyx mori* (NP_001037047); DPKD-1,2,3: *Danaus plexippus* (EHJ76238); NVKVD1 and NVKVD2: *Nasonia vitripennis* (NM_001161523 and NM_001170879).

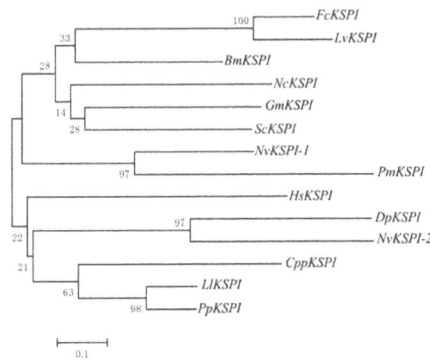

Figure 3. Phylogenetic analysis of *NvKSPI-1*, *NvKSPI-2*, and other KSPIs' amino acid sequences based on the neighbor-joining method. The origin of amino acid sequences and their GenBank accession numbers are as follows: *FcKSPI*: *Fenneropenaeus chinensis* (ABC33915); *LvKSPI*: *Litopenaeus vannamei* (AAT09421); *BmKSPI*: *Bombyx mori* (NP_001037047); *NcKSPI*: *Neospora caninum* (AAM29188); *GmKSPI*: *Glossina morsitans* (AFG28187); *ScKSPI*: *Stomoxys calcitrans* (AAY98015); *PmKSPI*: *Panstrongylus megistus* (ADF97836); *HsKSPI*: *Homo sapiens* (NP_115955); *DpKSPI*: *Danaus plexippus* (EHJ76238); *CppKSPI*: *Culex pipiens pallens* (AFN41343); *LlKSPI*: *Lutzomyia longipalpis* (ABV60319); *PpKSPI*: *Phlebotomus papatasi* (ABV44739); *NvKSPI-1* and *NvKSPI-2*: *Nasonia vitripennis* (NM_001161523 and NM_001170879).

2.2. Expression and Purification of Recombinant NvKSPI-1 and NvKSPI-2

Two recombinant proteins fused with glutathione S-transferase at their *N*-terminuses, namely GST-*NvKSPI-1* and GST-*NvKSPI-2*, were successfully detected by sodium dodecyl sulfate polyacrylamide gel electrophoresis (SDS-PAGE) with molecular weights of about 33 kDa and 36 kDa, respectively. The recombinant proteins were mainly detected in the supernatant and not in the precipitate. As soluble fusion proteins, GST-*NvKSPI-1* and GST-*NvKSPI-2* were purified with the GST•Bind™ Resin Kit (Novagen, Hilden, Germany) and the purified proteins were analyzed by SDS-PAGE (Figure 4). The protein concentrations of purified GST-*NvKSPI-1* and GST-*NvKSPI-2* were 1.8 mg/mL and 2.1mg/mL, respectively, and the concentration of the GST fusion proteins was 2.4 mg/mL.

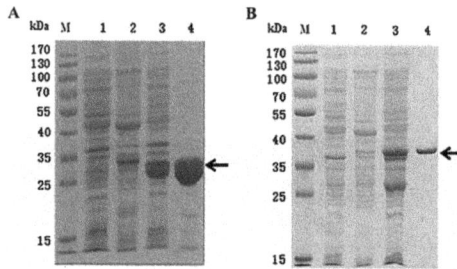

Figure 4. SDS-PAGE analysis of *NvKSPI-1* (**A**) and *NvKSPI-2* (**B**) recombinant proteins. M: Protein molecular weight marker; 1: Not induced by IPTG; 2: Precipitation after IPTG induction. 3: Supernatant after IPTG induction; 4: Purified proteins.

2.3. Transcriptional Profiles of NvKSPI-1 and NvKSPI-2 in Different Tissues and Developmental Stages

Levels of *NvKSPI-1* and *NvKSPI-2*, confirmed by genes by RT-qPCR using *18S rRNA* as the reference gene, showed similar transcript profiles (Figures 5 and 6). In tissues, *NvKSPI-1* and *NvKSPI-2* mRNAs were both expressed specifically in the venom apparatus, while they were detected at lower levels in the ovary and much lower amounts in other tissues tested. In detecting the transcript levels of *NvKSPI-1* and *NvKSPI-2* at different developmental stages (0–7 days post eclosion) in the venom apparatus of female adults, they were found to be lower at 0, 1 and 3 days post eclosion, but higher on the second day and highest on the fourth day post eclosion, after which they declined gradually. *NvKSPI-1* and *NvKSPI-2* had similar transcript profiles.

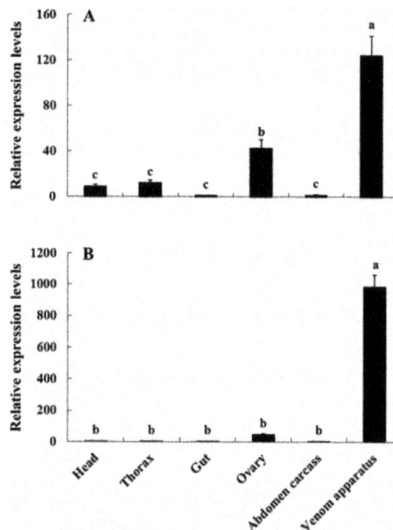

Figure 5. Tissue distribution of transcript levels of *NvKSPI-1* (**A**) and *NvKSPI-2* (**B**) in *N. vitripennis*. Abdomen carcass represents the rest of abdomen post dissection. Venom apparatus contains venom reservoir and gland, and venom released. All values in the figure are represented as mean ± standard deviation. Bars labeled with different letters are significantly different (one-way ANOVA followed by LSD test, $p < 0.05$).

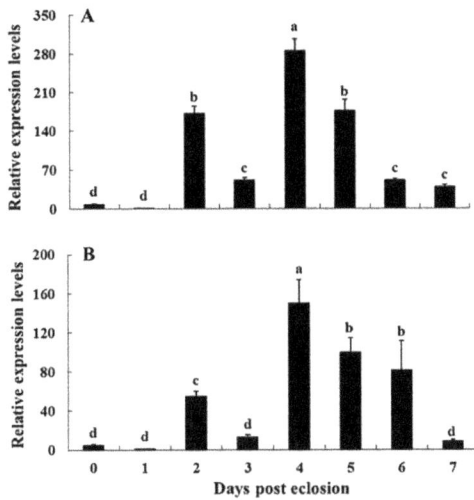

Figure 6. Developmental stages of transcript levels of *NvKSPI-1* (**A**) and *NvKSPI-2* (**B**) in venom apparatus of *N. vitripennis*. Female adults were sampled on days 0 to 7 post eclosion. All values are presented as mean ± standard deviation. Bars labeled with different letters are significantly different (one-way ANOVA followed by LSD test, $p < 0.05$).

2.4. Serine Protease Inhibition Activity of NvKSPIs

To define whether *NvKSPI-1* and *NvKSPI-2* can affect serine protease activity, we determined their enzyme activity inhibition spectrum to trypsin, chymotrypsin, and proteinase K with recombinant GST-*NvKSPI-1* and GST-*NvKSPI-2*. Results showed that only *NvKSPI-1* could inhibit the activity of trypsin, while *NvKSPI-2* did not inhibit the activity of any of the three serine proteases tested ($F = 4.159$, $p = 0.04$; $F = 0.247$, $p = 0.932$; $F = 0.552$, $p = 0.714$, respectively) (Figure 7).

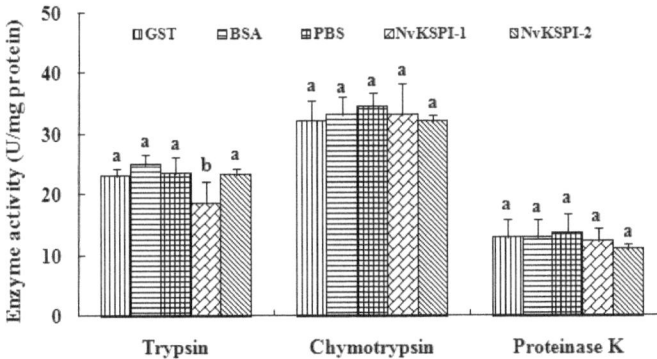

Figure 7. Effects of recombinant *NvKSPI-1* and *NvKSPI-2* on the activity of serine proteases. GST: GST-tag protein expressed by pGEX-4T-2; BSA and Buffer: treated by BSA and reaction buffer respectively; *NvKSPI-1* and *NvKSPI-2*: treated by recombination proteins respectively. All values in the figure are represented as mean ± standard deviation. Bars labeled with different letters are significantly different (one-way ANOVA followed by LSD test, $p < 0.05$).

2.5. Effects of NvKSPIs on Prophenoloxidase (PPO) Activation and Phenoloxidase (PO) Activity

To test whether *NvKSPI-1* and *NvKSPI-2* played a role in PPO activation and PO of hemolymph of host pupae, inhibition of PPO activation and PO activity were tested with recombinant GST-*NvKSPI-1* and GST-*NvKSPI-2* and compared with positive and negative controls. The PO activity between samples incubated with *NvKSPI-1*/*M. luteus* and *NvKSPI-2*/*M. luteus* were significantly different compared with the BSA and GST controls ($F = 73.255$, $p < 0.001$) (Figure 8A). The results indicated that recombinant GST-*NvKSPI-1* and GST-*NvKSPI-2* could inhibit PPO activation in the hemolymph of the host. When PPO in the hemolymph was pre-activated with *M. luteus*, recombinant GST-*NvKSPI-1* and GST-*NvKSPI-2* had no effect on PO activity ($F = 27.26$, $p < 0.001$) (Figure 8B). These results suggested that *NvKSPI-1* and *NvKSPI-2* could inhibit PPO activation in the host hemolymph but could not inhibit PO activity after PPO was activated.

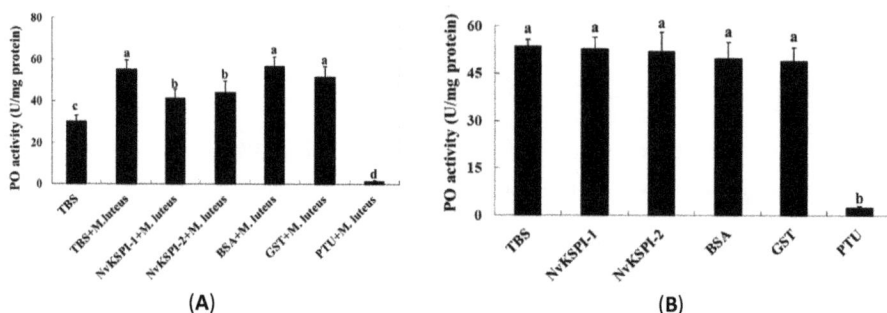

(A)   **(B)**

Figure 8. Effects of recombinant *NvKSPI-1* and *NvKSPI-2* on the PPO activation (**A**) and PO activity (**B**) of hemolymph of *M. domestica* pupae. (**A**) Screened hemolymph was incubated with TBS alone, or TBS/*M. luteus*, *NvKSPI-1*/*M. luteus*, *NvKSPI-2*/*M. luteus*, BSA/*M. luteus*, GST/*M. luteus*, PTU/*M. luteus* in TBS for 20 min at 25 °C. PPO activation was assayed using L-dopamine as a substrate, as described in Materials and Methods. (**B**) Screened hemolymph was incubated with *M. luteus* in TBS for 10 min at 25 °C, and then incubated with TBS, *NvKSPI-1*, *NvKSPI-2*, BSA, GST, or PTU for 10 min at 25 °C. PO activity was assayed using L-dopamine as a substrate, as described in Materials and Methods. All values in the figure are represented as mean ± standard deviation. Bars labeled with different letters are significantly different (one-way ANOVA followed by LSD test, $p < 0.05$).

3. Discussion

Structural studies of KSPIs have shown that they usually possess one to several conserved KDs consisting of six cysteine residues, but active sites of KSPIs have become highly variable due to long-term evolution [15]. Studies on *P. leniusculus* and *Penaeus monodon* (Crustacea: Decapoda) revealed that there were at least 26 and 20 different KDs from the hemocyte KSPIs of *P. leniusculus* and *P. monodon*, respectively. The position of the P1 site (usually the second amino acid residue after the second cysteine), a determinant for substrate specificity, varied highly [2,35]. In general, the domain with the P1 site of Arg or Lys inhibits trypsin; the P1 site of Tyr, Leu, Pro, Phe, or Met inhibits chymotrypsin; the P1 site of Ala or Ser inhibits elastase; and the P1 site of Thr or Asp inhibits proteinase K or subtilisin A.

Here, we characterized two KSPIs in *N. vitripennis* (*NvKSPI-1* and *NvKSPI-2*); both have a typical KD consisting of six cysteine residues. For *NvKSPI-1*, the P1 site is Arg, and it was supposed to inhibit the activity of trypsin but not chymotrypsin (or proteinase K). On the other hand, the P1 site of *NvKSPI-2* is Thr, and it was supposed not to inhibit the activity of our tested serine proteases (trypsin, chymotrypsin, and proteinase K). Our results with recombinant *NvKSPI-1* and *NvKSPI-2* were consistent with the general principle. As one of the most important families of protease inhibitors, KSPIs not on played a role in protecting the pancreas of vertebrates, but are also involved in many physiological processes in arthropods, such as dissolution, food digestion, blood coagulation, embryogenesis, ontogenesis, and inflammation and immune responses [16,36]. In two common shrimps, *Litopenaeus vannamei* (Crustacea: Decapoda) and *Fenneropenaeus chinensis* (Crustacea: Decapoda), KSPIs were cloned from their hemocytes' cDNA libraries, and the mRNA levels of the shrimp KSPIs were upregulated after injection of a Gram-negative marine bacterium, *Vibrio alginolyticus*, suggesting a probable role of KSPIs in the immune response [7,9]. In the black tiger shrimp, *P. monodon*, a specific KSPI, SPIPm2, did not function as a protease inhibitor but inhibited the regulatory function of WSV477. Interaction between SPIPm2 and viral protein WSV477 reduced the replication of the white spot syndrome virus [10]. In bivalves, a five-domain KSPI was identified from the pearl oyster *Pinctada fucata* (Mollusca: Bivalvia) that could inhibit chymotrypsin and trypsin activities [37]. Two KSPIs (MdSPI-1 and MdSPI-2) that are important humoral factors of innate immunity were identified from the surf clam *Mesodesma donacium* (Mollusca: Bivalvia). They were significantly upregulated at 2 and 8 h post infection with *V. anguillarum* [38]. In the coconut rhinoceros beetle, *Oryctes rhinoceros* (Coleoptera: Scarabaeidae), a single-domain KSPI was shown to play a key role in protection against bacterial infections [39]. In blood-sucking bugs *Rhodnius prolixus* (Hemiptera: Reduviidae) and *Dipetalogaster maximus* (Hemiptera: Reduviidae), synthesized double-headed KSPIs, which are highly specific for thrombin, could prevent the host blood coagulation [3,40]. In *Triatoma infestans* (Hemiptera: Reduviidae), one KSPI, factor XIIa, was found to participate in its meal acquisition and digestion, while another KSPI, Infestin 1R, was also demonstrated to be able to impair mammalian cell invasion by *Trypanosoma cruzi* [21]. In the mosquito, *Aedes aegypti* (Diptera: Culicidae), KSPIs not only acted as anticoagulants during blood feeding and digestive processes, but served other functions during the development of the mosquito [41]. In silk moths, *Bombyx mori* (Lepidoptera: Bombycidae), one three-domain KSPI was identified from the cDNA library of the pupae, and was speculated to inhibit the invasion of pathogenic microorganisms [11]. In Asian honey bees, *Apis cerana* (Hymenoptera: Apidae), a venom KSPI acting as a microbial serine protease inhibitor was identified and demonstrated to have inhibitory activity against subtilisin A and proteinase K [42]. In the desert locust, *Schistocerca gregaria* (Orthoptera: Acrididae), one KSPI specific for elastase, subtilisin, and chymotrypsin were identified from its ovary gland [12]. Therefore, KSPIs have diverse functions in insects, particularly related to insect immune responses.

Compared to other insects, few studies on KSPIs have been reported in parasitoid wasps. Only de Graaf *et al.* [30] and our group identified two cDNA sequences of KSPIs (*NvKSPI-1* and *NvKSPI-2*) from *N. vitripennis*, and our previous proteome determination of venom of *N. vitripennis* indicated that *NvKSPI-1* and *NvKSPI-2* could be secreted into the venom (unpublished), but no further studies were performed. Parasitoid wasps are important potential resources for biological control. *N. vitripennis* is an ectoparasitoid of flies and deposits its eggs along with the venom to ensure the development of the offspring by inhibiting the immune response or growth and development of the host [43–45]. Here, we presented a detailed study on the functions of *NvKSPI-1* and *NvKSPI-2* to host immunity. Our results showed that *NvKSPI-1* and *NvKSPI-2* were both expressed in *N. vitripennis* venom apparatus, in which their transcribed levels were hundreds of times higher than in other tissues tested, suggesting that *NvKSPI-1* and *NvKSPI-2* were possibly involved in inhibition of the host immune response. For the expression profiles of *NvKSPIs* in venom apparatus, the expression level of *NvKSPIs* declined at 3 days post eclosion compared to that of the second day; this might be due to the biological rhythm of female wasps. The first three days post eclosion were the main mating and oviposition periods; after oviposition peak, female wasps would re-express *NvKSPIs* for the subsequent parasitism. Of course, *NvKSPIs* might be involved in other biological functions during the eclosion period. PPO and PO activation assays of hemolymph from *M. domestica* (host) pupae showed that *NvKSPI-1* and *NvKSPI-2* were inhibitors of PPO. The PPO activation system is an important component of the immune system in arthropods, and activation of zymogen PPO to active PO involves a serine protease cascade [46,47] and is tightly regulated by serine proteases and serpins [48–50]. *N. vitripennis*, as an important ectoparasitoid, had two KSPIs (*NvKSPI-1* and *NvKSPI-2*) that were expressed specifically in venom apparatus and secreted into venom, suggesting that *NvKSPI-1* and *NvKSPI-2* played a role in repressing host melaninization through inhibition of trypsin activity and PPO activation. However, more studies are needed to investigate how *NvKSPI-1* and *NvKSPI-2* inhibit PPO activation and screen for their potential interaction factors.

4. Materials and Methods

4.1. Insect Rearing

Cultures of *Musca domestica* (Diptera: Muscidae) and *N. vitripennis* were collected from the experimental field of Zhejiang University, Hangzhou, China. The laboratory colonies of *N. vitripennis* and its host *M. domestica* were maintained as described previously [51] and used in all experiments. In brief, the host larvae were fed an artificial diet composed of 15% milk powder, 35% wheat bran, and 50% water, in 500 mL glass canning jars (at 25 ± 1 °C, light:dark = 10h:14h, relative humidity (R. H.) = 75%) until eclosion. *M. domestica* adults were maintained on a mixture of sugar and milk powder (10:1) and water, within a stainless steel-mesh cage (55 cm × 55 cm × 55 cm) under the same conditions just described. Freshly pupated hosts were exposed to mated female wasps (pupae:wasps = 1:10) in a 500 mL glass canning jar for 24 h. The parasitized pupae were maintained under the conditions just described. After emerging, the female wasps were collected and held in glass containers (also under the conditions just described) and fed *ad lib* on 20% (*v/v*) honey solution to lengthen the life span for 3–4 days until dissection of the venom reservoir and gland.

4.2. Sample Preparation

Female *N. vitripennis* from 0 to 7 days post eclosion were collected and paralyzed for 5 min at −70 °C. The head, thorax, gut, ovary, remaining abdomen carcass (the rest of the abdomen after dissection), and venom apparatus (containing venom reservoir and gland) were then collected on ice under a stereoscope (Lecia, Frankfurt, Germany). Samples were stored at −70 °C.

4.3. RNA Extraction and cDNA Cloning

The collected tissues were homogenized in liquid nitrogen, and total RNA were extracted using the TRizol reagent (Invitrogen, Carlsbad, California, USA) according to the manufacturer's instructions. RNA integrity was confirmed by ethidium bromide gel staining and RNA quantity was determined spectrophotometrically at $A_{260/280}$. Single-stranded cDNAs were synthesized by using PrimeScript™ One Step RT-PCR Kit (Takara, Dalian, China). Oligonucleotide primers (Table 1) were designed based on cDNA sequences of *N. vitripennis* KSPIs (GenBank accession numbers are NM_001161523 and NM_001170879). PCR conditions were as follows: an initial delay at 94 °C for 5 min; 34 cycles of denaturation at 94 °C for 30 min; annealing at 55 °C for 30s; and extending at 72 °C for 1 min. PCR products were fractionated in a 1% agarose gel by electrophoresis and purified with the DNA gel extraction kit (Aidlab, Beijing, China), and then cloned into the pGEM®-T easy cloning vector (Promega, Madison, Wisconsin, USA). Positive clones were selected by PCR and confirmed by sequencing at Invitrogen, Shanghai, China.

4.4. Sequence Analysis

The cDNAs and deduced amino acid sequences were analyzed by DNAStar software (version 5.02, DNAStar, Madison, Wisconsin, USA) and online Blast [52]. The signal peptides were analyzed by Signal P [53]. Multiple sequence alignments were carried out with DNAman software (Lynnon Biosoft, Quebec, QC, Canada) and Clustal W2 [54]. The phylogenetic tree was constructed by using MEGA 5.1 (Tokyo Metropolitan University, Tokyo, Japan) with the neighbor-joining (NJ) method.

Table 1. Primers used in this study.

Primers Function	Primer Name	Primer Sequence (5'-3')
	NvKSPI-1-F	TTGAGACGTGTCACCGAACA
Gene cloning	*NvKSPI* -1-R	ACTTGGCAATGGTAAGTTCTTG
	NvKSPI -2-F	ATGAAGTTAACCGTATTCCTCTTC
	NvKSPI -2-R	TTAAACAAGTCCGAAGTTTGCATC
	RT-*NvKSPI-1*-F	ACTACCAACCAGTATGTGAC
	RT-*NvKSPI-1*-R	TCACTGTACTTGGCAATGGT
RT-qPCR	RT-*NvKSPI-2*-F	CGGACGATTATGAAGAAGAAG
	RT-*NvKSPI-2*-R	ACTTGATAGCTTCCTCGTTAG
	RT-18S-F	TGGGCCGGTACGTTTACTTT
	RT-18S-R	CACCTCTAACGTCGCAATAC
	NvKSPI-1-F-B	CGCGGATCCTGCATTTGTCCAAGAAACTACC
Recombinant expression	*NvKSPI-1*-R-E	CCGGAATTCTTAACATTCACTGTACTTGGCAATG
	NvKSPI-2-F-B	CGCGGATCCCAACTTGAATCGGACGATTATGAAG
	NvKSPI-2-R-E	CCGGAATTCTTAAACAAGTCCGAAGTTTGCATCG

4.5. Protein Expression, Purification, and Antibody Preparation

The primers (Table 1) containing *BamH I* and *Xho I* restriction enzyme sites were designed to amplify the ORF cDNA sequences of *NvKSPI-1* and *NvKSPI-2* by PCR. PCR products and the pGEX-4T-2 vector were ligated after they were double digested by *BamH I* and *Xho I* (Takara, Dalian, China). The recombinant plasmids were confirmed by sequencing, transformed into *Escherichia coli* BL21 (DE3) cells (AxyGen, Shanghai, China), and induced by 1 mM isopropyl thiogalactoside (IPTG) for protein expression. The recombinant proteins were analyzed by 12% SDS-PAGE and then purified using GST•Bind™ Resin Kit (Novagen, Hilden, Germany) according to the protocols. Protein concentrations were determined by using the BCA Protein Assay Kit (Novagen, Hilden, Germany).

4.6. Real-Time Quantitative PCR (RT-qPCR)

RT-qPCR was used to determine the transcription profiles of *NvKSPI-1* and *NvKSPI-2 in different tissues and ages of N. vitripennis* females. The 18S rRNA gene (accession number GQ410677) was used as an internal control. Ten microliters of each first-strand cDNA product were diluted with 90 μL of sterilized water before use. The RT-qPCR system was performed with each 20 μL reaction containing 10 μL SsoFast EvaGreen SuperMix (Bio-Rad, Hercules, California, USA), 1 μL forward primer (200 nM), 1 μL reverse primer (200 nM), 1 μL diluted cDNA, and 7 μL sterile water. The thermal cycling conditions were 95 °C for 30 s, followed by 40 cycles of 95 °C for 5 s and 60 °C for 34 s. Amplification was monitored on the iCycleriQ ™ Real-Time PCR Detection System (Bio-Rad, Hercules, California, USA). The specificity of the SYBR-Green PCR signal was further confirmed by melting curve analysis. The experiments were repeated three times as independent biological replicates. The mRNA expression was quantified using the $2^{-\Delta\Delta Ct}$ method [55].

4.7. Serine Protease Inhibition Assays

Protease inhibition assays were performed with purified recombinant *NvKSPI-1* and *NvKSPI-2* proteins using the method described by Ling *et al.* [56]. Three typical serine proteases (bovine pancreatic trypsin, bovine pancreatic chymotrypsin, and proteinase K, 200 ng/mL) (Sigma, Taufkirchen, Germany) and their corresponding substrates (N-benzoyl-Val-Gly-Arg-*p*-nitroanilide, N-succinyl-Ala-Ala-Pro-Phe-*p*-nitroanilid, and N-succinyl-Ala-Ala-Pro-Phe-*p*-nitroanilid) (Sigma, Taufkirchen, Germany) were selected for determination of the spectrum of enzyme inhibition by recombinant *NvKSPI-1* and *NvKSPI-2* as follows. The recombinant *NvKSPI-1* and *NvKSPI-2* (1 μg each) were pre-incubated with a reaction buffer (100 mM Tris-HCl, 100 mM NaCl, 1 mM CaCl₂, pH 7.5) containing 200 ng/mL serine protease for 30 min at room temperature, and then 200 μL substrate was added (0.1 mM, 100 mM Tris-HCl, 100 mM NaCl, 1 mM CaCl₂, pH 7.5) before measuring the absorbance at 405 nm every min for 30 min. Substrates alone were used as blanks. Buffer and buffer with bovine serum albumin (BSA) were used as controls. One unit of enzyme activity was defined as an increase of absorbance by 0.001 per min. The experiments were repeated three times as independent biological replicates.

4.8. Prophenoloxidase (PPO) Activation and Phenoloxidase (PO) Activity Assays

PPO activation and PO activity assays were performed with purified recombinant *NvKSPI-1* and *NvKSPI-2* according to the method described by Ling *et al.* [48]. *M. domestica* in the white pupal stage were sterilized by 75% alcohol, rinsed with sterilized water and finally air-dried at room temperature before use. Hemolymph was collected from pupae by puncturing the pupal cuticle with a sterilized dissecting pin and rapidly transferred into a sterilized 1.5 mL Eppendorf tube. The collected hemolymph was then centrifuged at 3300 g for 5 min at 4 °C, and the supernatant was transferred into another 1.5 mL Eppendorf tube. Two microliters of each hemolymph sample were incubated with or without *Micrococcus luteus* (0.5 μg) in 10 μL of Tris buffered saline (TBS) at pH 7.4 in wells of a 96-well plate for 60 min at room temperature. Then 200 μL of L-dopamine (2 mM in TBS, pH 6.5) was added and absorbance at 470 nm was monitored. One unit of PO activity was defined as an increase of absorbance at 470 nm by 0.001 per min. Plasma samples with low PO activity when incubated alone but high PO activity after incubation with *M. luteus* were selected for subsequent PPO and PO activation assays.

For PPO activation assays, each 2 μL of prepared hemolymph was incubated with 10 μL TBS, 10 μL TBS/*M. luteus* (0.5 μg)/*NvKSPI-1* (1 μg) mixture, 10 μL TBS/*M. luteus* (0.5 μg)/*NvKSPI-2* (1 μg) mixture, 10 μL TBS/*M. luteus* (0.5 μg)/BSA (1 μg) mixture, 10 μL TBS/*M. luteus* (0.5 μg)/GST (1 μg) mixture, and 10 μL TBS/PTU (phenyl thiourea, saturated), respectively, for 20 min at 25 °C. Finally, 200 μL of L-dopamine substrate (2 mM) was added to each sample, and the PO activity was measured at 470 nm in a plate reader for 30 min. One unit of PO activity was defined as an increase of absorbance at 470 nm by 0.001 per minute.

For PO activity assays, each 2 μL of prepared hemolymph was incubated with 10 μL TBS/*M. luteus* (0.5 μg) mixture for 10 min at 25 °C. Then 2 μL TBS, 2 μL TBS, 2 μL *NvKSPI-1* (1 μg), 2 μL *NvKSPI-2* (1 μg), 2 μL BSA (1 μg), 2 μL GST (1 μg), and 2 μL PTU was added, respectively, and incubated for 10 min at 25 °C. Finally, 200 μL of L-dopamine substrate (2 mM) were added to each sample, and the PO activity was measured at 470 nm in a plate reader for 30 min. One unit of PO activity was defined as an increase of absorbance at 470 nm by 0.001 per min. All of the above experiments were repeated three times as independent biological replicates.

4.9. Statistical Analysis

All data were calculated as the mean ± standard deviation. Differences between samples were analyzed by one-way analysis of variance (ANOVA). Means were compared by least significant difference (LSD) tests. All statistical calculations were run using DPS software (version 8.01) [35] and statistical significance was set at $p < 0.05$.

5. Conclusions

In summary, we executed the molecular cloning and functional studies of *KSPI-1* and *KSPI-2* in *N. vitripennis*. We found that *KSPI-1* and *KSPI-2* are specifically expressed by *N. vitripennis* venom apparatus. The recombinant GST-*NvKSPI-1* and GST-*NvKSPI-2* can inhibit the PPO activation in host hemolymph and *NvKSPI-1* can inhibit the trypsin activity. Those findings suggest that *NvKSPI-1* and *NvKSPI-2* play a role in repressing host melaninization through inhibition of trypsin activity and PPO activation. However, more studies are needed to investigate the immune mechanism of KSPIs in insects.

Acknowledgments: This research was supported by grants from the National Program on Key Basic Research Projects (973 Program, 2013CB127600), the National Natural Science Foundation of China (Grant No. 31272098, 31472038), the Research Fund for the Doctoral Program of Higher Education of China (Grant Number: 2012010113004), the National Science Fund for Innovative Research Groups of Biological Control (Grant No. 31321063), the Zhejiang Provincial Natural Science Foundation of China (Grant Number: Y14C140006), and the Fundamental Research Funds for the Central Universities (Grant Number: 2014FZA6014).

Author Contributions: Gong-Yin Ye and Qi Fang conceived and designed the experiments; Cen Qian performed the experiments; Lei Wang analyzed the data; Cen Qian and Gong-Yin Ye wrote the paper.

Conflicts of Interest: The authors declare no conflict of interest. The founding sponsors had no role in the design of the study; in the collection, analyses, or interpretation of data; in the writing of the manuscript and in the decision to publish the results.

References

1. Kazal, L.A.; Spicer, D.S.; Brahinsky, R.A. Isolation of a crystalline trypsin inhibitor-anticoagulant protein from pancreas. *J. Am. Chem. Soc.* **1948**, *70*, 3034–3040. [CrossRef] [PubMed]

2. Laskowski, M., Jr.; Kato, I. Protein inhibitors of proteinases. *Annu. Rev. Biochem.* **1980**, *49*, 593–626. [CrossRef] [PubMed]

3. Friedrich, T.; Kroger, B.; Bialojan, S.; Lemaire, H.G.; Hoffken, H.W.; Reuschenbach, P.; Otte, M.; Dodt, J. A Kazal-type inhibitor with thrombin specificity from *Rhodnius prolixus*. *J. Biol. Chem.* **1993**, *268*, 16216–16222. [PubMed]

4. Johansson, M.W.; Keyser, P.; Soderhall, K. Purification and cDNA cloning of a four-domain Kazal proteinase inhibitor from crayfish blood cells. *Eur. J. Biochem.* **1994**, *223*, 389–394. [CrossRef] [PubMed]

5. Sommerhoff, C.P.; Sollner, C.; Mentele, R.; Piechottka, G.P.; Auerswald, E.A.; Fritz, H. A Kazal-type inhibitor of human mast cell tryptase: Isolation from the medical leech *Hirudo medicinalis*, characterization, and sequence analysis. *Biol. Chem. Hoppe. Seyler.* **1994**, *375*, 685–694. [CrossRef] [PubMed]

6. Vargas-Albores, F.; Villalpando, E. A new type of Kazal proteinase inhibitor related to shrimp *Penaeus* (Litopenaeus) *vannamei* immunity. *Fish Shellfish Immunol.* **2012**, *33*, 134–137. [CrossRef] [PubMed]

7. Jiménez-Vega, F.; Vargas-Albores, F. A four-Kazal domain protein in *Litopenaeus vannamei* hemocytes. *Dev. Comp. Immunol.* **2005**, *29*, 385–391. [CrossRef] [PubMed]

8. González, Y.; Pons, T.; Gil, J.; Besada, V.; Alonso-del-Rivero, M.; Tanaka, A.S.; Araujo, M.S.; Chávez, M.A. Characterization and comparative 3D modeling of CmPI-II, a novel "non-classical" Kazal-type inhibitor from the marine snail *Cenchritis muricatus* (Mollusca). *Biol. Chem.* **2007**, *388*, 1183–1194. [CrossRef] [PubMed]

9. Wang, Z.H.; Zhao, X.F.; Wang, J.X. Characterization, kinetics, and possible function of Kazal-type proteinase inhibitors of Chinese white shrimp, *Fenneropenaeus chinensis*. *Fish Shellfish Immunol.* **2009**, *26*, 885–897. [CrossRef] [PubMed]

10. Visetnan, S.; Donpudsa, S.; Supungul, P.; Tassanakajon, A.; Rimphanitchayakit, V. Domain 2 of a Kazal serine proteinase inhibitor SPIPm2 from *Penaeus monodon* possesses antiviral activity against WSSV. *Fish Shellfish Immunol.* **2014**, *41*, 526–530. [CrossRef] [PubMed]

11. Zheng, Q.L.; Chen, J.; Nie, Z.M.; Lv, Z.B.; Wang, D.; Zhang, Y.Z. Expression, purification and characterization of a three-domain Kazal-type inhibitor from silkworm pupae (*Bombyx mori*). *Comp. Biochem. Physiol. B Biochem. Mol. Biol.* **2007**, *146*, 234–240. [CrossRef] [PubMed]

12. Brillard-Bourdet, M.; Hamdaoui, A.; Hajjar, E.; Boudier, C.; Reuter, N.; Ehret-Sabatier, L.; Bieth, J.G.; Gauthier, F. A novel locust (*Schistocerca gregaria*) serine protease inhibitor with a high affinity for neutrophil elastase. *Biochem. J.* **2006**, *400*, 467–476. [PubMed]

13. Krowarsch, D.; Cierpicki, T.; Jelen, F.; Otlewski, J. Canonical protein inhibitors of serine proteases. *Cell Mol. Life Sci.* **2003**, *60*, 2427–2444. [CrossRef] [PubMed]

14. Christeller, J.T. Evolutionary mechanisms acting on proteinase inhibitor variability. *FEBS. J.* **2005**, *272*, 5710–5722. [CrossRef] [PubMed]

15. Rimphanitchayakit, V.; Tassanakajon, A. Structure and function of invertebrate Kazal-type serine proteinase inhibitors. *Dev. Comp. Immunol.* **2010**, *34*, 377–386. [CrossRef] [PubMed]

16. Zhu, L.; Song, L.; Chang, Y.; Xu, W.; Wu, L. Molecular cloning, characterization and expression of a novel serine proteinase inhibitor gene in bay scallops (*Argopecten irradians*, Lamarck 1819). *Fish Shellfish Immunol.* **2006**, *20*, 320–331. [CrossRef] [PubMed]

17. Kanost, M.R. Serine proteinase inhibitors in arthropod immunity. *Dev. Comp. Immunol.* **1999**, *23*, 291–301. [CrossRef]

18. Jiang, H.; Kanost, M.R. The clip-domain family of serine proteinases in arthropods. *Insect Biochem. Mol. Biol.* **2000**, *30*, 95–105. [CrossRef]

19. Di Cera, E. Serine proteases. *IUBMB Life* **2009**, *61*, 510–515. [CrossRef] [PubMed]

20. Campos, I.T.; Amino, R.; Sampaio, C.A.; Auerswald, E.A.; Friedrich, T.; Lemaire, H.G.; Schenkman, S.; Tanaka, A.S. Infestin, a thrombin inhibitor presents in *Triatoma infestans* midgut, a *Chagas* disease vector: Gene cloning, expression and characterization of the inhibitor. *Insect Biochem. Mol. Biol.* **2002**, *32*, 991–997. [CrossRef]

21. Campos, I.T.; Tanaka-Azevedo, A.M.; Tanaka, A.S. Identification and characterization of a novel factor XIIa inhibitor in the hematophagous insect, *Triatoma infestans* (Hemiptera: Reduviidae). *FEBS Lett.* **2004**, *577*, 512–516. [CrossRef] [PubMed]

22. Lovato, D.V.; de Campos, I.T.N.; Amino, R.; Tanaka, A.S. The full-length cDNA of anticoagulant protein infestin revealed a novel releasable Kazal domain, a neutrophil elastase inhibitor lacking anticoagulant activity. *Biochimie* **2006**, *88*, 673–681. [CrossRef] [PubMed]

23. Meiser, C.K.; Piechura, H.; Werner, T.; Dittmeyer-Schäfer, S.; Meyer, H.E.; Warscheid, B.; Schaub, G.A.; Balczun, C. Kazal-type inhibitors in the stomach of *Panstrongylus megistus* (Triatominae, Reduviidae). *Insect Biochem. Mol. Biol.* **2010**, *40*, 345–353. [CrossRef] [PubMed]

24. Zhu, J.Y.; Ye, G.Y.; Fang, Q.; Wu, M.L.; Hu, C. A pathogenic picorna-like from the endoparasitoid wasp, *Pteromalus puparum*: Initial discovery and partial genomic characterization. *Virus Res.* **2008**, *138*, 144–149. [CrossRef] [PubMed]

25. Dong, S.Z.; Ye, G.Y.; Zhu, J.Y.; Chen, Z.X.; Hu, C.; Liu, S. Vitellin of *Pteromalus puparum* (Hymenoptera: Pteromalidae), a pupal endoparasitoid of *Pieris rapae* (Lepidoptera: Pieridae): Biochemical characterization, temporal patterns of production and degradation. *J. Insect Physiol.* **2007**, *53*, 468–477. [CrossRef] [PubMed]

26. Dong, S.Z.; Ye, G.Y.; Hu, C. Roles of Ecdysteroid and juvenile hormone in vitellogenesis in an endoparasitic wasp, *Pteromalus puparum* (Hymenoptera: Pteromalidae). *Gen. Comp. Endoc.* **2009**, *160*, 102–108. [CrossRef] [PubMed]

27. Rivers, D.B.; Uçkan, F.; Ergin, E.; Keefer, D.A. Pathological and ultrastructural changes in cultured cells induced by venom from the ectoparasitic wasp *Nasonia vitripennis* (Walker) (Hymenoptera: Pteromalidae). *J. Insect Physiol.* **2010**, *56*, 1935–1948. [CrossRef] [PubMed]

28. Zhang, Z.; Ye, G.Y.; Cai, J.; Hu, C. Comparative venom toxicity between *Pteromalus puparum* and *Nasonia vitripennis* (Hymenoptera: Pteromalidae) toward the hemocytes of their natural hosts, non-target insects and cultured insect cells. *Toxicon* **2005**, *46*, 337–349. [CrossRef] [PubMed]

29. Dong, S.Z.; Ye, G.Y.; Yao, P.C.; Huang, Y.L.; Chen, X.X.; Shen, Z.C.; Hu, C. Effects of starvation on the vitellogenesis, ovarian development and fecundity in the ectoparasitoid, *Nasonia vitripennis* (Hymenoptera: Pteromalidae). *Insect Sci.* **2008**, *15*, 429–440. [CrossRef]

30. De Graaf, D.C.; Aerts, M.; Brunain, M.; Desjardins, C.A.; Jacobs, F.J.; Werren, J.H.; Devreese, B. Insight into the venom composition of the ectoparasitoid wasp *Nasonia vitripennis* from bioinformatics and proteomic studies. *Insect Mol. Biol.* **2010**, *19*, 11–26. [CrossRef] [PubMed]

31. Qian, C. Identification of Venom Proteins from *Nasonia vitripennis* and Functional Analysis of Pacifastin and Kazal Type Genes. Ph. D. Dissertation, Zhejiang University, Hangzhou, China, 2013.

32. Shuker, D.; Lynch, J.; Peire Morais, A. Moving from model to non-model organisms? Lessons from *Nasonia* wasps. *Bioessays* **2003**, *25*, 1247–1248. [CrossRef] [PubMed]

33. Wurm, Y.; Keller, L. Parasitoid wasps: from natural history to genomic studies. *Curr. Biol.* **2010**, *20*, R242–R244. [CrossRef] [PubMed]

34. Werren, J.H.; Richards, S.; Desjardins, C.A.; Niehuis, O.; Gadau, J.; Colbourne, J.K.; Desplan, C.; Beukeboom, L.W.; Elsik, C.G.; Grimmelikhuijzen, C.J.; *et al.* Functional and evolutionary insights from the genomes of three parasitoid *Nasonia* species. *Science* **2010**, *327*, 343–348. [CrossRef] [PubMed]

35. Lu, W.; Apostol, I.; Qasim, M.A.; Warne, N.; Wynn, R.; Zhang, W.L.; Anderson, S.; Chiang, Y.W.; Ogin, E.; Rothberg, I.; *et al.* Binding of amino acid side-chains to S1 cavities of serine proteinases. *J. Mol. Biol.* **1997**, *266*, 441–461. [CrossRef] [PubMed]

36. Van Hoef, V.; Breugelmans, B.; Spit, J.; Simonet, G.; Zels, S.; Vanden, B.J. Phylogenetic distribution of protease inhibitors of the Kazal-family within the Arthropoda. *Peptides* **2013**, *41*, 59–65. [CrossRef] [PubMed]

37. Zhang, D.; Ma, J.; Jiang, S. Molecular characterization, expression and function analysis of a five-domain Kazal-type serine proteinase inhibitor from pearl oyster *Pinctada fucata*. *Fish Shellfish Immunol.* **2014**, *37*, 115–121. [CrossRef] [PubMed]

38. Maldonado-Aguayo, W.; Núñez-Acuña, G.; Valenzuela-Muñoz, V.; Chávez-Mardones, J.; Gallardo-Escárate, C. Molecular characterization of two Kazal-type serine proteinase inhibitor genes in the surf clam *Mesodesma donacium* exposed to *Vibrio anguillarum*. *Fish Shellfish Immunol.* **2013**, *34*, 1448–1454. [CrossRef] [PubMed]

39. Horita, S.; Ishibashi, J.; Nagata, K.; Miyakawa, T.; Yamakawa, M.; Tanokura, M. Isolation, cDNA cloning, and structure-based functional characterization of oryctin, a hemolymph protein from the coconut rhinoceros beetle, *Oryctes rhinoceros*, as a novel serine protease inhibitor. *J. Biol. Chem.* **2010**, *285*, 30150–30158. [CrossRef] [PubMed]

40. Mende, K.; Petoukhova, O.; Koulitchkova, V.; Schaub, G.A.; Lange, U.; Kaufmann, R.; Nowak, G. Dipetalogastin, a potent thrombin inhibitor from the blood-sucking insect. *Dipetalogaster maximus* cDNA cloning, expression and characterization. *Eur. J. Biochem.* **1999**, *266*, 583–590. [CrossRef] [PubMed]

41. Watanabe, R.M.; Soares, T.S.; Morais-Zani, K.; Tanaka-Azevedo, A.M.; Maciel, C.; Capurro, M.L.; Torquato, R.J.; Tanaka, A.S. A novel trypsin Kazal-type inhibitor from *Aedes aegypti* with thrombin coagulant inhibitory activity. *Biochimie* **2010**, *92*, 933–939. [CrossRef] [PubMed]

42. Kim, B.Y.; Lee, K.S.; Zou, F.M.; Wan, H.; Choi, Y.S.; Yoon, H.J.; Kwon, H.W.; Je, Y.H.; Jin, B.R. Antimicrobial activity of a honeybee (*Apis cerana*) venom Kazal-type serine protease inhibitor. *Toxicon* **2013**, *76*, 110–117. [CrossRef] [PubMed]

43. Beckage, N.E.; Gelman, D.B. Wasp parasitoid disruption of host development: Implications for new biologically based strategies for insect control. *Annu. Rev. Entomol.* **2004**, *49*, 299–330. [CrossRef] [PubMed]

44. Asgari, S. Venom proteins from polydnavirus-producing endoparasitoids: Their role in host-parasite interactions. *Arch. Insect Biochem. Physiol.* **2006**, *61*, 146–156. [CrossRef] [PubMed]

45. Abt, M.; Rivers, D.B. Characterization of phenoloxidase activity in venom from the ectoparasitoid *Nasonia vitripennis* (Walker) (Hymenoptera: Pteromalidae). *J. Invertebr. Pathol.* **2007**, *94*, 108–118. [CrossRef] [PubMed]

46. Cerenius, L.; Lee, B.L.; Soderhall, K. The proPO-system: pros and cons for its role in invertebrate immunity. *Trends Immunol.* **2008**, *29*, 263–271. [CrossRef] [PubMed]

47. Cerenius, L.; Soderhall, K. The prophenoloxidase-activating system in invertebrates. *Immunol. Rev.* **2004**, *198*, 116–126. [CrossRef] [PubMed]

48. Tong, Y.; Jiang, H.; Kanost, M.R. Identification of plasma proteases inhibited by *Manduca sexta* serpin-4 and serpin-5 and their association with components of the prophenol oxidase activation pathway. *J. Biol. Chem.* **2005**, *280*, 14932–14942. [CrossRef] [PubMed]

49. Tong, Y.; Kanost, M.R. *Manduca sexta* serpin-4 and serpin-5 inhibit the prophenol oxidase activation pathway: cDNA cloning, protein expression, and characterization. *J. Biol. Chem.* **2005**, *280*, 14923–14931. [CrossRef] [PubMed]

50. Zhu, Y.; Wang, Y.; Gorman, M.J.; Jiang, H.; Kanost, M.R. *Manduca sexta* serpin-3 regulates prophenoloxidase activation in response to infection by inhibiting prophenoloxidase-activating proteinases. *J. Biol. Chem.* **2003**, *278*, 46556–46564. [CrossRef] [PubMed]

51. Ye, G.Y.; Dong, S.Z.; Dong, H.; Hu, C.; Shen, Z.C.; Cheng, J.A. Effects of host (Boettcherisca peregrina) copper exposure on development, reproduction and vitellogenesis of the ectoparasitic wasp, *Nasonia vitripennis*. *Insect Sci.* **2009**, *16*, 43–50. [CrossRef]

52. BLAST: Basic Local Alignment Search Tool. Available online: http://blast.ncbi.nlm.nih.gov/Blast.cgi (accessed on 30 July 2015).

53. CBS: Center of Biological Sequence Analysis. Available online: http://www.cbs.dtu.dk/services/SignalP/ (accessed on 30 July 2015).

54. ClustalW2. Available online: http://www.ebi.ac.uk/Tools/clustalw2/ (accessed on 30 July 2015).

55. Livak, K.J.; Schmittgen, T.D. Analysis of relative gene expression data using real-time quantitative PCR and the $2^{-\Delta\Delta Ct}$ Method. *Methods* **2001**, *25*, 402–408. [CrossRef] [PubMed]
56. Ling, E.J.; Rao, X.J.; Ao, J.Q.; Yu, X.Q. Purification and characterization of a small cationic protein from the tobacco hornworm *Manduca sexta*. *Insect Biochem. Mol. Biol.* **2009**, *39*, 263–271. [CrossRef] [PubMed]

© 2015 by the authors; licensee MDPI, Basel, Switzerland. This article is an open access article distributed under the terms and conditions of the Creative Commons Attribution (CC-BY) license (http://creativecommons.org/licenses/by/4.0/).

Article

Bee Venom Acupuncture Augments Anti-Inflammation in the Peripheral Organs of hSOD1^{G93A} Transgenic Mice

Sun-Hwa Lee [1], Sun-Mi Choi [2] and Eun Jin Yang [1,*]

[1]   Division of Clinical Research, Korea Institute of Oriental Medicine, Daejeon 305-811, Korea; enddlsh@kiom.re.kr

[2]   Executive Director of R&D, Korea Institute of Oriental Medicine, Daejeon 305-811, Korea; smchoi@kiom.re.kr

*   Correspondence: yej4823@hanmail.net; Tel.: +82-42-863-9497; Fax: +82-42-868-9339

Academic Editor: Sokcheon Pak

Received: 24 February 2015; Accepted: 6 July 2015; Published: 29 July 2015

Abstract: Amyotrophic lateral sclerosis (ALS) includes progressively degenerated motor neurons in the brainstem, motor cortex, and spinal cord. Recent reports demonstrate the dysfunction of multiple organs, including the lungs, spleen, and liver, in ALS animals and patients. Bee venom acupuncture (BVA) has been used for treating inflammatory diseases in Oriental Medicine. In a previous study, we demonstrated that BV prevented motor neuron death and increased anti-inflammation in the spinal cord of symptomatic hSOD1^{G93A} transgenic mice. In this study, we examined whether BVA's effects depend on acupuncture point (ST36) in the organs, including the liver, spleen and kidney, of hSOD1^{G93A} transgenic mice. We found that BV treatment at ST36 reduces inflammation in the liver, spleen, and kidney compared with saline-treatment at ST36 and BV injected intraperitoneally in symptomatic hSOD1^{G93A} transgenic mice. Those findings suggest that BV treatment combined with acupuncture stimulation is more effective at reducing inflammation and increasing immune responses compared with only BV treatment, at least in an ALS animal model.

Keywords: amyotrophic lateral sclerosis (ALS); bee venom acupuncture (BVA); anti-inflammation

1. Introduction

Amyotrophic lateral sclerosis (ALS) is characterized by progressive degeneration of motor neurons and muscle weakness. The death of ALS patients is caused by respiratory failure within 3–5 years of the diagnosis. ALS has two types: familial ALS (fALS) caused by genetic mutations, including superoxidase dismutase 1 (SOD1), alsin, senataxin, angiogenin, VAMP-associated protein B, dynactin, transactive response (TAR) DNA-binding protein 43 (TDP43), fused in sarcoma (FUS) and C9ORF72; and sporadic ALS (sALS), which includes ninety percent of all ALS cases and is induced by various environmental and genetic factors. The etiology of ALS is varied and there is no effective therapy for ALS patients. Riluzole, a glutamate release inhibitor approved by the FDA, is used only as a medical treatment for expansion of life by 3–5 months [1].

Bee venom (BV) is used for anti-inflammatory, anti-nociceptive, and anti-allergic effects in allergic rhinitis mice, complete Freund's adjuvant (CFA)-induced arthritis models, and neuropathic pain models [2–4]. In addition, BV treatment prevents the loss of dopaminergic neurons in 1-methyl-4-phenyl-1,2,3,6-tetrahydropyridine (MPTP)-induced Parkinson's disease (PD) and motor neurons in hSOD1^{G93A}-overexpressed ALS-mimic transgenic mice [5–7]. However, it is unclear whether BV's effects depend on acupuncture points or not. Therefore, the purpose of this study is to investigate whether BV treatment at ST36 is more effective than only BV treatment for the reduction of inflammation in the peripheral organs, including the liver, spleen, and kidney, in symptomatic hSOD1^{G93A} transgenic mice.

In this study, we found that BV treatment at ST36 reduced inflammation in the liver, spleen, and kidney compared with that of symptomatic hSOD1^{G93A} transgenic mice treated with saline at ST36 and those injected with BV intraperitoneally. Those findings suggest that BV treatment combined with acupuncture stimulation is more effective at reducing inflammation and increasing immune responses than is BV-only treatment, at least in an ALS animal model.

2. Results

2.1. BV Treatment at ST36 Reduces Inflammatory Proteins in the Liver of hSOD1^{G93A} Transgenic Mice

To investigate the effects of BV on inflammation specific to injection method, we conducted BV treatment two ways, namely at acupuncture point ST36 and intraperitoneally (i.p.), in 14-week-old hSOD1^{G93A} transgenic mice. As shown in Figure 1A, the expression level of Iba-1 in hepatocytes of the liver of hSOD1^{G93A} transgenic mice was increased by 3.4-fold compared with age-matched wild type (WT) (1 ± 0.24) mice. Furthermore, we found that BV treatment at ST36 significantly reduced Iba-1 by 2.8-fold compared with age-matched hSOD1^{G93A} transgenic mice. In addition, the anti-inflammatory effects of BV treatment at ST36 increased by 2.3-fold compared with BV injected by i.p. in the hSOD1^{G93A} mice. To confirm the anti-inflammatory effects of BV treatment at ST36, we studied the expression level of the cyclooxygenase 2 (COX2) protein in the liver of WT and hSOD1^{G93A} transgenic mice. COX2 positive hepatocytes were increased in the liver of hSOD1^{G93A} transgenic mice compared with WT (1 ± 0.55) mice, but BV treatment at ST36 significantly reduced them by 3.3-fold compared with saline-treated hSOD1^{G93A} transgenic mice. Those findings suggest that BV treatment at ST36 is effective at reducing inflammation in the liver of hSOD1^{G93A} transgenic mice.

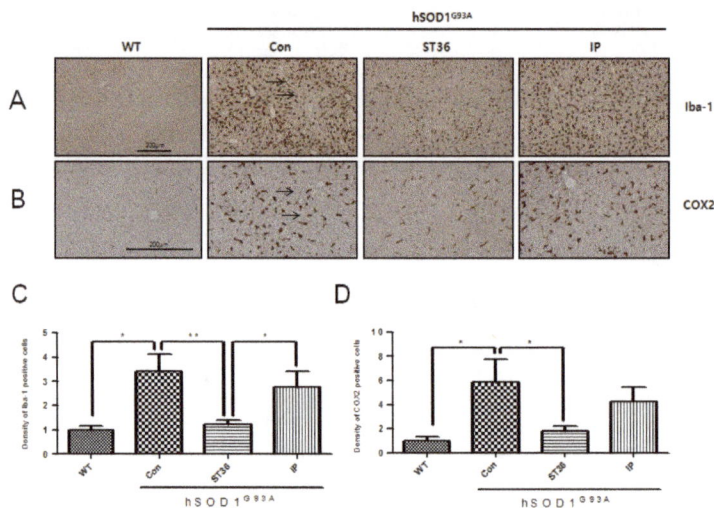

Figure 1. The effects of BV on inflammation in the liver of hSOD1^{G93A} mice. BV treatment at ST36 reduces inflammation in the liver in hSOD1^{G93A} mice. Immunohistochemical staining of paraffin-embedded sections of non-Tg mice (WT, $n = 3$), hSOD1^{G93A} mice (Con, $n = 3$), BV-treatment at Joksamli (ST36) acupuncture point in hSOD1^{G93A} mice (ST36, $n = 7$) and BV-treatment intraperitoneally in hSOD1^{G93A} mice (IP, $n = 4$). Representative images of immunohistochemistry with Iba-1 (**A**) and COX2 (**B**) in the liver of hSOD1^{G93A} mice or non-Tg mice. Quantification of Iba-1 (**C**) and COX2 (**D**) immunoreactivity (IR). It assigned the optical density of WT to one and analyzed relative optical density of Con, ST36, and IP. Data are expressed as the mean ± SEM. * $p < 0.05$, ** $p < 0.01$ from a one-way ANOVA with a Newman-Keuls test. Scale bar indicates 200 μm. WT: non-Tg; Con: saline-treatment at ST36; ST36: BV-treatment at ST36; IP: BV-injection intraperitoneally.

2.2. BV Treatment at ST36 Reduces Inflammation in the Spleen of hSOD1^{G93A} Transgenic Mice

To study the anti-inflammatory effects of BV treatment at ST36 in the spleen, which is involved in the immune response, we immunostained with Iba-1, COX2, and tumor necrosis factor (TNF)-α antibodies for the tissues of WT or hSOD1^{G93A} transgenic mice. As shown in Figure 2A, we found that the expression level of Iba-1 was greatly increased by 7.3-fold in the white pulp of the spleen of the hSOD1^{G93A} transgenic mice compared with WT (1 ± 0.59) mice. BV treatment at ST36 significantly reduced Iba-1 by 5.2-fold compared with saline-treated hSOD1^{G93A} transgenic mice. In addition, the expression level of Iba-1 in the white pulp of the spleen was decreased by 3.5-fold by BV treatment at ST36 compared with BV injected i.p. in the hSOD1^{G93A} transgenic mice. BV treatment at ST36 in hSOD1^{G93A} transgenic mice also significantly reduced COX2-immuno positive cells by 5.6-fold, which were increased by 11.2-fold compared with WT mice (1 ± 0.86). As a pro-inflammatory protein, TNF-α expression in the white pulp of the spleen of the hSOD1^{G93A} transgenic mice also increased by 11.7-fold compared with WT mice (1 ± 0.25), but BV treatment at ST36 significantly reduced its expression by 5.3-fold compared with hSOD1^{G93A} transgenic mice. BV injection by i.p. decreased the expression of TNF-α by 3.5-fold in the spleen of the hSOD1^{G93A} transgenic mice compared with saline-treated hSOD1^{G93A} transgenic mice. BV treatment at ST36 seems to be more effective at reducing the expression of the pro-inflammatory protein TNF-α in the spleen than BV-injected i.p., but it was not significantly different. These findings suggest that BV treatment at ST36 could augment immune responses by reducing the inflammatory proteins in the spleen of the hSOD1^{G93A} transgenic mice.

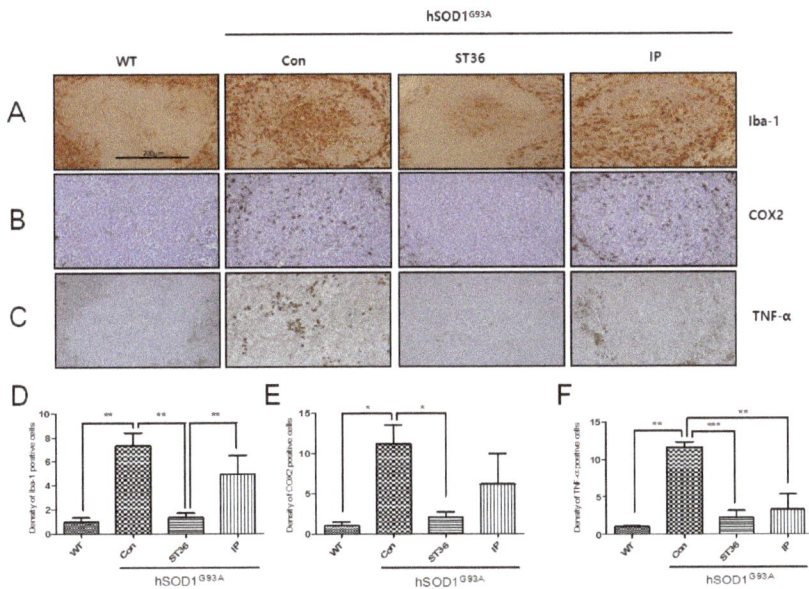

Figure 2. The effects of BV in the spleen of hSOD1^{G93A} mice. BV treatment decreases the expression of inflammatory proteins in the spleen of hSOD1^{G93A} mice. Representative images of the immunohistochemistry of inflammation-related proteins Iba-1 (**A**), COX2 (**B**), and TNF-α (**C**) in the spleen of three groups (Con, ST36, and IP) of hSOD1^{G93A} mice and WT mice. Quantification of Iba-1 (**D**), COX2 (**E**) and TNF-α (**F**) IR. It assigned the optical density of WT to 1 and analyzed relative optical density of Con, ST36, and IP. Data are shown as the mean ± SEM. * $p < 0.05$, ** $p < 0.01$, *** $p < 0.001$ from a one-way ANOVA with a Newman-Keuls test. Scale bar indicates 200 μm. WT: non-Tg; Con: saline-treatment at ST36; ST36: BV-treatment at ST36; IP: BV-injection intraperitoneally.

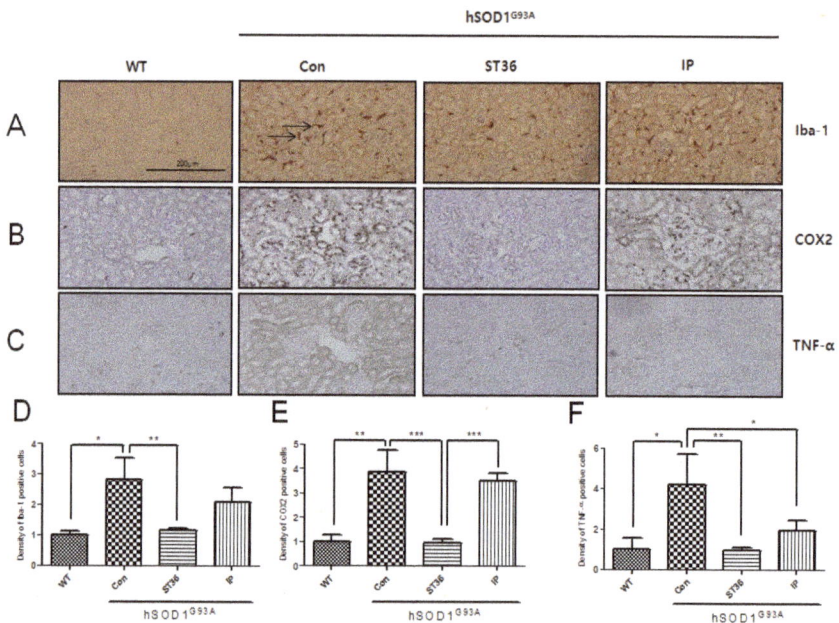

Figure 3. The effects of BV in the kidney of hSOD1^{G93A} mice. BV treatment at ST36 is more effective at reducing the expression of inflammatory proteins in the kidney of the hSOD1^{G93A} mice. Representative images of kidney tissue immunostained with Iba-1 (**A**), COX2 (**B**), and TNF-α (**C**) of three groups of hSOD1^{G93A} mice and WT mice. Quantification of immune-positive cells with Iba-1 (**D**), COX2 (**E**) and TNF-α (**F**) IR. It assigned the optical density of WT to 1 and analyzed relative optical density of Con, ST36, and IP. Data are shown as the mean ± SEM. * $p < 0.05$, ** $p < 0.01$, *** $p < 0.001$ from a one-way ANOVA with a Newman-Keuls test. Scale bar indicates 200 μm. WT: non-Tg; Con: saline-treatment at ST36; ST36: BV-treatment at ST36; IP: BV-injection intraperitoneally.

2.3. BV Treatment at ST36 Downregulates Inflammation in the Kidney of hSOD1^{G93A} Transgenic Mice

We examined the expression level of inflammatory proteins and the effects of BV in the kidney of the hSOD1^{G93A} transgenic mice. The expression of Iba-1 was increased by 2.8-fold in the kidney, as was shown in the liver and spleen, in the hSOD1^{G93A} transgenic mice compared with age-matched WT mice (1 ± 0.23) (Figure 3A). BV treatment at ST36 in the kidney of hSOD1^{G93A} transgenic mice reduced its expression by 2.3-fold compared with saline-treated hSOD1^{G93A} transgenic mice (Figure 3A). COX2 protein increased by 3.8-fold in the renal tubules and renal glomeruli of the kidney in symptomatic hSOD1^{G93A} transgenic mice compared with age-matched WT mice (1 ± 0.44) (Figure 3B). BV treatment at ST36 reduced COX2 expression in the kidney by 4.2-fold compared with saline-treated hSOD1^{G93A} transgenic mice. Furthermore, BV treatment at ST36 is more effective, by 4.2-fold, at decreasing COX2 expression compared with i.p. injection of BV in the hSOD1^{G93A} transgenic mice. TNF-α positive cells were increased by 4.2-fold in the renal tube of the kidney of hSOD1^{G93A} transgenic mice compared with WT mice (1 ± 0.93) (Figure 3C). However, BV treatment at ST36 significantly reduced, by 4.7-fold, TNF-α cytoplasmic staining cells in the renal tube of the kidney compared with saline-treatment at ST36 in hSOD1^{G93A} transgenic mice. BV injection i.p. in hSOD1^{G93A} transgenic mice also reduced, by 2.2-fold, the TNF-α expression in the kidney. BV treatment at ST36 showed more reduction of the expression of TNF-α in the renal tube of the kidney compared with that of i.p. injection of BV in hSOD1^{G93A} transgenic mice, but this difference was not significant. Those findings suggest that BV treatment at ST36 could improve kidney function by increasing the anti-inflammatory proteins in an ALS animal model.

3. Discussion

ALS causes the loss of motor neurons in the brainstem, cerebral cortex, and spinal cord and leads to irreversible paralysis of muscles and finally to respiratory impairment. There are several cellular and molecular pathological mechanisms involved, such as glutamate excitation, inflammatory events, oxidative stress, mitochondrial dysfunction, protein aggregation, and energy failure, as in other neurodegenerative diseases. However, it is insufficient to develop treatments and preventative measures for these biomarkers of ALS. Neuroinflammation in the central nervous system (CNS) from ALS contributes to the disease process and the immune system of sALS patients is altered by immune cells, including remarkable reductions in CD4+CD25+ T-regulatory (T-reg) cells as well as CD14+ monocytes [8]. This suggests that the reduction of T-reg cells in the blood affects the CNS immune system by involving activated microglia in ALS degeneration [9]. Several papers have reported anti-inflammatory therapy using Copaxone, Cyclosporine, and minocycline in animal models and clinical trials, but those have limitations for the treatment of ALS [10]. In a previous study, we demonstrated immune dysfunction of organs, including the lungs and spleen, in hSOD1^{G93A} transgenic mice and that electroacupuncture and melittin treatment enhanced anti-inflammation proteins [11,12]. Based on previous data, we investigated the effects of BV on inflammation of organs, including the spleen, liver, and kidney, of symptomatic hSOD1^{G93A} transgenic mice. We found that BV treatment at ST36 reduced inflammatory proteins, including Iba-1, COX2, and TNF-α, in the liver, spleen, and kidney of hSOD1^{G93A} transgenic mice compared with BV injected by i.p. This suggests that BV's effects may be more effective with treatment at an acupuncture point and it explains the synergistic effect of acupuncture combined with BV compared with BV treatment only.

Finkelstein *et al.* reported liver abnormalities and atrophy, and an increase of cytokines and hepatic lymphocytes in hSOD1^{G93A} transgenic mice [13]. In our study, we found an increase of Iba-1 and COX2 positive cells in the liver of symptomatic hSOD1^{G93A} transgenic mice, but BV treatment at ST36 significantly reduced the expression level of inflammatory proteins, including Iba-1 and COX2 compared with BV injected i.p. in symptomatic hSOD1^{G93A} transgenic mice. This suggests that BV treatment at the acupuncture point ST36 may reduce hepatotoxicity from inflammation and affect liver metabolism in ALS patients.

In hSOD1^{G93A} mice as an ALS animal model, the spleen is markedly reduced in size and weight compared with age-matched B6 wild type mice even though their spleen cell number is identical. In addition, splenic follicular architecture, T cell function, and the lymphoproliferative response are decreased in end-stage hSOD1^{G93A} transgenic mice. Immune dysregulation affecting both the adaptive and innate immune systems is a consistent hallmark in ALS [14]. In our study, we observed that the white pulp, the immune-related component of the spleen, was strongly immunostained with anti-Iba-1, anti-COX2, and anti-TNF-α in symptomatic hSOD1^{G93A} transgenic mice compared with age-matched non-transgenic mice. Furthermore, BV treatment at ST36 reduced the expression level of Iba-1, COX2, and TNF-α compared with saline-treatment at ST36 in symptomatic hSOD1^{G93A} transgenic mice. Those findings suggest that BV treatment at ST36 improves the immune regulation of hSOD1^{G93A} transgenic mice through a synergic effect of acupuncture combined with BV.

Jonsson *et al.* have reported that granular inclusion of mutant SOD1 protein is detected in the liver and kidney by immunohistochemical analysis of ALS patients [15]. We observed increased expression of inflammation-related proteins, including Iba-1, COX2, and TNF-α, in the renal tubules or renal glomeruli in symptomatic hSOD1^{G93A} transgenic mice. BV treatment at ST36 significantly increased anti-inflammation proteins in the kidney compared with saline-treatment at ST36 in hSOD1^{G93A} transgenic mice. This suggests that BV treatment at this acupuncture point could improve kidney dysfunction in an ALS animal model. In addition, we suggest that the mechanism of BV treatment at ST36 may occur by combined effect of some components of BV suppressing inflammatory signaling and the activation of the endogenous modulatory systems by acupoint stimulation. Therefore, it should be required to confirm the activity of individual components of BV. Future studies should investigate whether BV's effects at ST36 are specific or generalized to other acupuncture points. In addition, the relationship between acupuncture point stimulation and meridian, and the mechanism of BV's synergistic effects when combined with acupuncture compared with BV alone need to be explored.

4. Materials and Methods

4.1. Animals

All mice were handled in accordance with the United States National Institutes of Health guidelines, and all procedures were approved by the Institutional Animal Care and Use Committees of the Korea Institute of Oriental Medicine (Protocol number: #13-109). Hemizygous transgenic B6SJL mice carrying the mutant human SOD1 gene, which has a glycine-to-alanine base pair mutation at the 93rd codon of the cytosolic Cu/Zn superoxide dismutase ($hSOD1^{G93A}$), were originally obtained from Jackson Laboratory (Bar Harbor, ME, USA).

Transgenic mice were identified using polymerase chain reaction (PCR) as described previously [16]. All of the mice were kept in standard housing with free access to water and standard rodent chow purchased from Orient Bio (Orient, Seongnam-si, Gyeonggi-do, Korea).

4.2. Bee Venom Treatment

Bee venom (BV) was purchased from Sigma (St. Louis, MO, USA) and diluted with saline. At a dose of 0.1 µg/g, bee venom was injected bilaterally at the Joksamli (ST36) acupuncture point ($n = 7$) or intraperitoneally (IP; $n = 4$) in 14 week-old $hSOD1^{G93A}$ transgenic mice. The mice were treated with BV once every other day for two weeks. According to the human acupuncture point landmark and a mouse anatomical reference [17], the ST36 acupuncture point is anatomically located at 5 mm below and lateral to the anterior tubercle of the tibia. Non-transgenic (WT; $n = 3$) and transgenic $hSOD1^{G93A}$ mice (Con; $n = 3$) were injected at the ST36 acupuncture point with normal saline of an equal volume.

4.3. Tissue Preparation and Immunohistochemistry

$hSOD1^{G93A}$ mice were anesthetized with pentobarbital and perfused with phosphate-buffered saline (PBS). The liver, spleen and kidney were removed and fixed in 4% paraformaldehyde for three days at 4 °C After three days, the liver, spleen and kidney were embedded in paraffin. The tissues were 5 µm-thick sections and were mounted on glass slides. The tissue sections were prepared for immunostaining through xylene treatment and gradual rehydration with 95%–75% ethanol. Following de-paraffinization, the slides were treated with 3% hydrogen peroxide (H_2O_2) for 15 min to inactivate endogenous peroxidases and then blocked in 5% bovine serum albumin (BSA) in 0.01% PBS-Triton X–100 (Sigma-Aldrich, Oakville, ON, Canada) for 1 h at room temperature. The sections were then incubated with various primary antibodies, including anti-Iba-1 (Wako, Osaka, Japan), anti-TNF-α (Abcam, Cambridge, UK), and anti-COX2 (Epitomics, Burlingame, CA, USA), overnight. Next, the sections were incubated with the secondary antibody for 1 h. For visualizing, the ABC kit and 3,3'-diaminobenzidine (DAB)/H_2O_2 substrate were used with a hematoxylin counterstain. After rinsing, the sections were dehydrated in ethanol, cleared in xylene, and coverslipped. Immunostained tissues were observed with a light microscope (Olympus, Tokyo, Japan) and analyzed by Image J 1.46j software (NIH) (GraphPad Software, San Diego, CA, USA).

4.4. Statistical Analysis

All data were analyzed using GraphPad Prism 5.0 (GraphPad Software, San Diego, CA, USA) and are presented as the mean ± standard error of the mean (SEM) where indicated. The results of immunohistochemistry and Western blots were analyzed using one-way ANOVAs followed by Newman-Keuls tests. Statistical significance was set at $p < 0.05$.

5. Conclusions

In summary, we examined whether BVA's effects depend on acupuncture point (ST36) in the organs, including the liver, spleen and kidney, of hSOD1^{G93A} transgenic mice. We found that BV treatment at ST36 reduces inflammation-related proteins including Iba-1, COX2, TNF-α in the liver, spleen, and kidney compared with saline-treatment at ST36 and BV injected i.p. in symptomatic hSOD1^{G93A} transgenic mice. Those findings suggest that BV treatment combined with acupuncture stimulation is more effective at reducing inflammation and increasing immune responses compared with only BV treatment, at least in an ALS animal model.

Acknowledgments: We would like to thank Mu-Dan Cai for animal maintenance. This research was supported by grants (K13010, K14010) from the Korea Institute of Oriental Medicine (KIOM).

Author Contributions: E.J.Y. designed the experiments and analyzed the data as well as wrote the manuscript. S.H.L. executed all the experiments and S.M.C. discussed with the manuscript. All authors have read and approved the final manuscript.

Conflicts of Interest: The authors declare no conflict of interest.

References

1. Miller, R.G.; Mitchell, J.D.; Moore, D.H. Riluzole for amyotrophic lateral sclerosis (ALS)/motor neuron disease (MND). *Cochrane Database Syst. Rev.* **2012**, *3*, CD001447. [PubMed]
2. Suh, S.J.; Kim, K.S.; Kim, M.J.; Chang, Y.C.; Lee, S.D.; Kim, M.S.; Kwon, D.Y.; Kim, C.H. Effects of bee venom on protease activities and free radical damages in synovial fluid from type II collagen-induced rheumatoid arthritis rats. *Toxicol. In Vitro* **2006**, *20*, 1465–1471. [CrossRef] [PubMed]
3. Roh, D.H.; Kwon, Y.B.; Kim, H.W.; Ham, T.W.; Yoon, S.Y.; Kang, S.Y.; Han, H.J.; Lee, H.J.; Beitz, A.J.; Lee, J.H. Acupoint stimulation with diluted bee venom (apipuncture) alleviates thermal hyperalgesia in a rodent neuropathic pain model: Involvement of spinal alpha 2-adrenoceptors. *J. Pain* **2004**, *5*, 297–303. [CrossRef] [PubMed]
4. Shin, S.H.; Kim, Y.H.; Kim, J.K.; Park, K.K. Anti-allergic effect of bee venom in an allergic rhinitis mouse model. *Biol. Pharm. Bull.* **2014**, *37*, 1295–1300. [CrossRef] [PubMed]
5. Alvarez-Fischer, D.; Noelker, C.; Vulinović, F.; Grünewald, A.; Chevarin, C.; Klein, C.; Oertel, W.H.; Hirsch, E.C.; Michel, P.P.; Hartmann, A. Bee venom and its component apamin as neuroprotective agents in a Parkinson disease mouse model. *PLoS ONE* **2013**, *8*, e61700. [CrossRef] [PubMed]
6. Doo, A.R.; Kim, S.T.; Kim, S.N.; Moon, W.; Yin, C.S.; Chae, Y.; Park, H.K.; Lee, H.; Park, H.J. Neuroprotective effects of bee venom pharmaceutical acupuncture in acute 1-methyl-4-phenyl-1,2,3,6-tetrahydropyridine-induced mouse model of Parkinson's disease. *Neurol. Res.* **2010**, *32*, 88–91. [CrossRef] [PubMed]
7. Yang, E.J.; Jiang, J.H.; Lee, S.M.; Yang, S.C.; Hwang, H.S.; Lee, M.S.; Choi, S.M. Bee venom attenuates neuroinflammatory events and extends survival in amyotrophic lateral sclerosis models. *J. Neuroinflammation* **2010**, *7*, 69. [CrossRef] [PubMed]
8. Mantovani, S.; Garbelli, S.; Pasini, A.; Alimonti, D.; Perotti, C.; Melazzini, M.; Bendotti, C.; Mora, G. Immune system alterations in sporadic amyotrophic lateral sclerosis patients suggest an ongoing neuroinflammatory process. *J. Neuroimmunol.* **2009**, *210*, 73–79. [CrossRef] [PubMed]
9. Kipnis, J.; Avidan, H.; Caspi, R.R.; Schwartz, M. Dual effect of CD4+CD25+ regulatory T cells in neurodegeneration: A dialogue with microglia. *Proc. Natl. Acad. Sci. USA* **2004**, *101*, 14663–14669. [CrossRef] [PubMed]
10. Phani, S.; Re, D.B.; Przedborski, S. The Role of the Innate Immune System in ALS. *Front. Pharmacol.* **2012**, *3*, 150. [CrossRef] [PubMed]
11. Jiang, J.H.; Yang, E.J.; Baek, M.G.; Kim, S.H.; Lee, S.M.; Choi, S.M. Anti-inflammatory effects of electroacupuncture in the respiratory system of a symptomatic amyotrophic lateral sclerosis animal model. *Neurodegener. Dis.* **2011**, *8*, 504–514. [CrossRef] [PubMed]
12. Lee, S.H.; Choi, S.M.; Yang, E.J. Melittin ameliorates the inflammation of organs in an amyotrophic lateral sclerosis animal model. *Exp. Neurobiol.* **2014**, *23*, 86–92. [CrossRef] [PubMed]

13. Finkelstein, A.; Kunis, G.; Seksenyan, A.; Ronen, A.; Berkutzki, T.; Azoulay, D.; Koronyo-Hamaoui, M.; Schwartz, M. Abnormal changes in NKT cells, the IGF-1 axis, and liver pathology in an animal model of ALS. *PLoS ONE* **2011**, *6*, e22374. [CrossRef] [PubMed]
14. McGeer, P.L.; McGeer, E.G. Inflammatory processes in amyotrophic lateral sclerosis. *Muscle Nerve* **2002**, *26*, 459–470. [CrossRef] [PubMed]
15. Jonsson, P.A.; Bergemalm, D.; Andersen, P.M.; Gredal, O.; Brännström, T.; Marklund, S.L. Inclusions of amyotrophic lateral sclerosis-linked superoxide dismutase in ventral horns, liver, and kidney. *Ann. Neurol.* **2008**, *63*, 671–675. [CrossRef] [PubMed]
16. Rosen, D.R.; Siddique, T.; Patterson, D.; Figlewicz, D.A.; Sapp, P.; Hentati, A.; Donaldson, D.; Goto, J.; O'Regan, J.P.; Deng, H.X.; *et al.* Mutations in Cu/Zn superoxide dismutase gene are associated with familial amyotrophic lateral sclerosis. *Nature* **1993**, *362*, 59–62. [CrossRef] [PubMed]
17. Yin, C.S.; Jeong, H.S.; Park, H.J.; Baik, Y.; Yoon, M.H.; Choi, C.B.; Koh, H.G. A proposed transpositional acupoint system in a mouse and rat model. *Res. Vet. Sci.* **2008**, *84*, 159–165. [CrossRef] [PubMed]

© 2015 by the authors; licensee MDPI, Basel, Switzerland. This article is an open access article distributed under the terms and conditions of the Creative Commons Attribution (CC-BY) license (http://creativecommons.org/licenses/by/4.0/).

toxins

MDPI

Article

Repetitive Treatment with Diluted Bee Venom Attenuates the Induction of Below-Level Neuropathic Pain Behaviors in a Rat Spinal Cord Injury Model

Suk-Yun Kang [1], Dae-Hyun Roh [2], Jung-Wan Choi [1], Yeonhee Ryu [1,*] and Jang-Hern Lee [3,*]

[1]   KM Fundamental Research Division, Korea Institute of Oriental Medicine, Daejeon 305-811, Korea;
     sy8974@kiom.re.kr (S.-Y.K.); cjw214@kiom.re.kr (J.-W.C.)

[2]   Department of Oral Physiology and Research Center for Tooth and Periodontal Tissue Regeneration,
     School of Dentistry, Kyung Hee University, Seoul 130-701, Korea; dhroh@khu.ac.kr

[3]   Department of Veterinary Physiology, College of Veterinary Medicine and BK21 Program for Veterinary
     Science, Seoul National University, Seoul 151-742, Korea

*   Correspondence: yhryu@kiom.re.kr (Y.R.); jhl1101@snu.ac.kr (J.-H.L.); Tel.: +82-2-880-1272 (J.-H.L.);
     Fax: +82-2-885-2732 (J.-H.L.)

Academic Editor: Ren Lai
Received: 11 May 2015; Accepted: 7 July 2015; Published: 10 July 2015

Abstract: The administration of diluted bee venom (DBV) into an acupuncture point has been utilized traditionally in Eastern medicine to treat chronic pain. We demonstrated previously that DBV has a potent anti-nociceptive efficacy in several rodent pain models. The present study was designed to examine the potential anti-nociceptive effect of repetitive DBV treatment in the development of below-level neuropathic pain in spinal cord injury (SCI) rats. DBV was applied into the Joksamli acupoint during the induction and maintenance phase following thoracic 13 (T13) spinal hemisection. We examined the effect of repetitive DBV stimulation on SCI-induced bilateral pain behaviors, glia expression and motor function recovery. Repetitive DBV stimulation during the induction period, but not the maintenance, suppressed pain behavior in the ipsilateral hind paw. Moreover, SCI-induced increase in spinal glia expression was also suppressed by repetitive DBV treatment in the ipsilateral dorsal spinal cord. Finally, DBV injection facilitated motor function recovery as indicated by the Basso–Beattie–Bresnahan rating score. These results indicate that the repetitive application of DBV during the induction phase not only decreased neuropathic pain behavior and glia expression, but also enhanced locomotor functional recovery after SCI. This study suggests that DBV acupuncture can be a potential clinical therapy for SCI management.

Keywords: bee venom; spinal cord injury; mechanical allodynia; thermal hyperalgesia; glia; acupuncture

---

1. Introduction

One of the pain therapies is the use of chemical stimulation into an acupuncture point to produce an analgesic effect and to reduce pain severity. In this regard, the injection of diluted bee venom (DBV) into an acupuncture point, termed apipuncture, has been used clinically in traditional Korean medicine to produce a significant analgesic effect in human patients [1,2]. Many experimental studies have demonstrated that injecting DBV into the Joksamli (ST36) acupuncture point produces a robust anti-nociceptive effect in various pain animal models, such as the writhing test, the formalin test, the carrageenan-induced inflammatory pain test and arthritis models [3–6]. Furthermore, we demonstrated that this DBV-induced anti-nociceptive effect is associated with the activation of descending coeruleospinal noradrenergic pathways, which subsequently activate spinal alpha-2 adrenoceptors [3,7,8]. DBV stimulation of ST36 also inhibits the activation of spinal astrocytes in a mouse formalin test [3]. In particular, we showed that a single injection of DBV

(0.25 mg/kg) into ST36 temporarily alleviated thermal hyperalgesia [9] and that repetitive stimulation using DBV significantly alleviated neuropathic pain-induced mechanical and cold allodynia in a sciatic nerve chronic constrictive injury (CCI) model of rats [7,8]. However, the precise roles of repetitive DBV treatment in the induction and maintenance phases of central neuropathic pain have not been examined.

Spinal cord injury (SCI), which is caused by direct traumatic damage to the spinal cord, has been related to many clinical complications, including functional disability, urinary tract problems, autonomic dysreflexia, altered sensations and pain [10–12]. These patients often have experiences of several types of pain; central chronic pain syndrome, which exhibits mechanical allodynia and thermal hyperalgesia, is one of the most common causes for a reduced quality of life [13,14]. Especially, below-level pain after SCI represents a clinically-significant symptom of central neuropathic pain that is very difficult to treat effectively [15,16]. Several experimental models of SCI have been developed to determine the detailed mechanisms and therapeutic strategies to treat SCI. The most widely-used models are rat SCI contusion, excitotoxic and hemisection models [14,17–19]. In this study, the rat SCI hemisection model was chosen, because this model is widely used to verify the mechanism behind SCI-induced chronic pain development.

Recently, a number of studies have reported the potential role of spinal astrocytes and microglia in both postoperative pain and neuropathic pain [20–24]. Moreover, intrathecal treatment with glia inhibitors, such as minocycline (a microglia inhibitor) and propentofylline (a glia modulating agent), reduced below-level neuropathic pain behaviors in SCI rats [25–27]. However, although there is some evidence that glial cells are activated during the development of SCI-induced neuropathic pain, the precise mechanisms underlying glial activation, particularly in lumbar segments distant from the spinal cord injury site, are poorly understood.

Based on the above-mentioned studies, we hypothesized that repetitive DBV treatment into an acupoint reduces SCI-induced mechanical allodynia and thermal hyperalgesia and that this reduction is mediated by the suppression of spinal astrocyte or microglia activation. Thus, the present study was designed to examine the following: (1) whether repetitive DBV acupuncture point treatment for five days during the induction and maintenance phases following thoracic 13 (T13) spinal cord hemisection would produce a more potent and prolonged analgesic effect compared to controls that received repetitive injections of vehicle; (2) whether the anti-nociceptive effect of DBV is mediated by the modulation of spinal astrocyte and microglia activation; and (3) whether repetitive DBV treatment affects motor functional recovery in SCI rats.

2. Results

2.1. Effect of Repetitive DBV Treatment during the Maintenance Phase of Spinal Cord Injury-Induced Pain

Spinal cord hemisection produced prominent mechanical allodynia and thermal hyperalgesia, as shown in Figure 1. During the maintenance phase, repetitive DBV treatment was administered twice a day from 15 to 20 days post-surgery. Repetitive DBV treatment significantly increased the decrease in the paw withdrawal threshold in the ipsilateral hind paw by SCI surgery at three and five days after DBV treatment (* $p < 0.05$, compared with saline-treated groups) (Figure 1A); however, mechanical allodynia of the contralateral paw did not show any change compared to saline-treated animals (Figure 1B).

As shown in Figure 1C, repetitive DBV treatment during the maintenance phase increased paw withdrawal latency to noxious thermal stimulus only Day 1 after DBV treatment compared with the repetitive saline-treated group (* $p < 0.05$). In the contralateral paw, repetitive DBV treatment had no effect on SCI-induced thermal hyperalgesia (Figure 1D).

2.2. Effect of Repetitive DBV Treatment during the Induction Phase of Spinal Cord Injury-Induced Pain

The withdrawal response threshold to innocuous mechanical stimuli and withdrawal response latency to noxious thermal stimuli were measured in repetitive DBV and vehicle treatment groups during the induction phase (twice a day from one to five days post-surgery, Figure 2). Groups treated with saline (vehicle, $n = 8$) in the ipsilateral paw showed an approximately 6 to 7-g threshold for mechanical allodynic behaviors in both paws by five days post-surgery. The peak was reached at Day 14 (10 days after the termination of injection). Repetitive DBV-treated groups (DBV, $n = 8$) displayed potently suppressed pain induction in the ipsilateral paw as early as five days post-surgery (* $p < 0.05$, ** $p < 0.01$ and *** $p < 0.001$, compared with saline-treated groups; Figure 2A). In the contralateral paw, DBV-treated groups did not display significantly altered SCI-induced mechanical allodynia compared to the saline-treated control group (Figure 2B). Repetitive DBV injection for five consecutive days during the induction phase significantly increased the SCI-induced decrease in the paw withdrawal latency to noxious thermal stimulus beginning seven days post-SCI surgery compared with the repetitive saline-treated group (* $p < 0.05$; Figure 2C). In the contralateral paw, groups treated with repetitive DBV for five days showed a tendency toward increasing the paw withdrawal latency after DBV treatment; however, this increase was not significant (Figure 2D).

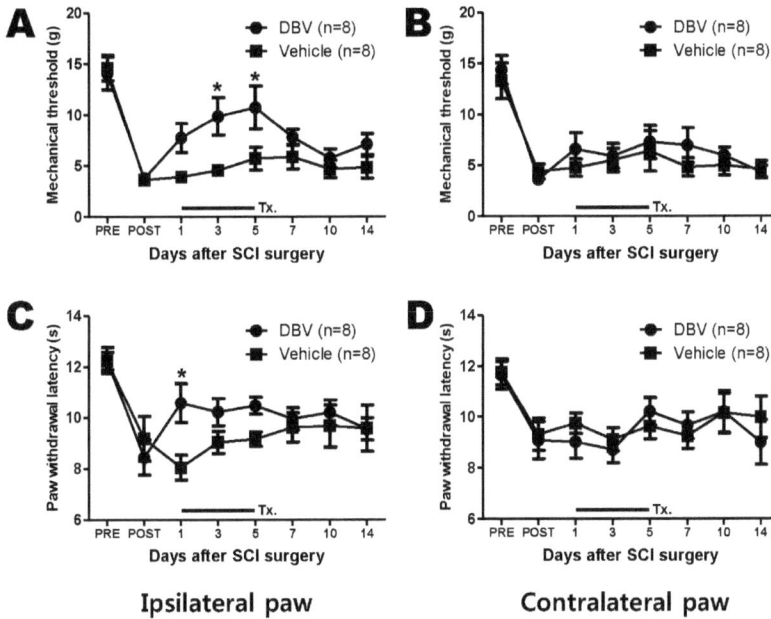

Figure 1. Graphs illustrating the effects of repetitive diluted bee venom (DBV) or vehicle on the maintenance phase of mechanical allodynia (**A,B**) and thermal hyperalgesia (**C,D**) in spinal cord injury animals. (**A**) Repetitive daily treatment with DBV (from 15 to 20 days post-surgery, twice a day) increased the paw withdrawal threshold by mechanical stimuli for the period of DBV treatment in the ipsilateral hind paw (* $p < 0.05$ compared to the vehicle-treated group); (**B**) whereas the contralateral paw did not display any differences compared to the vehicle-treated group; (**C**) repetitive DBV treatment reversed the spinal cord injury (SCI)-induced decrease in the paw withdrawal latency (s) to noxious thermal stimuli compared to the vehicle-treated group (* $p < 0.05$); (**D**) no significant difference in the paw withdrawal latency was observed in the contralateral paw between the DBV and vehicle-treated groups. Two-way ANOVA followed by Bonferroni's test. PRE; one day before SCI surgery, POST; 15 days after SCI surgery. Tx.: DBV or vehicle treatment.

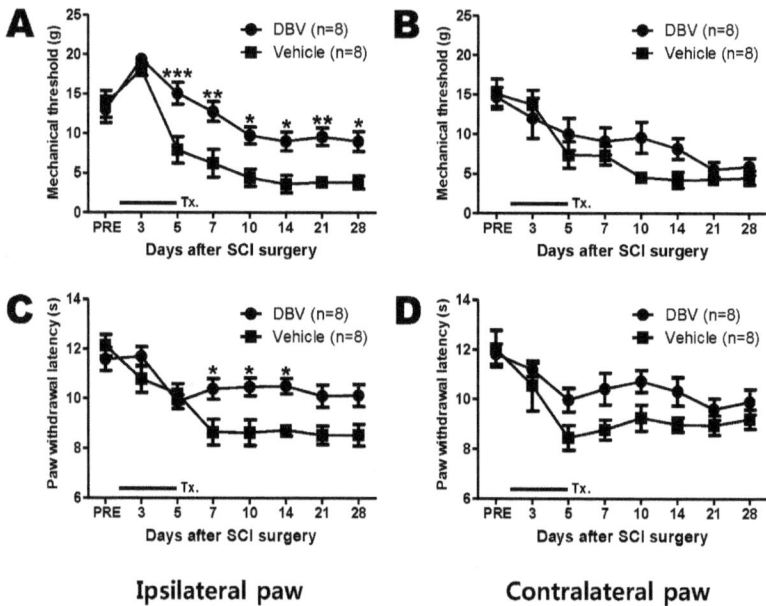

Figure 2. Graphs illustrating the effects of repetitive DBV or vehicle treatment during the induction phase on mechanical allodynia (**A,B**) and thermal hyperalgesia (**C,D**) in spinal cord injury animals. (**A**) Repetitive daily treatment with DBV (twice a day from one to five days post-surgery) suppressed the induction of SCI-induced mechanical allodynia in the ipsilateral hind paw compared with vehicle-treated rats (* $p < 0.05$, ** $p < 0.01$, *** $p < 0.001$); (**B**) whereas the contralateral paw did not display any differences; (**C**) repetitive treatment with DBV reversed the SCI-induced decrease in the paw withdrawal latency (s) to noxious thermal stimuli compared to the vehicle-treated group (* $p < 0.05$); (**D**) no significant difference in the paw withdrawal latency was observed in the contralateral paw between the DBV- and vehicle-treated groups. Two-way ANOVA followed by Bonferroni's test. PRE: one day before SCI surgery; Tx.: DBV or vehicle treatment.

2.3. Effect of Repetitive DBV Treatment during the Induction Period on Glia Expression after Spinal Cord Injury

To determine how repetitive DBV treatment might affect glia, astrocyte and microglia expression, the Western blot assay was performed on the lumbar spinal cord dorsal horn at 14 days after SCI surgery. Astrocytes can respond quickly to various pathological stimuli, and this response is related to an increase in GFAP. Repetitive saline injection during the induction phase significantly increased GFAP expression in the spinal dorsal horn compared with that of normal animals (** $p < 0.01$), and repetitive DBV treatment at ST36 suppressed SCI-enhanced GFAP expression (# $p < 0.05$), suggesting that DBV treatment has a potent anti-nociceptive effect on SCI-induced central neuropathic pain. However, the contralateral spinal cord dorsal horn did not reproduce this suppressive effect of DBV (Figure 3A). In Figure 3B, repetitive saline injection also showed a significant increase in Iba-1 expression in the ipsilateral spinal cord dorsal horns compared to that of normal animals (*** $p < 0.001$), and DBV-treated rats displayed significantly decreased Iba-1 expression compared to the vehicle-treated group (## $p < 0.01$).

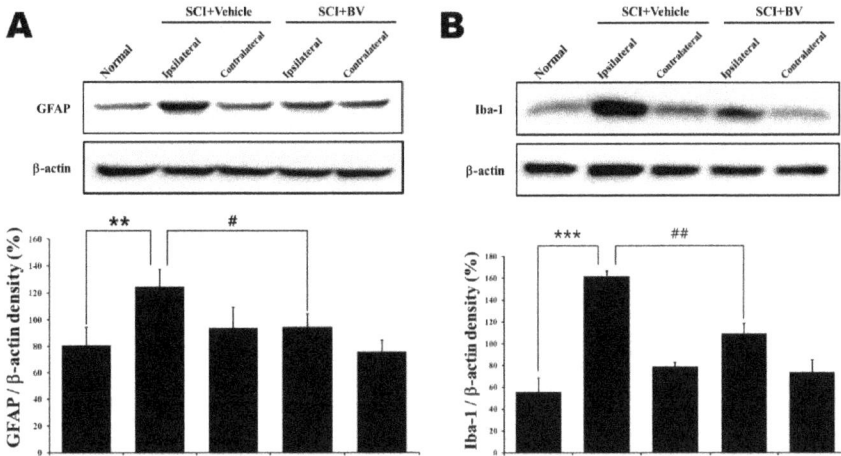

Figure 3. Graphs illustrating the effects of repetitive DBV or vehicle treatment during the induction phase on astrocyte (**A**) and microglia (**B**) expression in the lumbar spinal cord of SCI rats. (A) Western blot data confirmed the effect of repetitive DBV administration on GFAP (astrocyte marker) expression in the spinal cord dorsal horn 14 days after SCI surgery. Vehicle-treated rats displayed significantly increased GFAP expression in the spinal cord compared to normal rats (** $p < 0.01$), and repetitive DBV-treated rats displayed significantly decreased GFAP expression in the spinal cord compared to the vehicle-treated group (# $p < 0.05$). (**B**) In the ipsilateral spinal cord, Iba-1(microglia marker) expression increased following SCI surgery compared to the Iba-1 expression level in normal rats (*** $p < 0.001$), and repetitive DBV-treated rats displayed significantly decreased Iba-1 expression compared to the vehicle-treated group (## $p < 0.01$).

2.4. Effect of Repetitive DBV Treatment on Motor Function Recovery after Spinal Cord Injury

Before hemisection, the Basso, Beattie and Bresnahan (BBB) scores were 21 in the repetitive DBV and saline groups (Figure 4). Immediately upon emerging from anesthesia, hemisected animals showed a dramatic loss of ipsilateral hindlimb function as indicated by BBB scores of zero for each group. From Days 1 to 5 after hemisection, rats treated with DBV during the induction phase displayed faster functional recovery rates throughout the four-week period following SCI surgery than those treated with vehicle (** $p < 0.01$ and *** $p < 0.001$ compared with saline-treated groups).

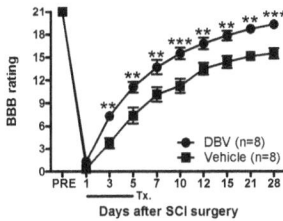

Figure 4. Graphs illustrating the effects of repetitive DBV or vehicle treatment during the induction phase on the recovery of motor function in SCI rats. The Basso, Beattie and Bresnahan (BBB) scores were 21 in all SCI groups before hemisection. Immediately after recovering from anesthesia, hemisected animals appeared to show a loss of ipsilateral hindlimb function as indicated by BBB scores of zero for each group. Animals treated repeatedly with DBV from Days 1 to 5 after hemisection showed faster functional recovery rates than those treated with vehicle (** $p < 0.01$, *** $p < 0.001$). PRE: one day before SCI surgery; Tx.: DBV or vehicle treatment.

3. Discussion

This study demonstrated that repetitive injections of DBV into the Joksamli acupuncture point during the induction phase (one to five days after SCI) of below-level neuropathic pain significantly produce a more potent and prolonged anti-nociceptive effect compared to repetitive DBV treatment during the maintenance phase (15 to 20 days after SCI) or repetitive injections of the vehicle. Importantly, repetitive treatment with DBV had less of an effect when administered during the maintenance phase. Acupuncture therapy, including manual acupuncture, electro-acupuncture and DBV therapy, produces a gradually increasing anti-nociceptive effect in chronic pain patients when injected repetitively over the course of several days, weeks or months [28]. Huang *et al.* found that high-frequency electro-acupuncture treatment twice a week for four weeks produced a significant reduction in mechanical hyperalgesia by the third and fourth weeks of treatment, whereas it caused no effect on thermal hyperalgesia in a chronic inflammatory pain model of rats [29]. In general, bee venom (BV) contains a number of potential pain-related substances, including melittin, histamine and phospholipase A2, and this mixture of biologically-active substances is able to induce toxic effects, contributing to certain clinical signs or symptoms of envenomation. Human responses to BV include small edema, redness, extensive local swelling, anaphylaxis, systemic toxic reaction and pain [30]. Thus, the use of BV always requires great care.

By contrast, BV has been also used in Oriental and Korean medicine to reduce pain and inflammation. We demonstrated previously that repetitive DBV injection into the Joksamli acupuncture point twice a day for two weeks could significantly decrease mechanical and cold allodynia and thermal hyperalgesia in CCI-induced neuropathic rats [7,8], whereas single DBV injection into the acupuncture point temporarily suppressed thermal hyperalgesia (up to 45 min after DBV injection), but not mechanical allodynia in CCI rats [9]. Our results demonstrate that repetitive injections of DBV into the acupuncture point play an important role during early induction, but not during maintenance of pain behaviors associated with central neuropathic pain conditions. To exclude a possible influence of the temporary anti-nociceptive effect of DBV (observed immediately after each daily DBV injection) on the long-term effects of repetitive DBV treatment, we performed the pain behavioral tests in the afternoon between 6 and 10 h after DBV injection. During the sixth to 10th hour after DBV injection, the temporary anti-nociceptive effect was no longer shown, thus the collected anti-nociceptive data indicate the net long-term effect produced by repetitive DBV treatment. The present study also demonstrates the significance of the appropriate time point for drug administration in SCI patients. Preemptive or initiatory medication in the spinal cord level has not been widely examined in SCI patients, because administering drug preemptively in these patients is almost impossible because of the unpredictable clinical occurrence of chronic pain [31]. However, it might be important to detect situations where the possibility for the development of below-level neuropathic pain is high, and the ability to alter this situation would be of considerable clinical value. Although predicting which patients suffering from SCI will go on to develop chronic central neuropathic pain is impossible, our results demonstrate that a critical time window exists in which early treatment with DBV would be effective. Recently, Tan *et al.* suggested that inhibition of early neuroimmune events could have a critical impact on the induction of long-term pain phenomena after SCI [32]. Marchand *et al.* also demonstrated that early treatment with etanercept, a tumor-necrosis-factor inhibitor, caused the reduction of mechanical allodynia after SCI, whereas delayed treatment of etanercept had no significant effect [27]. These results are consistent with the time-dependent effect of DBV observed in our present study. Collectively, these findings, including the present results, imply that repetitive DBV acupuncture therapy during the induction phase is able to produce a powerful analgesic effect on chronic central neuropathic pain and suggest the clinical use of repetitive DBV treatment as a potential novel strategy in the early management of SCI-induced neuropathic pain.

Moreover, the findings of this study demonstrate that the suppression of glial cell activation in ipsilateral, but not contralateral, spinal cord dorsal horn is closely related to the anti-allodynic and anti-hyperalgesic effects of repetitive DBV treatment during the induction phase in SCI rats.

Glial cells, in particular astrocytes and microglia, have been known as important modulators or key factors of nociception. Although glia have been traditionally recognized to have simple functions that are necessary for neuronal communication in normal conditions, they are now recognized as key modulators of plasticity changes in pathophysiological conditions. Furthermore, glia can interact directly with neurons, and then, they also play important neuromodulatory and/or neuroimmune roles in the CNS [33,34]. Recently, several studies demonstrated the involvement of glia activation in chronic pain conditions, including inflammatory pain, peripheral and central neuropathic pain [20,21,23,35]. Direct metabolic inhibitors of glia activation, like minocycline and propentofylline, have been shown to have an anti-nociceptive effect in SCI rats [25–27,32]. Moreover, the blockade of astrocyte gap junctions by the intrathecal injection of carbenoxolone during the induction period of SCI-induced neuropathic pain reduced the development of below-level mechanical allodynia and thermal hyperalgesia and suppressed astrocytic activation in spinal cord [36]. Thus, unsurprisingly, astrocyte activation may contribute to the induction of central neuropathic pain in SCI rats. DBV stimulation of the ST36 acupuncture point also suppressed the activation of spinal cord astrocytes and reduced nociceptive behaviors in the mouse formalin test [3]. Our results showed that GFAP and Iba-1 expression in the ipsilateral lumbar 4 (L4) to L6 segments significantly increased 14 days after SCI in vehicle-treated SCI rats. However, interestingly, the increase in GFAP and Iba-1 expression was significantly decreased by repetitive DBV treatment during the induction phase. In the contralateral L4 to L6 segments, GFAP and Iba-1 expression in vehicle-treated SCI rats did not differ from that in normal animals, and repetitive DBV treatment during the induction phase did not modify GFAP and Iba-1 expression in contralateral dorsal horn examined in the present study. This result indicates that glia activation in the ipsilateral lumbar spinal dorsal horn could be caused by the damage in a spinal cord injured segment (T13) and that the mechanism underlying this remote activation of glia could ultimately lead to the development of below-level neuropathic pain. Thus, these findings suggest that the early activation of astrocytes and microglia can initiate the induction of below-level neuropathic pain.

Finally, the BBB open field locomotor test was used to examine the functional recovery by repetitive DBV acupuncture point treatment during the induction phase in SCI rats. Because the ipsilateral hindlimb was operated on in the hemisection model, we only recorded the motility of the ipsilateral hindlimb based on the locomotor rating set by Basso *et al.* [37]. Repetitive DBV treatment during the induction phase evoked a significant and rapid recovery of motor function. Thus, early repetitive DBV treatment presents the advantage of motor function recovery, because DBV-treated rats appeared to present facilitated motor functional recovery from Day 3 after SCI. Significant functional recovery was observed after repetitive DBV treatment during the induction period; however, hindlimb deficits in the saline control group were relatively prolonged for 28 days. The demyelination of axons in the injured spinal cord is a known cause of motor function deficits, and remyelination or regeneration by natural formation are extremely limited due to glial scarring and growth inhibitors contained in the environment [38]. Glia scarring is the prominent factor that inhibits axonal regeneration in the central nervous system. By limiting the formation of astrocyte scars, we can facilitate axonal regeneration physically and biochemically [39]. Another possibility is the rerouting or plasticity of injured spinal cord. Iwashita *et al.* reported that a partial recovery is available due to a rerouting mechanism in untreated SCI animals [40]. Moreover, neuroplastic changes of the CNS in response to injury have been shown to be highly susceptible to intervention during the post-injury phase [41]. The observed functional recovery here might have also been partially evoked by the reintroduction of afferent feedback signals into the injured spinal cord by the rerouted nerve. One report indicated that animals under tactile stimulation, such as direct mechanical disturbance or electrical stimulation, resulted in greater locomotion restoration [42]. Therefore, the DBV-induced constant chemical stimulation at an acupuncture point may have facilitated locomotor function recovery. Collectively, we suggest that the repetitive DBV treatment during the induction phase can facilitate motor function recovery by enhancing sensory stimulation and by suppressing secondary injury development.

In conclusion, the present study demonstrates that repetitive DBV acupuncture therapy in spinal cord-injured rats can reduce the development of below-level mechanical allodynia and thermal hyperalgesia and can prevent glia activation in the ipsilateral spinal cord dorsal horn. In contrast, DBV treatment during the maintenance period after SCI did not modify glia expression in the spinal dorsal horn, nor below-level mechanical allodynia and thermal hyperalgesia previously established following SCI. In addition, the facilitation of motor function recovery occurred by repetitive DBV treatment. These results suggest that the repetitive application of DBV acupuncture therapy suppressed SCI-induced central neuropathic pain syndrome development and might be a potential clinical therapy for the management of SCI.

4. Materials and Methods

4.1. Animals

All experiments were performed on Sprague–Dawley male rats weighing 180 to 200 g. Animals were obtained from Orient Bio (Sungnam, Korea). The rats were housed in cages with free access to food and water. For 1 week before the study, they were also maintained in temperature- and light-controlled rooms ($24 \pm 2\,^{\circ}\text{C}$, 12/12 h light/dark cycle with lights on at 07:00 h). All experimental procedures used in this study were reviewed and approved by the Animal Care and Use Committee at Korea Institute of Oriental Medicine and performed as in the NIH guidelines (NIH Publication No. 86–23, revised 1985). We made an effort to minimize animal distress and to reduce the number of animals used in this study.

4.2. Spinal Cord Hemisection Surgery

Spinal cord hemisection surgery was performed according to the method described by Christensen *et al.* [13]. Briefly, rats were transiently anesthetized with a combination of 2.5 mg of Zoletil 50 (Virbac Laboratories) and 0.47 mg of Rompun (Bayer Korea) in saline to reduce handling-induced stress and then mounted on the surgical field. Then, the dorsal surface was palpated to locate the cranial borders of the sacrum and the spinous processes of the lower thoracic and lumbar vertebrae. The thoracic 11 to 12 (T11 to T12) vertebrae were recognized by counting spinous processes from the sacrum. A laminectomy was performed between the T11 to T12 vertebral segments, and the lumbar enlargement region was identified with the accompanying dorsal vessel; then, the spinal cord was hemisected directly cranial to the lumbar 1 (L1) dorsal root entry zone with a No. 15 scalpel blade. We tried not to damage the major dorsal vessel or its vascular branches. All surgical procedures were performed under visual guidance using an operation microscope. Then, the musculature and the fascia were sutured, and the skin was finally apposed. After the hemisected animals recovered in a temperature-controlled incubation chamber, they were housed individually in a cage with a thick layer of sawdust and were monitored.

4.3. Bee Venom Treatment and Experimental Groups

First, whole bee venom (Sigma, St. Louis, MO, USA; 0.25 mg/kg) was dissolved in a 50-μL volume of saline. The apposed solution was subcutaneously administered into the Joksamli (ST36) acupuncture point on the same side as the SCI surgery (ipsilateral side). The Joksamli point was located 5 mm below and lateral to the anterior tubercle of the tibia. Previously, we reported that this dose was effective in producing anti-nociception when injected into an acupuncture point, and thus, we chose the dose for evaluating the possible anti-nociceptive effects of peripheral injection [9]. Repetitive DBV or saline injections during the induction phase were initiated on the first day post-SCI surgery and were then applied twice a day (at 8 a.m. and 8 p.m., respectively) for 5 consecutive days. During the maintenance phase, repetitive DBV or saline was administered to SCI rats from Days 15 to 20 after surgery. Although previous data suggest that repetitive DBV injection does not induce pathological changes at the site of injection [8], we examined all animals receiving DBV injections into the ST 36 acupuncture point for the appearance of edema and possible infection. In addition, we massaged the injection site area daily immediately after DBV treatment in order to prevent the accumulation of DBV in the tissues.

4.4. Mechanical Allodynia Test

All behavioral assessments were performed under the ethical guidelines set forth by the International Association for the Study of Pain (IASP). Pain behavior assessments were performed one day before hemisection surgery to obtain baseline values of withdrawal responses to mechanical and heat stimuli. Then, rats were assigned randomly to each treatment group, and behavioral testing was subsequently performed blindly. During the experimental period, all behavioral tests were performed at the following time points after surgery: 3, 5, 7, 10, 14, 21 and 28 days. These tests were conducted at the same time of the day to reduce errors in relation to diurnal rhythm. Animals were placed on a metal mesh grid under a plastic chamber, and the tactile threshold was measured by applying a von Frey filament (North Coast Medical) to the mid-plantar surface of the hind paw until a positive response for withdrawal behavior was elicited. Nine calibrated fine von Frey filaments (0.40, 0.70, 1.20, 2.00, 3.63, 5.50, 8.50, 15.1 and 21.0 g) were used. They were presented serially to the hind paw in ascending order of strength with sufficient force to evoke slight bending against the paw. A brisk paw withdrawal response was considered as a positive response, for which the next filament was tested. If there was no response, the next filament was the next greater force. When animals did not respond at 21 g of pressure, the animal was recognized as being at the cut-off value. The 50% withdrawal response threshold was determined using the up-down method.

4.5. Thermal Hyperalgesia Test

To determine nociceptive responses to heat stimuli, paw withdrawal response latency (WRL) was measured using a previously described procedure [43]. Briefly, animals were placed in a plastic chamber (15 cm in diameter and 20 cm in length) on a glass floor and allowed to acclimatize for 10 min before thermal hyperalgesia testing. A radiant heat source was positioned under the glass floor beneath each hind paw, and paw withdrawal latency was measured to the nearest 0.1 s using a plantar analgesia meter (IITC Life Science Inc., Woodland Hills, CA, USA). The intensity of the light source was calibrated to produce a paw withdrawal response between 10 and 12 s in naive animals. The test was examined twice on both the ipsilateral and contralateral hind paws, and the mean withdrawal latency in each hind paw was calculated. The cutoff time was set at 20 s.

4.6. Motor Function Recovery

After the rats underwent spinal cord hemisection surgery, they were tested for motor function or coordination in an open-field test space using the BBB locomotor rating scale [36]. Briefly, the BBB scale ranges from 0 (no discernible hindlimb movement) to 21 (normal movement, including coordinated gait with parallel paw placement of the hindlimb and consistent trunk stability). Scores from 0 to 7 showed the recovery of isolated movements in the three joints (hip, knee and ankle). Scores from 8 to 13 indicated the intermediate recovery phase showing stepping, paw placement and forelimb-hindlimb coordination. In addition, scores from 14 to 21 mainly showed the late phase of recovery with toe clearance during every step phase. Only the scores of the ipsilateral hind limb on the hemisected side were examined, because there was no significant difference in locomotor function of the contralateral hind limb.

4.7. Western Blot Assay

All procedures for the Western blot assay were followed as described in our previous report [44]. After the mice were anesthetized by injecting a combination of 2.5 mg of Zoletil 50 with 0.47 mg of Rompun in saline, the spinal cord was obtained using the pressure expulsion method into a cooled saline-filled glass dish and was frozen quickly in liquid nitrogen. To investigate the functional changes of the L4-6 spinal cord segments, we verified the attachment site of spinal nerves in anesthetized rats. In addition, the extracted spinal segments were divided into ipsilateral and contralateral halves under a neurosurgical microscope. Subsequently, the ipsilateral and contralateral spinal dorsal horns were used for Western blot analysis. The spinal cords were homogenized with RIPA buffer (cell signaling, Beverly, MA, USA) containing protease inhibitor, phosphatase inhibitor and 0.1% SDS (sodium dodecyl sulfate). In addition, insoluble materials were removed by centrifugation at 12,000 g for 20 min at 4 °C. The sample protein concentrations were determined using Bradford reagents (Bio-Rad Laboratories, Hercules, CA, USA), and spinal cord lysates were separated by 10% or 15% SDS-PAGE (SDS-polyacrylamide gel electrophoresis). Subsequently, lysates were transferred to a nitrocellulose membrane. Non-specific binding was pre-blocked with 5% non-fat milk (Becton, Dickinson & Company, Franklin Lakes, NY, USA) in T-TBS and 8% bovine serum albumin (MP Biomedical) for 30 min at room temperature. Then, the membrane was incubated overnight at 4 °C with mouse anti-β-actin (1:1000, Sigma, St. Louis, MO, USA), mouse anti-GFAP antibody (1:1000, Millipore, Billerica, MA, USA) or rabbit anti-Iba1 antibody (1:1000, Abcam, Cambridge, UK) in 5% non-fat milk solution. The membrane was washed three times with T-TBS for 10 min each time and incubated with goat anti-mouse IgG horseradish peroxidase (1:5000; Calbiochem, Darmstadt, Germany) or goat anti-rabbit IgG horseradish peroxidase (1:5000; Calbiochem, Darmstadt, Germany) for 1 h at room temperature. After the membrane was washed three times with T-TBS, antibody reactive expressions were visualized using a chemiluminescence assay kit (Pharmacia-Amersham, Freiburg, Germany). The intensity of protein bands was analyzed by Image J software (Graph Pad Software, Stapleton, NY, USA, 2010).

4.8. Statistical Analysis

All data were expressed as the mean ± standard error of the mean (SEM) and analyzed statistically using the Prism 5.0 program (Graph Pad Software). Data from behavior studies were tested using two-way analysis of variance (ANOVA) in order to determine the significant effect of the repetitive DBV treatment. Bonferroni's multiple comparison test as *post hoc* analysis was also performed to determine the *p*-value among experimental groups. For Western blotting analysis, column analysis was examined by Student's *t*-test for comparisons between two mean values. $p < 0.05$ was considered statistically significant.

Acknowledgments: This research was supported by a grant (K15070) from the Korea Institute of Oriental Medicine (Daejeon, South Korea). This research was also supported by National Research Foundation (NRF) Grant (2014R1A2A2A01007695) funded by the Korean Government (Ministry of Science, ICT and Future Planning), Republic of Korea.

Author Contributions: S.Y.K. and D.H.R. designed the experiments and analyzed the data, as well as wrote the manuscript. J.W.C. partially executed the experiments, and Y.R. and J.H.L. discussed the manuscript. All authors have read and approved the final manuscript.

Conflicts of Interest: The authors declare no conflict of interest.

References

1. Somerfield, S.D.; Brandwein, S. Bee venom and adjuvant arthritis. *J. Rheumatol.* **1988**, *15*, 1878. [PubMed]
2. Billingham, M.E.; Morley, J.; Hanson, J.M.; Shipolini, R.A.; Vernon, C.A. Letter: An anti-inflammatory peptide from bee venom. *Nature* **1973**, *245*, 163–164. [CrossRef] [PubMed]
3. Kang, S.Y.; Kim, C.Y.; Roh, D.H.; Yoon, S.Y.; Park, J.H.; Lee, H.J.; Beitz, A.J.; Lee, J.H. Chemical stimulation of the ST36 acupoint reduces both formalin-induced nociceptive behaviors and spinal astrocyte activation via spinal alpha-2 adrenoceptors. *Brain Res. Bull.* **2011**, *86*, 412–421. [CrossRef] [PubMed]
4. Kwon, Y.B.; Ham, T.W.; Kim, H.W.; Roh, D.H.; Yoon, S.Y.; Han, H.J.; Yang, I.S.; Kim, K.W.; Beitz, A.J.; Lee, J.H. Water soluble fraction (<10 kDa) from bee venom reduces visceral pain behavior through spinal alpha 2-adrenergic activity in mice. *Pharmacol. Biochem. Behav.* **2005**, *80*, 181–187. [PubMed]

5. Lee, J.H.; Kwon, Y.B.; Han, H.J.; Mar, W.C.; Lee, H.J.; Yang, I.S.; Beitz, A.J.; Kang, S.K. Bee venom pretreatment has both an antinociceptive and anti-inflammatory effect on carrageenan-induced inflammation. *J. Vet. Med. Sci.* **2001**, *63*, 251–259. [CrossRef] [PubMed]

6. Kwon, Y.B.; Lee, J.D.; Lee, H.J.; Han, H.J.; Mar, W.C.; Kang, S.K.; Beitz, A.J.; Lee, J.H. Bee venom injection into an acupuncture point reduces arthritis associated edema and nociceptive responses. *Pain* **2001**, *90*, 271–280. [CrossRef]

7. Kang, S.Y.; Roh, D.H.; Yoon, S.Y.; Moon, J.Y.; Kim, H.W.; Lee, H.J.; Beitz, A.J.; Lee, J.H. Repetitive treatment with diluted bee venom reduces neuropathic pain via potentiation of locus coeruleus noradrenergic neuronal activity and modulation of spinal NR1 phosphorylation in rats. *J. Pain* **2012**, *13*, 155–166. [CrossRef] [PubMed]

8. Kang, S.Y.; Roh, D.H.; Park, J.H.; Lee, H.J.; Lee, J.H. Activation of spinal α2-adrenoceptors using diluted bee venom stimulation reduces cold allodynia in neuropathic pain rats. *Evid. Based Complement. Alternat. Med.* **2012**, *2012*, 784713. [CrossRef] [PubMed]

9. Roh, D.H.; Kwon, Y.B.; Kim, H.W.; Ham, T.W.; Yoon, S.Y.; Kang, S.Y.; Han, H.J.; Lee, H.J.; Beitz, A.J.; Lee, J.H. Acupoint stimulation with diluted bee venom (apipuncture) alleviates thermal hyperalgesia in a rodent neuropathic pain model: Involvement of spinal alpha 2-adrenoceptors. *J. Pain* **2004**, *5*, 297–303. [CrossRef] [PubMed]

10. Jones, D.G.; Anderson, E.R.; Galvin, K.A. Spinal cord regeneration: Moving tentatively towards new perspectives. *NeuroRehabilitation* **2003**, *18*, 339–351. [PubMed]

11. Nesic, O.; Lee, J.; Johnson, K.M.; Ye, Z.; Xu, G.Y.; Unabia, G.C.; Wood, T.G.; McAdoo, D.J.; Westlund, K.N.; Hulsebosch, C.E.; Regino Perez-Polo, J. Transcriptional profiling of spinal cord injury-induced central neuropathic pain. *J. Neurochem* **2005**, *95*, 998–1014. [CrossRef] [PubMed]

12. Samson, G.; Cardenas, D.D. Neurogenic bladder in spinal cord injury. *Phys. Med. Rehabil. Clin. N. Am.* **2007**, *18*, 255–274. [CrossRef] [PubMed]

13. Christensen, M.D.; Everhart, A.W.; Pickelman, J.T.; Hulsebosch, C.E. Mechanical and thermal allodynia in chronic central pain following spinal cord injury. *Pain* **1996**, *68*, 97–107. [CrossRef]

14. Siddall, P.J.; Loeser, J.D. Pain following spinal cord injury. *Spinal Cord* **2001**, *39*, 63–73. [CrossRef] [PubMed]

15. Finnerup, N.B.; Jensen, T.S. Spinal cord injury pain–mechanisms and treatment. *Eur. J. Neurol.* **2004**, *11*, 73–82. [CrossRef] [PubMed]

16. Peng, X.M.; Zhou, Z.G.; Glorioso, J.C.; Fink, D.J.; Mata, M. Tumor necrosis factor-alpha contributes to below-level neuropathic pain after spinal cord injury. *Ann. Neurol.* **2006**, *59*, 843–851. [CrossRef] [PubMed]

17. Christensen, M.D.; Hulsebosch, C.E. Chronic central pain after spinal cord injury. *J. Neurotrauma* **1997**, *14*, 517–537. [CrossRef] [PubMed]

18. Tator, C.H.; Fehlings, M.G. Review of the secondary injury theory of acute spinal cord trauma with emphasis on vascular mechanisms. *J. Neurosurg.* **1991**, *75*, 15–26. [CrossRef] [PubMed]

19. Hains, B.C.; Yucra, J.A.; Eaton, M.J.; Hulsebosch, C.E. Intralesion transplantation of serotonergic precursors enhances locomotor recovery but has no effect on development of chronic central pain following hemisection injury in rats. *Neurosci. Lett.* **2002**, *324*, 222–226. [CrossRef]

20. Hashizume, H.; de Leo, J.A.; Colburn, R.W.; Weinstein, J.N. Spinal glial activation and cytokine expression after lumbar root injury in the rat. *Spine (Phila Pa 1976)* **2000**, *25*, 1206–1217. [CrossRef]

21. Gordh, T.; Chu, H.; Sharma, H.S. Spinal nerve lesion alters blood-spinal cord barrier function and activates astrocytes in the rat. *Pain* **2006**, *124*, 211–221. [CrossRef] [PubMed]

22. Obata, H.; Eisenach, J.C.; Hussain, H.; Bynum, T.; Vincler, M. Spinal glial activation contributes to postoperative mechanical hypersensitivity in the rat. *J. Pain* **2006**, *7*, 816–822. [CrossRef] [PubMed]

23. Hulsebosch, C.E. Gliopathy ensures persistent inflammation and chronic pain after spinal cord injury. *Exp. Neurol.* **2008**, *214*, 6–9. [CrossRef] [PubMed]

24. Carlton, S.M.; Du, J.; Tan, H.Y.; Nesic, O.; Hargett, G.L.; Bopp, A.C.; Yamani, A.; Lin, Q.; Willis, W.D.; Hulsebosch, C.E. Peripheral and central sensitization in remote spinal cord regions contribute to central neuropathic pain after spinal cord injury. *Pain* **2009**, *147*, 265–276. [CrossRef] [PubMed]

25. Gwak, Y.S.; Crown, E.D.; Unabia, G.C.; Hulsebosch, C.E. Propentofylline attenuates allodynia, glial activation and modulates GABAergic tone after spinal cord injury in the rat. *Pain* **2008**, *138*, 410–422. [CrossRef] [PubMed]

26. Gwak, Y.S.; Hulsebosch, C.E. Remote astrocytic and microglial activation modulates neuronal hyperexcitability and below-level neuropathic pain after spinal injury in rat. *Neuroscience* **2009**, *161*, 895–903. [CrossRef] [PubMed]

27. Marchand, F.; Tsantoulas, C.; Singh, D.; Grist, J.; Clark, A.K.; Bradbury, E.J.; McMahon, S.B. Effects of Etanercept and Minocycline in a rat model of spinal cord injury. *Eur. J. Pain* **2009**, *13*, 673–681. [CrossRef] [PubMed]

28. Guerra de Hoyos, J.A.; Andrés Martín Mdel, C.; Bassas y Baena de Leon, E.; Vigára Lopez, M.; Molina López, T.; Verdugo Morilla, F.A.; González Moreno, M.J. Randomised trial of long term effect of acupuncture for shoulder pain. *Pain* **2004**, *112*, 289–298. [CrossRef] [PubMed]

29. Huang, C.; Hu, Z.P.; Long, H.; Shi, Y.S.; Han, J.S.; Wan, Y. Attenuation of mechanical but not thermal hyperalgesia by electroacupuncture with the involvement of opioids in rat model of chronic inflammatory pain. *Brain Res. Bull.* **2004**, *63*, 99–103. [CrossRef] [PubMed]

30. Hoffman, D.R. Hymenoptera venom allergens. *Clin. Rev. allErgy Immunol.* **2006**, *30*, 109–128. [CrossRef]

31. Burton, A.W.; Lee, D.H.; Saab, C.; Chung, J.M. Preemptive intrathecal ketamine injection produces a long-lasting decrease in neuropathic pain behaviors in a rat model. *Reg. Anesth. Pain Med.* **1999**, *24*, 208–213. [PubMed]

32. Tan, A.M.; Zhao, P.; Waxman, S.G.; Hains, B.C. Early microglial inhibition preemptively mitigates chronic pain development after experimental spinal cord injury. *J. Rehabil. Res. Dev.* **2009**, *46*, 123–133. [CrossRef] [PubMed]

33. Romero-Sandoval, E.A.; Horvath, R.J.; de Leo, J.A. Neuroimmune interactions and pain: Focus on glial-modulating targets. *Curr. Opin. Investig. Drugs* **2008**, *9*, 726–734. [PubMed]

34. Milligan, E.D.; Watkins, L.R. Pathological and protective roles of glia in chronic pain. *Nat. Rev. Neurosci.* **2009**, *10*, 23–36. [CrossRef] [PubMed]

35. Zhang, T.; Zhang, J.; Shi, J.; Feng, Y.; Sun, Z.S.; Li, H. Antinociceptive synergistic effect of spinal mGluR2/3 antagonist and glial cells inhibitor on peripheral inflammation-induced mechanical hypersensitivity. *Brain Res. Bull.* **2009**, *79*, 219–223. [CrossRef] [PubMed]

36. Roh, D.H.; Yoon, S.Y.; Seo, H.S.; Kang, S.Y.; Han, H.J.; Beitz, A.J.; Lee, J.H. Intrathecal injection of carbenoxolone, a gap junction decoupler, attenuates the induction of below-level neuropathic pain after spinal cord injury in rats. *Exp. Neurol.* **2010**, *224*, 123–132. [CrossRef] [PubMed]

37. Basso, D.M.; Beattie, M.S.; Bresnahan, J.C. A sensitive and reliable locomotor rating scale for open field testing in rats. *J. Neurotrauma* **1995**, *12*, 1–21. [CrossRef] [PubMed]

38. Cheng, H.; Cao, Y.; Olson, L. Spinal cord repair in adult paraplegic rats: Partial restoration of hind limb function. *Science* **1996**, *273*, 510–513. [CrossRef] [PubMed]

39. Yiu, G.; He, Z. Glial inhibition of CNS axon regeneration. *Nat. Rev. Neurosci.* **2006**, *7*, 617–627. [CrossRef] [PubMed]

40. Iwashita, Y.; Kawaguchi, S.; Murata, M. Restoration of function by replacement of spinal cord segments in the rat. *Nature* **1994**, *367*, 167–170. [CrossRef] [PubMed]

41. Edgerton, V.R.; Tillakaratne, N.J.; Bigbee, A.J.; de Leon, R.D.; Roy, R.R. Plasticity of the spinal neural circuitry after injury. *Annu. Rev. Neurosci.* **2004**, *27*, 145–167. [CrossRef] [PubMed]

42. Marqueste, T.; Decherchi, P.; Messan, F.; Kipson, N.; Grélot, L.; Jammes, Y. Eccentric exercise alters muscle sensory motor control through the release of inflammatory mediators. *Brain Res.* **2004**, *1023*, 222–230. [CrossRef] [PubMed]

43. Hargreaves, K.; Dubner, R.; Brown, F.; Flores, C.; Joris, J. A new and sensitive method for measuring thermal nociception in cutaneous hyperalgesia. *Pain* **1988**, *32*, 77–88. [CrossRef]

44. Patterson, W.B. Surgical issues in geriatric oncology. *Semin. Oncol.* **1989**, *16*, 57–65. [PubMed]

© 2015 by the authors; licensee MDPI, Basel, Switzerland. This article is an open access article distributed under the terms and conditions of the Creative Commons Attribution (CC-BY) license (http://creativecommons.org/licenses/by/4.0/).

toxins

MDPI

Article

Analgesic Effects of Bee Venom Derived Phospholipase A$_2$ in a Mouse Model of Oxaliplatin-Induced Neuropathic Pain

Dongxing Li [1], Younju Lee [1], Woojin Kim [2], Kyungjin Lee [3], Hyunsu Bae [1] and Sun Kwang Kim [1,2,*]

[1] Department of Physiology, College of Korean Medicine, Kyung Hee University, 26 Kyungheedae-ro, Dongdamoon-gu, Seoul 130-701, Korea; leedongxing@naver.com (D.L.); eyounju@naver.com (Y.L.); hbae@khu.ac.kr (H.B.)

[2] Department of East-West Medicine, Graduate School, Kyung Hee University, 26 Kyungheedae-ro, Dongdamoon-gu, Seoul 130-701, Korea; thasnow@gmail.com

[3] Department of Herbology, College of Korean Medicine, Kyung Hee University, 26 Kyungheedae-ro, Dongdamoon-gu, Seoul 130-701, Korea; niceday@khu.ac.kr

* Correspondence: skkim77@khu.ac.kr; Tel.: +82-2-961-0491; Fax: +82-7-4194-9316

Academic Editor: Sokcheon Pak

Received: 15 May 2015; Accepted: 23 June 2015; Published: 29 June 2015

Abstract: A single infusion of oxaliplatin, which is widely used to treat metastatic colorectal cancer, induces specific sensory neurotoxicity signs that are triggered or aggravated when exposed to cold or mechanical stimuli. Bee Venom (BV) has been traditionally used in Korea to treat various pain symptoms. Our recent study demonstrated that BV alleviates oxaliplatin-induced cold allodynia in rats, via noradrenergic and serotonergic analgesic pathways. In this study, we have further investigated whether BV derived phospholipase A$_2$ (bvPLA$_2$) attenuates oxaliplatin-induced cold and mechanical allodynia in mice and its mechanism. The behavioral signs of cold and mechanical allodynia were evaluated by acetone and a von Frey hair test on the hind paw, respectively. The significant allodynia signs were observed from one day after an oxaliplatin injection (6 mg/kg, i.p.). Daily administration of bvPLA$_2$ (0.2 mg/kg, i.p.) for five consecutive days markedly attenuated cold and mechanical allodynia, which was more potent than the effect of BV (1 mg/kg, i.p.). The depletion of noradrenaline by an injection of *N*-(2-chloroethyl)-*N*-ethyl-2-bromobenzylamine hydrochloride (DSP4, 50 mg/kg, i.p.) blocked the analgesic effect of bvPLA$_2$, whereas the depletion of serotonin by injecting DL-*p*-chlorophenylalanine (PCPA, 150 mg/kg, i.p.) for three successive days did not. Furthermore, idazoxan (α2-adrenegic receptor antagonist, 1 mg/kg, i.p.) completely blocked bvPLA$_2$-induced anti-allodynic action, whereas prazosin (α1-adrenegic antagonist, 10 mg/kg, i.p.) did not. These results suggest that bvPLA$_2$ treatment strongly alleviates oxaliplatin-induced acute cold and mechanical allodynia in mice through the activation of the noradrenergic system, via α2-adrenegic receptors, but not via the serotonergic system.

Keywords: phospholipase A$_2$; Bee Venom; oxaliplatin; neuropathic pain; noradrenergic

1. Introduction

Oxaliplatin is an effective platinum derivative, which is widely used in the treatment of colorectal carcinoma [1,2], but causes neurotoxicity predominantly within the peripheral nervous system [3,4]. Two different types of oxaliplatin-induced peripheral neuropathy have been described hitherto, *i.e.*, cold and mechanical hypersensitivity [5,6]. However, effective treatment for oxaliplatin-induced cold and mechanical hypersensitivity still remains to be elucidated. Hence, it is required to discover therapeutic options for the management of oxaliplatin-induced neuropathic pain.

Bee Venom (BV) has been traditionally used in Korea to relieve pain and to treat chronic inflammatory diseases [7–12]. Previous studies have demonstrated that the analgesic effects of BV in various pain models are mediated mainly by activation of α2-adrenergic and/or serotonergic receptors [12–16]. Rho and his colleagues have reported that subcutaneous injections of BV attenuated heat hyperalgesia and cold and mechanical allodynia in the rats with nerve injury-induced neuropathic pain through the activation of the endogenous noradrenergic system [17,18]. In a rat model of oxaliplatin-induced neuropathic pain, we found that the anti-allodynic effect of BV is at least partially mediated by the noradrenergic and serotonergic system, but not by the opioid system [9,16].

Phospholipase A_2 from BV (bvPLA$_2$), a prototypic group III enzyme that hydrolyzes fatty acids in membrane phospholipids, is one of the major active components of BV [19,20]. Several studies have shown that this bvPLA$_2$ prevents neuronal cell death and spinal cord injury [21,22]. Our previous study demonstrated that BV mitigates cisplatin-induced nephrotoxicity [23] and found that bvPLA$_2$ can reduce such nephrotoxicity more potently than BV [24]. However, the effect of PLA$_2$ on oxaliplatin-induced neuropathic pain and its mechanism have not been studied yet.

The aim of this study was to evaluate and compare the analgesic effect of BV and bvPLA$_2$ on oxaliplatin-induced cold and mechanical allodynia in mice. In addition, we examined whether the anti-allodynic effect of bvPLA$_2$ is mediated by the serotonergic or noradrenergic pain inhibitory system.

2. Results

2.1. Effects of BV and bvPLA$_2$ on Oxaliplatin-Induced Cold and Mechanical Hypersensitivity

First, we investigated the effects of a single administration of oxaliplatin (6 mg/kg, i.p.) on behavioral sensitivity to cold and mechanical stimuli in mice (Figure 1). The administration significantly increased the frequency of licking and shaking of the hind paw in response to cold acetone stimuli. A significant cold allodynia was observed at day 1, peaked at day 3 and lasted for at least seven days after the oxaliplatin administration, compared to the vehicle group (Figure 1a). Similarly, an administration of oxaliplatin significantly increased the withdrawal responses of the hind paw to von Frey filament applications (as expressed % of withdrawal response) at day 1, peaked at day 3–4 and maintained up to day 5 (Figure 1b).

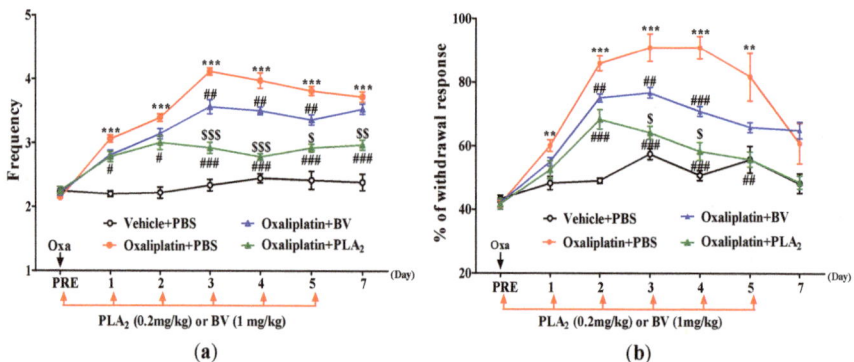

Figure 1. Effects of BV and bvPLA$_2$ on oxaliplatin-induced cold and mechanical allodynia in mice. The behavioral tests for cold (**a**) and mechanical (**b**) allodynia were performed before (PRE) and after the administration of oxaliplatin (6 mg/kg, i.p.). Vehicle (5% glucose) + PBS, Oxaliplatin + PBS, Oxaliplatin + BV, and Oxaliplatin + bvPLA$_2$ groups received daily injection of PBS, BV (1 mg/kg, i.p.) or bvPLA$_2$ (0.2 mg/kg, i.p.) for five days after an oxaliplatin or vehicle administration. Results are expressed as mean $\pm$ SEM; n = 6 mice/group; The data was analyzed with one-way ANOVA followed by the Tukey's multiple comparison test. ** $p < 0.01$, *** $p < 0.001$, *vs.* Vehicle + PBS; # $p < 0.05$, ## $p < 0.01$, ### $p < 0.001$ *vs.* Oxaliplatin + PBS; $ $p < 0.05$, $$ $p < 0.01$, $$$ $p < 0.001$, *vs.* Oxaliplatin + BV.

Daily BV treatment (1 mg/kg, i.p.) for five consecutive days significantly reduced the cold allodynia from three days after the oxaliplatin administration and such analgesic effect last up to day 5. In addition, bvPLA$_2$ treatment (0.2 mg/kg, i.p.) significantly attenuated the cold allodynia from day 1 after the oxaliplatin injection and such effect endured at least for the following six days (Figure 1a). BV treatment also significantly attenuated mechanical allodynia from day 2 after the oxaliplatin injection and this BV analgesia was continued up to day 4. Moreover, bvPLA$_2$ treatment showed a significant reduction in mechanical allodynia from day 2 after the oxaliplatin injection and such effect lasted up to day 5 (Figure 1b). The relieving effects of bvPLA$_2$ on oxaliplatin-induced cold and mechanical allodynia were significantly more potent than those of BV (Figure 1).

2.2. Effects of BvPLA$_2$ on Oxaliplatin-Induced Cold and Mechanical Allodynia in Serotonin Depleted Mice

We investigated the effects of bvPLA$_2$ on oxaliplatin-induced cold and mechanical allodynia in serotonin depleted mice by injecting DL-*p*-chlorophenylalanine (PCPA, 150 mg/kg, i.p.) for three successive days [25,26]. PCPA pretreatment itself did not affect the behavioral signs of cold and mechanical allodynia induced by oxaliplatin ($p > 0.05$, Oxaliplatin + PBS + PCPA [$n = 6$] *vs.* Oxaliplatin + PBS + NS [$n = 4$], Cold (frequency): 1.83 ± 0.04 *vs.* 1.92 ± 0.05 at 0d, 4.61 ± 0.07 *vs.* 4.67 ± 0.07 at 3d, 3.83 ± 0.06 *vs.* 3.79 ± 0.08 at 5d, 3.25 ± 0.07 *vs.* 3.04 ± 0.10 at 7d; Mechanical (%): 30.00 ± 1.83 *vs.* 30.00 ± 0.00 at 0d, 94.17 ± 0.83 *vs.* 93.75 ± 1.25 at 3d, 72.03 ± 1.54 *vs.* 68.75 ± 1.25 at 5d, 50 ± 2.24 *vs.* 51.25 ± 1.25 at 7d). Thus, we pooled the data from the two groups as a control group (Oxaliplatin + PBS + PCPA/NS, Figure 2). Compared to this control group, bvPLA$_2$ treatment in mice without serotonin depletion (Oxaliplatin + PLA$_2$ + NS group) significantly attenuated the cold and mechanical hypersensitivity ($p < 0.01$ at days 3 and 5). Such anti-allodynic effects of bvPLA$_2$ were not blocked by PCPA pretreatment (Figure 2), indicating that the serotonergic mechanism is not involved in the analgesic effect of bvPLA$_2$ on oxaliplatin-induced neuropathic pain.

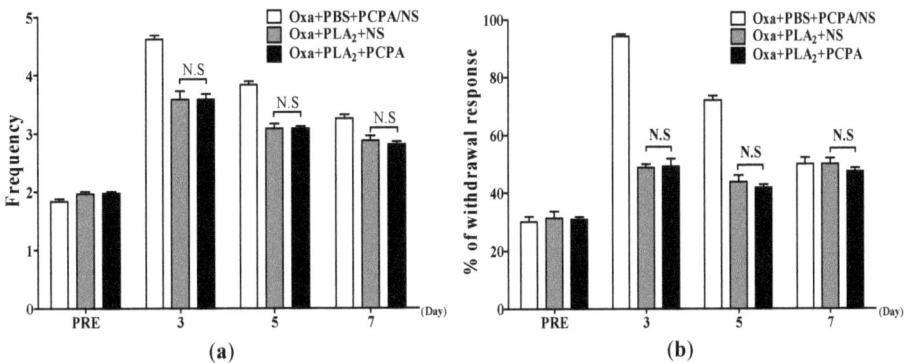

Figure 2. Effects of bvPLA$_2$ on oxaliplatin-induced cold and mechanical allodynia in serotonin depleted mice. The behavioral tests for cold (**a**) and mechanical (**b**) allodynia were performed before and after an administration of oxaliplatin (Oxa, 6 mg/kg, i.p.). Serotonin was depleted by daily injections of PCPA (150 mg/kg, i.p.) for three consecutive days prior to an oxaliplatin administration. Oxa + PBS + PCPA/NS, $n = 10$; Other groups, $n = 6$ mice/group; Results are expressed as mean $\pm$ SEM; N.S, no significance ($p > 0.05$), The data were analyzed with one-way ANOVA followed by the Tukey's multiple comparison test.

2.3. Noradrenergic Mechanism of the Anti-Allodynic Effects of BvPLA$_2$ in Oxaliplatin-Administered Mice

We evaluated the effects of bvPLA$_2$ on oxaliplatin-induced allodynia in noradrenaline depleted mice by a pretreatment of DSP4 [27]. The anti-allodynic effects of bvPLA$_2$ (Oxa + PLA$_2$ + NS group, $p < 0.01$ *vs.* control Oxa + PBS + DSP4/NS group at days 3, 5 and 7) were significantly blocked by DSP4 pretreatment (Figure 3), unlike the aforementioned PCPA pretreatment. These results suggest that activation of the noradrenergic pain inhibitory pathway at least partially mediates the bvPLA$_2$-induced anti-allodynic action in oxaliplatin-administered mice.

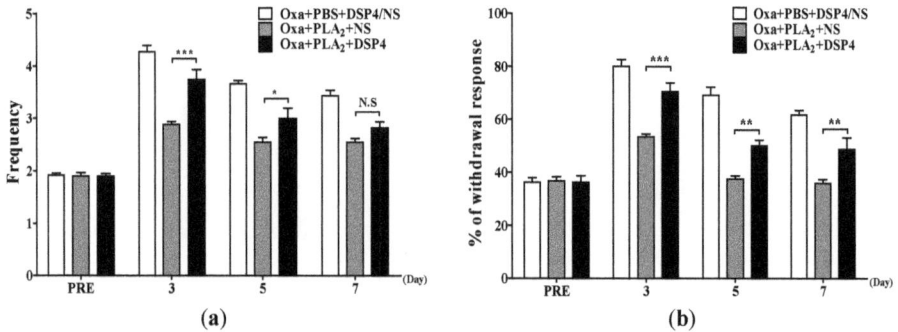

Figure 3. Effects of bvPLA$_2$ on oxaliplatin-induced cold and mechanical allodynia in noradrenaline depleted mice. The behavioral tests for cold (**a**) and mechanical (**b**) allodynia were performed before and after an administration of oxaliplatin (6 mg/kg, i.p.). Noradrenaline was depleted by an injection of DSP4 (50 mg/kg, i.p.) at a day before an oxaliplatin administration. Since DSP4 pretreatment itself did not affect the cold and mechanical allodynia signs induced by oxaliplatin ($p > 0.05$, Oxa + PBS + DSP4 [$n = 6$] *vs.* Oxa + PBS + NS [$n = 6$]), we pooled the data from the two groups as a control (Oxa + PBS + DSP4/NS, n = 12). Other groups, $n = 6$ mice/group; Results are expressed as mean ± SEM; N.S, no significance ($p > 0.05$), * $p < 0.05$, ** $p < 0.01$, *** $p < 0.001$, The data was analyzed with one-way ANOVA followed by the Tukey's multiple comparison test.

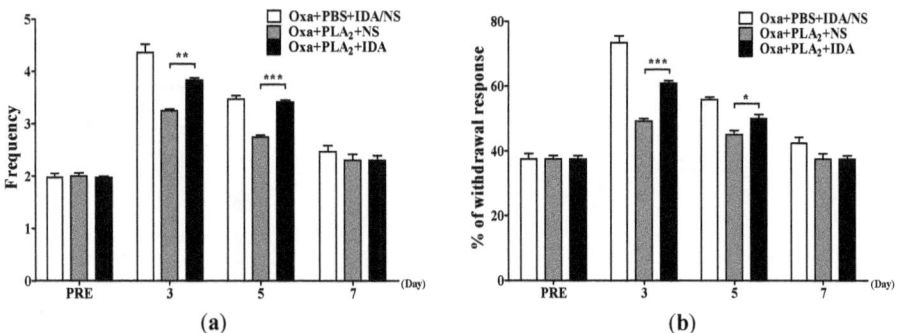

Figure 4. Effect of α2-adrenergic receptor antagonist, idazoxan, on bvPLA$_2$-induced anti-allodynic action in oxaliplatin-administered mice. The behavioral tests for cold (**a**) and mechanical (**b**) allodynia were performed before and after an administration of oxaliplatin (6 mg/kg, i.p.). Idazoxan (IDA, 1 mg/kg, i.p.) was administered 30 min prior to bvPLA$_2$ injection. Since IDA pretreatment itself did not affect the cold and mechanical allodynia signs induced by oxaliplatin ($p > 0.05$, Oxa + PBS + IDA [$n = 6$] *vs.* Oxa + PBS + NS [$n = 6$]), we pooled the data from the two groups as a control (Oxa + PBS + IDA/NS, $n = 12$). Other groups, $n = 6$ mice/group; Results are expressed as mean ± SEM; * $p < 0.05$, ** $p < 0.01$, *** $p < 0.001$, The data was analyzed with one-way ANOVA followed by the Tukey's multiple comparison test.

To identify which adrenergic receptor subtype mediates the analgesic effects of bvPLA$_2$ on oxaliplatin-induced neuropathic pain in mice, we examined the effect of prazosin (PRA, α1-adrenergic receptor antagonist) or idazoxan (IDA, α2-adrenergic receptor antagonist) on the bvPLA$_2$-induced anti-allodynic action. As shown in Figure 4, IDA (1 mg/kg, i.p.) significantly blocked the relieving effect of bvPLA$_2$ on oxaliplatin-induced cold and mechanical allodynia. However, PRA (10 mg/kg, i.p.) did not affect the bvPLA$_2$ effect, because there were no significant differences in cold and mechanical sensitivity between the Oxa + PLA$_2$ + NS and Oxa + PLA$_2$ + PRA groups ($p > 0.05$, Table 1). These results indicate that bvPLA$_2$ treatment alleviates oxaliplatin-induced acute cold and mechanical allodynia in mice via activation of α2-adrenergic receptors, but not α1-adrenergic receptors.

Table 1. Effect of α1-adrenergic receptor antagonist, prazosin (PRA) on bvPLA$_2$-induced anti-allodynic action in oxaliplatin-administered mice.

Post-Oxaliplatin Day	D0	D3	D5	D7
Acetone test		Frequency		
Oxa + PBS + PRA/NS	2.1 ± 0.04	4.4 ± 0.08	3.6 ± 0.15	3.3 ± 0.09
Oxa + PLA$_2$ + NS	2.0 ± 0.06	3.5 ± 0.15 ***	3.0 ± 0.10 **	2.6 ± 0.10 ***
Oxa + PLA$_2$ + PRA	2.1 ± 0.06	3.1 ± 0.12 ***	3.1 ± 0.09 *	2.8 ± 0.07 **
von Frey test <0.4 g>		% withdrawal	response	
Oxa + PBS + PRA/NS	40.8 ± 0.83	76.7 ± 1.67	55.8 ± 1.54	44.2 ± 1.54
Oxa + PLA$_2$ + NS	40.0 ± 1.83	59.2 ± 2.39 ***	48.1 ± 1.29 **	41.5 ± 1.12
Oxa + PLA$_2$ + PRA	40.8 ± 2.01	57.5 ± 1.71 ***	47.5 ± 1.71 **	40.33 ± 1.67

Oxa + PBS + PRA/NS, $n = 12$; Other groups, $n = 6$ mice/group; Results are expressed as mean ± SEM; * $p < 0.05$, ** $p < 0.01$, *** $p < 0.001$, *vs.* Oxa + PBS + PRA/NS, one-way ANOVA followed by the Tukey's multiple comparison test.

3. Discussion

Oxaliplatin-induced neuropathic pain represents a major obstacle to successful cancer treatment as it restricts both individual and cumulative dosages. However, despite these limitations, it is widely used and many patients suffer from the development of long-lasting consequences (*i.e.*, peripheral neuropathy) [28,29]. Patients may also experience cold-induced paresthesias, throat and jaw tightness, and occasionally focal weakness [30]. Oxaliplatin is structurally similar to other platinum based chemotherapy drugs, such as cisplatin and carboplatin. They all have neurotoxicity, however oxaliplatin has little nephrotoxicity and hematotoxicity [31]. It has been shown that oxaliplatin-induced acute neuropathy is characterized by a specific somatosensory profile, *i.e.*, cold and mechanical hypersensitivity [32]. Therefore, cold and mechanical hypersensitivity is a hallmark of oxaliplatin-induced neuropathy. Indeed, in this study, a single intraperitoneal administration of oxaliplatin (6 mg/kg) significantly increased the cold and mechanical sensitivity in mice, corroborating the previous reports using rats [33,34]. This mouse model might help in exploring the molecular and genetic mechanism of oxaliplatin-induced neuropathic pain in the future studies, since transgenic and gene knock-out/knock-in animals have been developed primarily in the mouse [35–37].

There have been few reports about the effective treatment and prevention of oxaliplatin-induced neuropathic pain. However, our previous study has shown that BV has a beneficial role in reliving oxaliplatin-induced neuropathic pain symptoms in rats, suggesting that BV could be an alternative therapeutic option for the management of oxaliplatin-induced peripheral neuropathy [9,16]. Interestingly, we recently found that bvPLA$_2$ mitigates cisplatin-induced nephrotoxicity and acetaminophen-induced hepatotoxicity, which was more potent than BV [24,38]. In the present study, we investigated for the first time whether bvPLA$_2$ has an analgesic effect on oxaliplain-induced cold and mechanical allodynia. Our data showed that the treatment of bvPLA$_2$ significantly alleviated the allodynia in oxaliplatin-administered mice and such bvPLA$_2$ effect was superior to the BV effect. This powerful analgesic effect of bvPLA$_2$ led us to investigate the underlying mechanisms.

The extensive data support a role for the monoamine neurotransmitters (*i.e.*, serotonin and noradrenaline) and opioids, in the modulation of pain [39,40]. Serotonin and noradrenaline have been implicated as principal mediators of endogenous analgesic mechanisms in the descending pain pathways [39]. Our previous study suggested that the anti-allodynic effect of BV on oxaliplatin-induced cold allodynia in rats involves the noradrenergic, but not opioid, system [9]. In addition, spinal 5-HT$_3$ receptors play an important role in the BV-induced anti-allodynic action in oxaliplatin-injected rats [16]. In contrast, the present study clearly showed that serotonin depletion by pretreatment of PCPA did not significantly affect the anti-allodynic effects of bvPLA$_2$ in oxaliplatin-administered mice. This result suggests that the serotonergic inhibitory pathway is not involved in the analgesic effect of bvPLA$_2$ on oxaliplatin-induced cold and mechanical hypersensitivity, unlike the case of BV. Other active components of BV, such as apamin [41], might be responsible for the serotonergic mechanism of BV-induced analgesia. However, in noradrenaline-depleted mice, the suppressive effect of bvPLA$_2$ on oxaliplatin-induced cold and mechanical allodynia was significantly prevented. These results indicate that the noradrenergic analgesic system is at least partially involved in the analgesic effects of bvPLA$_2$ in oxaliplatin-administered mice.

α-adrenergic receptors have been demonstrated to play an important role in the modulation of cold allodynia via the noradrenergic pain inhibitory system [42]. Previous studies showed that either α1- [43,44] or α2-adrenergic receptors [45,46] are responsible for the adrenergic sensitivity of nerve injury-induced neuropathic rats. A dual contribution of α1- and α2-adrenergic receptors to neuropathic pain was also suggested [47,48]. Also other articles have elucidated that the analgesic effects of BV are mediated mainly by activation of α2-adrenergic receptors in various pain models, such as nerve injury-induced neuropathic pain, acetic acid-induced visceral pain and inflammatory pain [12–15]. Our recent study also showed that α2-adrenergic receptors mediate the anti-allodynic effect of BV in oxaliplatin-induced neuropathic pain in rats [9]. In the present study, we further examined which adrenergic receptor subtype mediates the effects of bvPLA$_2$ on cold and mechanical allodynia in oxaliplatin-administered mice. The current results showed that IDA (α2-adrenergic antagonist) was able to completely block the anti-allodynic effect of bvPLA$_2$ on oxaliplatin-induced neuropathic pain in mice, whereas PRA (α1-adrenergic antagonist) did not. These results indicate that the noradrenergic mechanism of the analgesic effect of bvPLA$_2$ on oxaliplatin-induced neuropathic pain is mediated by activation of α2-adrenergic, but not α1-adrenergic, receptors.

In this study, we have clearly shown that bvPLA$_2$ treatment strongly alleviates oxaliplatin-induced acute cold and mechanical allodynia in mice through the activation of the noradrenergic system, via α2-adrenegic receptors. Besides such action through the neurochemical mechanism, bvPLA$_2$ is known to have a potent immune modulatory effect. The major pathway of the bvPLA$_2$-induced immune modulation is to increase peripheral regulatory T cells, which play a key role in the maintenance of tolerance in the immune system. Our recent studies showed that such strategies using bvPLA$_2$ could be successful in the prevention of cisplatin-induced acute kidney and acetaminophen-induced acute liver injury, by suppressing immune response via the modulation of regulatory T cells [24,38]. Another previous study also demonstrated that regulatory T cells attenuate peripheral nerve injury-induced neuropathic pain in rats [49]. Thus, it would be of high interest to see if bvPLA$_2$ treatment before an oxaliplatin administration prevents the development of neuropathic cold and mechanical allodynia by regulating peripheral immune response. To our best knowledge, there are no clinical trials for bvPLA$_2$ treatment. Further research is required in this field. In addition, our current work is limited to behavioral and pharmacological approaches. Molecular and genetic studies using the advantage of the mouse model are now in progress to elucidate more detailed mechanism of bvPLA$_2$-induced analgesia.

4. Experimental Section

4.1. Animals

Male C57BL/6 mice (6–8 weeks old) were purchased from the Daehan Biolink (Chungbuk, Korea). They were kept under specific pathogen-free conditions with air conditioning and a 12 h light/dark cycle. The mice had free access to food and water during the experiments. The study was approved by the Kyung Hee University Animal Care and Use Committee (KHUASP(SE)-15-024).

4.2. Behavioral Tests

Behavioral tests representing different sensory components of neuropathic pain were conducted before and after an oxaliplatin administration. The mice were habituated to handling by investigators and to all testing procedures for a week before the start of the experiments. The experimenters were blind to oxaliplatin and any treatments.

Cold sensitivity was measured by an acetone test [50]. Mice were placed in a clear plastic box (12 × 8 × 6 cm) with a wire mesh floor and allowed to habituate for 30 min prior to the testing. Acetone (10 μL, Reagents Chemical Ltd., Gyonggi-do, Korea) was sprayed onto the plantar skin of each hind paw 3 times and the frequency of licking and shaking of the affected paw was counted after the acetone spray for 30 s. The advantage of acetone test is that it is quite simple, economical, and that the assessment for multiple mice can be made in a short period of time. The disadvantage is that acetone can also induce some behavioral response in naïve mice. However, mice having allodynia (e.g., oxaliplatin-administered mice) show a significant increase in the level of response to acetone when compared to the control mice.

Mechanical sensitivity was measured by the von Frey hair test [51]. Mice were placed in a clear plastic box (12 × 8 × 6 cm) with a wire mesh floor and allowed to habituate for 30 min before the testing. A von Frey filament (Linton Instrumentation, Norfolk, UK) with a bending force of 0.4 g were applied to the midplantar skin (avoiding the base of the tori) of each hind paw 10 times, with each application held for 3 s [52]. The number of withdrawal responses to the von Frey filament applications from both hind paws were counted and then expressed as an overall percentage response.

4.3. Oxaliplatin Administration and BV/bvPLA₂ Treatment

Oxaliplatin (Sigma, St. Louis, MO, USA) was dissolved in a 5% glucose solution at a concentration of 2 mg/mL and was intraperitoneally (i.p.) injected at 6 mg/kg [9,33]. The vehicle control group received the same volume of a 5% glucose solution through the same injection route.

BV (1 mg/kg, i.p., Sigma) or bvPLA₂ (0.2 mg/kg, i.p., Sigma) [9,24,38] dissolved in phosphate buffered saline (PBS) was injected in oxaliplatin-administered mice for a continuous five days. Cold and mechanical sensitivity were measured by acetone test and von Frey hair test, respectively, before each bvPLA₂ or BV treatment. Control group was treated intraperitoneally with PBS.

4.4. Depletion of Serotonin or Noradrenaline

DL-*p*-chlorophenylalanine (PCPA, Sigma, an inhibitor of serotonin synthesis, 150 mg/kg/day) or vehicle (normal saline, NS) was injected intraperitoneally to mice prior to oxaliplatin administration for three days. The dosage and treatment course of PCPA have been widely used to deplete 5-HT stores [25,26]. After 5-HT depletion with PCPA, oxaliplatin and bvPLA2 were administered as aforementioned.

N-(2-Chloroethyl)-N-ethyl-2-bromobenzylamine hydrochloride (DSP4, TOCRIS, 50 mg/kg) or vehicle (NS) was injected intraperitoneally to mice a day before an oxaliplatin administration. DSP4 at a concentration of 50 mg/kg has been shown to be an effective dose for maximal NE depletion [53], with the advantage that mice did not require special care following injection as no adverse effects could be observed.

4.5. α1- or α2-Adrenergic Receptor Antagonist

To test which adrenergic receptor subtype mediates the anti-allodynic effects of bvPLA₂ in oxaliplatin-administered mice, specific antagonists were administered intraperitoneally 30 min prior to bvPLA₂ treatment for five days: α1-adrenergic receptor antagonist (prazosin, 10 mg/kg, Sigma), α2-adrenergic receptor antagonist (idazoxan, 1 mg/kg, Sigma). The dose of each antagonist was determined based on the previous studies showing the selective and effective antagonistic action against adrenergic receptor-mediated responses [54–56].

4.6. Statistical Analyses

The data are presented as mean ± SEM and were analyzed by the unpaired t-test or one-way ANOVA followed by the Tukey's multiple comparison test to determine the statistical differences among the groups. $p < 0.05$ was considered as statistically significant.

5. Conclusions

In conclusion, our findings reveal that cold and mechanical sensitivity were significantly increased after a single injection of oxaliplatin in mice. BV and bvPLA$_2$ can exert significant relieving effects on oxaliplatin-induced cold and mechanical hypersensitivity, in which bvPLA$_2$ is more potent than BV. The serotonergic mechanism is not involved in the analgesic effect of bvPLA$_2$ on oxaliplatin-induced neuropathic pain, whereas the noradrenergic pain inhibitory system at least partially mediates the bvPLA$_2$ effect. Finally, we demonstrated that bvPLA$_2$ treatment alleviates oxaliplatin-induced acute cold and mechanical allodynia in mice via activation of α2-adrenergic receptors. These findings may provide clinically useful evidence for the application of bvPLA$_2$ in the management of peripheral neuropathic pain that occurs after the oxaliplatin administration.

Acknowledgments: This research was supported by a grant of the Korea Health Technology R&D Project through the Korea Health Industry Development Institute (KHIDI), funded by the Ministry of Health & Welfare, Republic of Korea (grant number: HI14 C0738).

Author Contributions: Hyunsu Bae and Sun Kwang Kim conceived and designed the study. Dongxing Li and Younju Lee performed the experiments. Dongxing Li, Woojin Kim, Kyungjin Lee and Sun Kwang Kim analyzed the data. Dongxing Li, Younju Lee, Woojin Kim, Kyungjin Lee and Sun Kwang Kim wrote the manuscript. All authors have read and approved the final manuscript.

Conflicts of Interest: The authors declare no conflict of interest.

References

1. Andre, T.; Boni, C.; Mounedji-Boudiaf, L.; Navarro, M.; Tabernero, J.; Hickish, T.; Topham, C.; Zaninelli, M.; Clingan, P.; Bridgewater, J.; *et al.* Oxaliplatin, fluorouracil, and leucovorin as adjuvant treatment for colon cancer. *N. Engl. J. Med.* **2004**, *350*, 2343–2351. [CrossRef] [PubMed]

2. De Gramont, A.; Figer, A.; Seymour, M.; Homerin, M.; Hmissi, A.; Cassidy, J.; Boni, C.; Cortes-Funes, H.; Cervantes, A.; Freyer, G.; *et al.* Leucovorin and fluorouracil with or without oxaliplatin as first-line treatment in advanced colorectal cancer. *J. Clin. Oncol.* **2000**, *18*, 2938–2947. [PubMed]

3. Cassidy, J.; Misset, J.L. Oxaliplatin-related side effects: Characteristics and management. *Semin. Oncol.* **2002**, *29*, 11–20. [CrossRef] [PubMed]

4. Extra, J.M.; Marty, M.; Brienza, S.; Misset, J.L. Pharmacokinetics and safety profile of oxaliplatin. *Semin. Oncol.* **1998**, *25*, 13–22. [PubMed]

5. Pasetto, L.M.; D'Andrea, M.R.; Rossi, E.; Monfardini, S. Oxaliplatin-related neurotoxicity: How and why? *Crit. Rev. Oncol. Hematol.* **2006**, *59*, 159–168. [CrossRef] [PubMed]

6. Grothey, A. Oxaliplatin-safety profile: Neurotoxicity. *Semin. Oncol.* **2003**, *30*, 5–13. [CrossRef]

7. Billingham, M.E.; Morley, J.; Hanson, J.M.; Shipolini, R.A.; Vernon, C.A. Letter: An anti-inflammatory peptide from bee venom. *Nature* **1973**, *245*, 163–164. [CrossRef] [PubMed]

8. Choi, M.S.; Park, S.; Choi, T.; Lee, G.; Haam, K.K.; Hong, M.C.; Min, B.I.; Bae, H. Bee venom ameliorates ovalbumin induced allergic asthma via modulating CD4+CD25+ regulatory T cells in mice. *Cytokine* **2013**, *61*, 256–265. [CrossRef] [PubMed]

9. Lim, B.S.; Moon, H.J.; Li, D.X.; Gil, M.; Min, J.K.; Lee, G.; Bae, H.; Kim, S.K.; Min, B.I. Effect of bee venom acupuncture on oxaliplatin-induced cold allodynia in rats. *Evid. Based Complement. Alternat. Med.* **2013**, *2013*, 369324. [CrossRef] [PubMed]

10. Yoon, S.Y.; Yeo, J.H.; Han, S.D.; Bong, D.J.; Oh, B.; Roh, D.H. Diluted bee venom injection reduces ipsilateral mechanical allodynia in oxaliplatin-induced neuropathic mice. *Biol. Pharm. Bull.* **2013**, *36*, 1787–1793. [CrossRef] [PubMed]

11. Somerfield, S.D.; Brandwein, S. Bee venom and adjuvant arthritis. *J. Rheumatol.* **1988**, *15*, 1878. [PubMed]

12. Roh, D.H.; Kwon, Y.B.; Kim, H.W.; Ham, T.W.; Yoon, S.Y.; Kang, S.Y.; Han, H.J.; Lee, H.J.; Beitz, A.J.; Lee, J.H. Acupoint stimulation with diluted bee venom (apipuncture) alleviates thermal hyperalgesia in a rodent neuropathic pain model: Involvement of spinal alpha 2-adrenoceptors. *J. Pain* **2004**, *5*, 297–303. [CrossRef] [PubMed]

13. Baek, Y.H.; Huh, J.E.; Lee, J.D.; Choi do, Y.; Park, D.S. Antinociceptive effect and the mechanism of bee venom acupuncture (apipuncture) on inflammatory pain in the rat model of collagen-induced arthritis: Mediation by alpha2-adrenoceptors. *Brain Res.* **2006**, *1073–1074*, 305–310. [CrossRef] [PubMed]

14. Kwon, Y.B.; Kang, M.S.; Han, H.J.; Beitz, A.J.; Lee, J.H. Visceral antinociception produced by bee venom stimulation of the Zhongwan acupuncture point in mice: Role of alpha(2) adrenoceptors. *Neurosci. Lett.* **2001**, *308*, 133–137. [CrossRef]

15. Kim, H.W.; Kwon, Y.B.; Han, H.J.; Yang, I.S.; Beitz, A.J.; Lee, J.H. Antinociceptive mechanisms associated with diluted bee venom acupuncture (apipuncture) in the rat formalin test: Involvement of descending adrenergic and serotonergic pathways. *Pharmacol. Res.* **2005**, *51*, 183–188. [CrossRef] [PubMed]

16. Lee, J.H.; Li, D.X.; Yoon, H.; Go, D.; Quan, F.S.; Min, B.I.; Kim, S.K. Serotonergic mechanism of the relieving effect of bee venom acupuncture on oxaliplatin-induced neuropathic cold allodynia in rats. *BMC Complement. Altern. Med.* **2014**, *14*, 471. [CrossRef] [PubMed]

17. Kang, S.Y.; Roh, D.H.; Park, J.H.; Lee, H.J.; Lee, J.H. Activation of spinal alpha2-adrenoceptors using diluted bee venom stimulation reduces cold allodynia in neuropathic pain rats. *Evid. Based Complement. Altern. Med.* **2012**, *2012*, 784713. [CrossRef] [PubMed]

18. Kang, S.Y.; Roh, D.H.; Yoon, S.Y.; Moon, J.Y.; Kim, H.W.; Lee, H.J.; Beitz, A.J.; Lee, J.H. Repetitive treatment with diluted bee venom reduces neuropathic pain via potentiation of locus coeruleus noradrenergic neuronal activity and modulation of spinal NR1 phosphorylation in rats. *J. Pain* **2012**, *13*, 155–166. [CrossRef] [PubMed]

19. Monti, M.C.; Casapullo, A.; Santomauro, C.; D'Auria, M.V.; Riccio, R.; Gomez-Paloma, L. The molecular mechanism of bee venom phospholipase A2 inactivation by bolinaquinone. *Chembiochem* **2006**, *7*, 971–980. [CrossRef] [PubMed]

20. Zhao, H.; Kinnunen, P.K. Modulation of the activity of secretory phospholipase A2 by antimicrobial peptides. *Antimicrob. Agents Chemother.* **2003**, *47*, 965–971. [CrossRef] [PubMed]

21. Jeong, J.K.; Moon, M.H.; Bae, B.C.; Lee, Y.J.; Seol, J.W.; Park, S.Y. Bee venom phospholipase A2 prevents prion peptide induced-cell death in neuronal cells. *Int. J. Mol. Med.* **2011**, *28*, 867–873. [PubMed]

22. Lopez-Vales, R.; Ghasemlou, N.; Redensek, A.; Kerr, B.J.; Barbayianni, E.; Antonopoulou, G.; Baskakis, C.; Rathore, K.I.; Constantinou-Kokotou, V.; Stephens, D.; *et al.* Phospholipase A2 superfamily members play divergent roles after spinal cord injury. *FASEB J.* **2011**, *25*, 4240–4252. [CrossRef] [PubMed]

23. Burrage, P.S.; Mix, K.S.; Brinckerhoff, C.E. Matrix metalloproteinases: Role in arthritis. *Front. Biosci.* **2006**, *11*, 529–543. [CrossRef] [PubMed]

24. Kim, H.; Lee, H.; Lee, G.; Jang, H.; Kim, S.-S.; Yoon, H.; Kang, G.-H.; Hwang, D.-S.; Kim, S.K.; Chung, H.-S.; *et al.* Phospholipase A2 inhibits cisplatin-induced acute kidney injury by modulating regulatory T cells via CD206 mannose receptor. *Kidney Int.* **2015**, *10*. [CrossRef]

25. Zhu, J.X.; Zhu, X.Y.; Owyang, C.; Li, Y. Intestinal serotonin acts as a paracrine substance to mediate vagal signal transmission evoked by luminal factors in the rat. *J. Physiol.* **2001**, *530*, 431–442. [CrossRef] [PubMed]

26. Maleki, N.; Nayebi, A.M.; Garjani, A. Effects of central and peripheral depletion of serotonergic system on carrageenan-induced paw oedema. *Int. Immunopharmacol.* **2005**, *5*, 1723–1730. [CrossRef] [PubMed]

27. Kudo, T.; Kushikata, T.; Kudo, M.; Hirota, K. Antinociceptive effects of neurotropin in a rat model of central neuropathic pain: DSP-4 induced noradrenergic lesion. *Neurosci. Lett.* **2011**, *503*, 20–22. [CrossRef] [PubMed]

28. Farquhar-Smith, P. Chemotherapy-induced neuropathic pain. *Curr. Opin. Support. Palliat. Care* **2011**, *5*, 1–7. [CrossRef] [PubMed]

29. Kannarkat, G.; Lasher, E.E.; Schiff, D. Neurologic complications of chemotherapy agents. *Curr. Opin. Neurol.* **2007**, *20*, 719–725. [CrossRef] [PubMed]

30. Lehky, T.J.; Leonard, G.D.; Wilson, R.H.; Grem, J.L.; Floeter, M.K. Oxaliplatin-induced neurotoxicity: Acute hyperexcitability and chronic neuropathy. *Muscle Nerve* **2004**, *29*, 387–392. [CrossRef] [PubMed]

31. Desoize, B.; Madoulet, C. Particular aspects of platinum compounds used at present in cancer treatment. *Crit. Rev. Oncol. Hematol.* **2002**, *42*, 317–325. [CrossRef]

32. Binder, A.; Stengel, M.; Maag, R.; Wasner, G.; Schoch, R.; Moosig, F.; Schommer, B.; Baron, R. Pain in oxaliplatin-induced neuropathy—Sensitisation in the peripheral and central nociceptive system. *Eur. J. Cancer* **2007**, *43*, 2658–2663. [CrossRef] [PubMed]

33. Ling, B.; Coudore-Civiale, M.A.; Balayssac, D.; Eschalier, A.; Coudore, F.; Authier, N. Behavioral and immunohistological assessment of painful neuropathy induced by a single oxaliplatin injection in the rat. *Toxicology* **2007**, *234*, 176–184. [CrossRef] [PubMed]

34. Ling, B.; Coudore, F.; Decalonne, L.; Eschalier, A.; Authier, N. Comparative antiallodynic activity of morphine, pregabalin and lidocaine in a rat model of neuropathic pain produced by one oxaliplatin injection. *Neuropharmacology* **2008**, *55*, 724–728. [CrossRef] [PubMed]

35. De Felipe, C.; Herrero, J.F.; O'Brien, J.A.; Palmer, J.A.; Doyle, C.A.; Smith, A.J.; Laird, J.M.; Belmonte, C.; Cervero, F.; Hunt, S.P. Altered nociception, analgesia and aggression in mice lacking the receptor for substance P. *Nature* **1998**, *392*, 394–397. [PubMed]

36. Honore, P.; Rogers, S.D.; Schwei, M.J.; Salak-Johnson, J.L.; Luger, N.M.; Sabino, M.C.; Clohisy, D.R.; Mantyh, P.W. Murine models of inflammatory, neuropathic and cancer pain each generates a unique set of neurochemical changes in the spinal cord and sensory neurons. *Neuroscience* **2000**, *98*, 585–598. [CrossRef]

37. Back, S.K.; Sung, B.; Hong, S.K.; Na, H.S. A mouse model for peripheral neuropathy produced by a partial injury of the nerve supplying the tail. *Neurosci. Lett.* **2002**, *322*, 153–156. [CrossRef]

38. Kim, H.; Keum, D.J.; Kwak, J.; Chung, H.S.; Bae, H. Bee venom phospholipase A2 protects against acetaminophen-induced acute liver injury by modulating regulatory T cells and IL-10 in mice. *PLoS ONE* **2014**, *9*, e114726. [CrossRef] [PubMed]

39. Lamont, L.A.; Tranquilli, W.J.; Grimm, K.A. Physiology of pain. *Vet. Clin. N. Am. Small Anim. Pract.* **2000**, *30*, 703–728. [CrossRef]

40. Ossipov, M.H.; Morimura, K.; Porreca, F. Descending pain modulation and chronification of pain. *Curr. Opin. Support. Palliat. Care* **2014**, *8*, 143–151. [PubMed]

41. Crespi, F. Apamin increases 5-HT cell firing in raphe dorsalis and extracellular 5-HT levels in amygdala: A concomitant *in vivo* study in anesthetized rats. *Brain Res.* **2009**, *1281*, 35–46. [CrossRef] [PubMed]

42. Millan, M.J. Descending control of pain. *Prog. Neurobiol.* **2002**, *66*, 355–474. [CrossRef]

43. Korenman, E.M.; Devor, M. Ectopic adrenergic sensitivity in damaged peripheral nerve axons in the rat. *Exp. Neurol.* **1981**, *72*, 63–81. [CrossRef]

44. Lee, D.H.; Liu, X.; Kim, H.T.; Chung, K.; Chung, J.M. Receptor subtype mediating the adrenergic sensitivity of pain behavior and ectopic discharges in neuropathic Lewis rats. *J. Neurophysiol.* **1999**, *81*, 2226–2233. [PubMed]

45. Leem, J.W.; Gwak, Y.S.; Nam, T.S.; Paik, K.S. Involvement of alpha2-adrenoceptors in mediating sympathetic excitation of injured dorsal root ganglion neurons in rats with spinal nerve ligation. *Neurosci. Lett.* **1997**, *234*, 39–42. [CrossRef]

46. Zhang, J.M.; Song, X.J.; LaMotte, R.H. An *in vitro* study of ectopic discharge generation and adrenergic sensitivity in the intact, nerve-injured rat dorsal root ganglion. *Pain* **1997**, *72*, 51–57. [CrossRef]

47. Hord, A.H.; Denson, D.D.; Stowe, B.; Haygood, R.M. Alpha-1 and alpha-2 adrenergic antagonists relieve thermal hyperalgesia in experimental mononeuropathy from chronic constriction injury. *Anesth. Analg.* **2001**, *92*, 1558–1562. [CrossRef] [PubMed]

48. Tracey, D.J.; Cunningham, J.E.; Romm, M.A. Peripheral hyperalgesia in experimental neuropathy: Mediation by alpha 2-adrenoreceptors on post-ganglionic sympathetic terminals. *Pain* **1995**, *60*, 317–327. [CrossRef]

49. Austin, P.J.; Kim, C.F.; Perera, C.J.; Moalem-Taylor, G. Regulatory T cells attenuate neuropathic pain following peripheral nerve injury and experimental autoimmune neuritis. *Pain* **2012**, *153*, 1916–1931. [CrossRef] [PubMed]

50. Flatters, S.J.; Bennett, G.J. Ethosuximide reverses paclitaxel- and vincristine-induced painful peripheral neuropathy. *Pain* **2004**, *109*, 150–161. [CrossRef] [PubMed]

51. Joseph, E.K.; Levine, J.D. Comparison of oxaliplatin- and cisplatin-induced painful peripheral neuropathy in the rat. *J. Pain* **2009**, *10*, 534–541. [CrossRef] [PubMed]

52. Shibata, K.; Sugawara, T.; Fujishita, K.; Shinozaki, Y.; Matsukawa, T.; Suzuki, T.; Koizumi, S. The astrocyte-targeted therapy by bushi for the neuropathic pain in mice. *PLoS ONE* **2011**, *6*, e23510. [CrossRef] [PubMed]

53. Scullion, G.A.; Kendall, D.A.; Sunter, D.; Marsden, C.A.; Pardon, M.C. Central noradrenergic depletion by DSP-4 prevents stress-induced memory impairments in the object recognition task. *Neuroscience* **2009**, *164*, 415–423. [CrossRef] [PubMed]

54. Johnson, J.D.; Campisi, J.; Sharkey, C.M.; Kennedy, S.L.; Nickerson, M.; Greenwood, B.N.; Fleshner, M. Catecholamines mediate stress-induced increases in peripheral and central inflammatory cytokines. *Neuroscience* **2005**, *135*, 1295–1307. [CrossRef] [PubMed]

55. Nelson, L.E.; Lu, J.; Guo, T.; Saper, C.B.; Franks, N.P.; Maze, M. The alpha2-adrenoceptor agonist dexmedetomidine converges on an endogenous sleep-promoting pathway to exert its sedative effects. *Anesthesiology* **2003**, *98*, 428–436. [CrossRef] [PubMed]

56. Zarrindast, M.R.; Homayoun, H.; Khavandgar, S.; Fayaz-Dastgerdi, M. The effects of simultaneous administration of alpha(2)-adrenergic agents with L-NAME or L-arginine on the development and expression of morphine dependence in mice. *Behav. Pharmacol.* **2002**, *13*, 117–125. [CrossRef] [PubMed]

© 2015 by the authors; licensee MDPI, Basel, Switzerland. This article is an open access article distributed under the terms and conditions of the Creative Commons Attribution (CC-BY) license (http://creativecommons.org/licenses/by/4.0/).

toxins

MDPI

Article

Exploring the Potential of Venom from *Nasonia vitripennis* as Therapeutic Agent with High-Throughput Screening Tools

Ellen L. Danneels *, Ellen M. Formesyn and Dirk C. de Graaf

Laboratory of Molecular Entomology and Bee Pathology, Ghent University, S2 Krijgslaan 281, B-9000 Ghent, Belgium; ellen.formesyn@gmail.com (E.M.F.); dirk.degraaf@ugent.be (D.C.G.)

* Correspondence: ellen.danneels@ugent.be; Tel.: +32-92-645-151; Fax: +32-92-645-242

Academic Editor: Sokcheon Pak
Received: 17 April 2015; Accepted: 29 May 2015; Published: 3 June 2015

Abstract: The venom from the ectoparasitoid wasp *Nasonia vitripennis* (Hymenoptera: Pteromalidae) contains at least 80 different proteins and possibly even more peptides or other small chemical compounds, demonstrating its appealing therapeutic application. To better understand the dynamics of the venom in mammalian cells, two high-throughput screening tools were performed. The venom induced pathways related to an early stress response and activated reporters that suggest the involvement of steroids. Whether these steroids reside from the venom itself or show an induced release/production caused by the venom, still remains unsolved. The proinflammatory cytokine IL-1β was found to be down-regulated after venom and LPS co-treatment, confirming the anti-inflammatory action of *N. vitripennis* venom. When analyzing the expression levels of the NF-κB target genes, potentially not only the canonical but also the alternative NF-κB pathway can be affected, possibly explaining some counterintuitive results. It is proposed that next to an NF-κB binding site, the promoter of the genes tested by the PCR array may also contain binding sites for other transcription factors, resulting in a complex puzzle to connect the induced target gene with its respective transcription factor. Interestingly, *Nasonia* venom altered the expression of some drug targets, presenting the venom with an exciting therapeutical potential.

Keywords: *Nasonia*; venom; therapeutic; reporter array; PCR array; NF-κB

1. Introduction

Hymenopteran parasitoids develop at the expense of other arthropods, ultimately killing their host. The ectoparasitoid wasp *Nasonia vitripennis* (Hymenoptera: Pteromalidae) preferably parasitizes pupae from flesh flies (Sarcophagidae) and blow flies (Calliphoridae). After locating a suitable host, the female wasp injects venom inside the fly pupa and lays her eggs in the space between the pupa and the puparium. The injection of this complex mixture of venom compounds prepares the host to present the best possible environment for the wasp offspring to survive. Host physiology is altered, in which the host development is arrested, its immune system is suppressed, and host metabolism is modified so that it is synchronized with the development of parasitoid larvae.

The venom from *N. vitripennis* is known to contain (at least) 80 different proteins [1,2], and possibly even peptides and other bio-molecules. Over the past century, natural products (NPs) have been the source of inspiration for the majority of FDA approved drugs. This is highlighted by the fact that nearly 50% of all drugs in clinical use are of natural product origin [3]. These interesting chemicals are derived from the phenomenon of biodiversity in which the interactions among organisms and their environment formulate the diverse complex chemical entities within the organisms that enhance their survival and competitiveness [4]. The therapeutic areas of infectious diseases and oncology have benefited from the complex molecular scaffolds found in NPs of which the chemical diversity is unmatched by synthetic molecules. Animal venoms are a rich source of NPs that have evolved high affinity and selectivity for a diverse range of biological targets, especially membrane proteins such as ion channels, receptors, and transporters. Therefore, venomics has emerged as an important addition to modern drug discovery efforts [5]. Snake venom is a treasure house of toxins that contributes significantly to the treatment of many medical conditions and presents a great potential as an anti-tumor agent [6]. The venom and its constituents from honey bees have many therapeutic applications ranging from anti-arthritis and pain-releasing to anti-cancer effects [7]. Venoms from parasitoid wasps contain a staggering amount of toxins, and because they can manipulate cell physiology in diverse ways [8–10], their therapeutic potential is interesting to investigate.

Although the natural hosts of *N. vitripennis* wasps are insect pupae, one might wonder how the venom-induced physiological alterations would translate to a mammalian system. The concept "bugs as drugs" emphasizes the interest in mining insects for medicinal drugs [11]. With the intent to explore a possible medicinal future, we performed a wide screening of the effects of this *Nasonia* venom on mammalian cellular signaling with high-throughput arrays that are designed for use with mammalian cells. However, to further investigate the specific interaction mechanism, *Drosophila melanogaster* could be used as a model system, which is less of a leap from the *Nasonia*-host insect system.

Cell-based assays provide a high-performance tool due to their exceptional sensitivity, reproducibility, specificity, and signal-to-noise ratio, for assessing the functions of natural products under physiological cellular conditions. By screening multiple pathway activities simultaneously, relevant pathways for further analysis can quickly be identified. Therefore, possible regulation by the complete venom mixture of 45 reporters that represent transcription factors (TFs) that play a central role in regulating gene expression, orchestrating a host of cellular processes, and are associated with many human diseases were investigated using a reporter array (Cignal™ 45-Pathway Reporter Array, SABiosciences, Frederick, MD, USA). Of the 45 pathways, four main research areas were targeted, including cancer, immunity, development, and toxicology. By reverse transfecting human embryonic kidney (HEK293T) cells into multi-pathway reporter arrays, the activity of 45 pathways was screened upon *N. vitripennis* venom treatment. Transcriptional activity is monitored by the dual luciferase technology that allows for quantification of the degree of activation of each particular signaling pathway in a 96 well format By analyzing the effects of the venom on multiple cellular signaling pathways, new directions for further investigations with possible biomedical application could be appointed.

Recently, *N. vitripennis* venom was shown to exert a suppressive action on the nuclear factor kappa-light-chain-enhancer of activated B cells (NF-κB) pathway in murine macrophages [8]. This important TF regulates a large number of target genes involved in multiple cellular processes including inflammation, immunity, and stress responses [12]. Dysregulation of this signal transduction pathway has been associated with inflammatory or autoimmune diseases [13] and cancer [14]. Previous investigations showed that lipopolysaccharide (LPS)-induced NF-κB activation in Raw264.7 macrophages resulted in an inhibition of the inflammatory response when the cells were incubated with *N. vitripennis* venom. By further investigating the interruption of this crucial immune pathway by the venom, it appeared that next to the suppression of the NF-κB cascade also the mitogen-activated protein kinase (MAPK) and glucocorticoid receptor (GR) signaling pathways were affected. Therefore, in order to fully explore venom activities on intracellular signaling after an immune activation, TNFα-induced HEK293T cells were incubated with *N. vitripennis* venom and also analyzed with the reporter array.

In 1999, Pahl listed over 150 target genes known to be expressed by the active NF-κB transcription factor [15]. To date, this list has been extended by more than 250 extra investigated NF-κB target genes and even more than 300 genes are predicted by computer-based methods to have composite NF-κB regulatory sites [16]. The majority of proteins encoded by NF-κB target genes participate in the host inflammatory and immune responses, which include cytokines and chemokines, as well as receptors required for immune recognition, proteins involved in antigen presentation, acute phase proteins, and cell adhesion molecules. Many of them are induced by exposure to a wide variety of bacteria, as well as hosts of viruses and their respective products. NF-κB, however, is involved in the control of the transcription of many genes whose functions extend beyond the immune response, but are involved in more general stress responses [17]. Various physiological stress conditions such as liver regeneration and hemorrhagic shock can activate NF-κB [18,19]. Also physical stress, in the form of irradiation as well as oxidative stress to cells, induces NF-κB, that in turn activates a large variety of stress response genes [20]. In fact, NF-κB relays the information of an imminent stress and at the same time enacts a response by promoting the transcription of genes whose products alleviate the stress condition. The human body is also exposed to environmental hazards and therapeutic drugs, activating NF-κB that in turn activates its target genes including many cell surface receptors [21]. Several stimuli, among them the cytokine TNFα, can lead to NF-κB activation that exerts anti-apoptotic activities. On the other hand, there is ample evidence for apoptosis-promoting functions of NF-κB as well [22,23]. The nature of the apoptotic stimulus determines the pro- or anti-apoptotic function of NF-κB. Activation of NF-κB can lead to the transcriptional induction of various TF genes, even members of their own Rel/NF-κB/IκB family.

The enormous amount of NF-κB target genes can be categorized in the different groups mentioned above, which has been done for the 84 tested genes in the NF-κB Signaling Targets PCR Array. To gain broader understanding of the effects of *N. vitripennis* venom on the immune response, the expression of these key genes responsive to NF-κB signal transduction were analyzed after incubation of Raw264.7 macrophages with this venom mixture. Additionally, alterations of specific NF-κB signaling target genes with a role in inflammatory diseases and oncology could hint at potential therapeutic lead compounds present in the venom.

2. Results and Discussion

2.1. Effect of Venom on Mammalian Intracellular Signaling

To investigate the effects of *N. vitripennis* venom on mammalian intracellular signaling, we screened a wide range of signaling pathways for their regulation after 8 h incubation with venom. We utilized a commercially available array (Cignal™ 45-Pathway Reporter Array, SABiosciences, Frederick, MD, USA) on HEK293T cells. These cells are easily and efficiently transfected with PEI transfection reagent. By comparison of two reporter constructs, the pathway-focused TF-responsive Firefly luciferase reporter and the constitutively expressing Renilla luciferase construct, activated pathways are identified. This dual luciferase technology allows for quantification of the degree of activation of each particular signaling pathway in a 96 well format. Since the venom from *N. vitripennis* was found to suppress the NF-κB pathway in fibrosarcoma cells when induced with tumor necrosis factor alpha (TNFα) [8], we also used TNFα-induced HEK293T cells in this reporter array to investigate the effects of the venom on intracellular signaling when the immune system in these cells is activated. All conditions were performed in four replicates and internal positive and negative controls on all plates were fulfilled.

The significant values with a fold change higher than 2-fold are summarized in Table 1 for the following three comparisons: venom-treated *versus* untreated cells, TNFα-treated *versus* untreated cells, and venom and TNFα co-treated *versus* TNFα-treated cells. The incubation of cells with *N. vitripennis* venom caused an increased activation of four reporters compared to control cells: the AARE, LXR, MEF2 and RXR reporters. The reporter that showed the highest up-regulation was the amino acid deprivation (AARE) reporter which is known to be an early response upon stress [24]. Malnutrition, various pathological situations, and xenobiotic toxins are able to alter amino acid availability which can result in a deficit of certain amino acids and increased uncharged transfer RNAs (tRNAs). Following amino acid deprivation, the general control non-derepressible-2 (GCN2) kinase is activated upon detection of these accumulated free tRNAs and phosphorylates the translation initiation factor (eukaryotic initiation factor 2a), thereby attenuating protein synthesis. Since the endoplasmic reticulum (ER) regulates the production and oxidative folding of proteins, this response may prevent further accumulation of misfolded proteins and alterations in redox state [25]. Furthermore, phosphorylated eIF2a enhances the translation of several mRNAs, including ATF4 which is rapidly induced under cell-stress conditions, such as glutathione depletion and oxidative stress. ATF4 is an important regulator of several ER stress target genes, amino acid transporters and antioxidants thereby preventing further accumulation of reactive oxygen species (ROS) [26]. Interestingly, free amino acids (FAA) were found to be up-regulated in the natural host after parasitization by *N. vitripennis* [27]. They suggested that the larval parasitoids use these FAA as a source of direct nutrition. Whether both processes concerning the elevation of amino acids in mammalian cells and the insect hosts can be linked together still needs to be elucidated.

Table 1. Effect of *N. vitripennis* venom in HEK293T cells, either induced with TNFα or not, on the transcriptional activity of reporters of 45 different pathways. Fold regulation (FR) of all tested reporters are presented for 3 different comparisons. When $p > 0.05$, insignificant values are between brackets; when $|FR| > 2$, values are in bold.

Pathway reporters	FR TNFα-treated *versus* untreated	FR Venom-treated *versus* untreated	FR TNFα- and venom-treated *versus* TNFα-treated
AARE reporter	1.601	**4.081**	**4.555**
AR reporter	**−2.613**	(1.291)	**7.814**
C/EBP reporter	(1.274)	1.925	**2.507**
CRE reporter	1.779	(1.457)	**2.020**
E2F reporter	1.001	(1.012)	**3.560**
p53 reporter	(2.569)	(3.867)	**4.177**
EGR1 reporter	(−0.911)	(1.548)	**2.228**
HSR reporter	(−0.901)	(1.311)	**3.999**
GLI reporter	**−3.073**	(−0.240)	**7.556**
IRF1 reporter	−1.188	(1.287)	**2.568**
LXR reporter	(−0.342)	**2.201**	1.622
MEF2 reporter	−1.192	**2.023**	**2.849**
NF-κB reporter	**100.640**	(−1.272)	(1.145)
Oct4 reporter	**−2.336**	(−1.436)	**3.007**
PR reporter	**10.676**	(−0.126)	(1.718)
RARE reporter	(−0.702)	(−0.048)	**2.3887**
RXR reporter	(−0.370)	**2.539**	(1.379)

Two other reporters up-regulated by the venom are the LXR and the RXR reporter, that measure the transcriptional activity of the Liver X receptor and the Retinoid X receptor, respectively. The natural ligands of LXRs are oxygenated forms of cholesterol, while RXRs are activated upon binding with vitamin A, derivatives, and rexinoids. The regulation of both receptors through binding with small chemical compounds that can pass biological membranes, could imply that compounds other than proteins are present in the venom, which can migrate through membranes and bind directly to TFs. Today, nothing is known about such compounds in the venom of *N. vitripennis*, presenting us with an unexplored and unexploited source of potentially useful compounds for medicine to further investigate.

When HEK293T cells were induced for 8 h with TNFα, the activation of two reporters was significantly up-regulated, while three were down-regulated compared to control cells. The NF-κB reporter was no less than 100 times up-regulated, showing the strong activation of this immune cascade. When cells were co-incubated with TNFα together with venom from *N. vitripennis*, no significant alteration in transcriptional activity could be observed compared to cells simply induced with TNFα. In L929sA cells however, a significant inhibition of the NF-κB activation was noted by the venom [8]. This can possibly be explained by the use of different cell lines, HEK293T cells instead of L929sA cells, or other induction times, 8 h instead of 6 h.

When cells were incubated with TNFα together with venom, no less than 13 of the 45 tested pathways showed significant differential transcriptional activity, compared to cells that were only induced with TNFα. Interestingly, the AR and the GLI reporter, that show down-regulated transcriptional activity after TNFα treatment, have the highest up-regulated activity (more than seven times higher) when co-incubated with *N. vitripennis* venom. The androgen receptor (AR) is a nuclear receptor that is activated by binding androgenic hormones, testosterone, or dihydrotestosterone, again hinting at possible presence of small chemical compounds in the venom. The GLI reporter is designed to measure hedgehog signaling activity. Interestingly, the sonic hedgehog signaling pathway plays an important role in the development of cancer, specifically brain and skin cancer [28]. Analysis of the reporter array also revealed that several reporters, like the p53 and HSR reporter involved in stress responses were affected by the venom when co-incubated with TNFα for 8 h. In addition, microarray on parasitized insect hosts also revealed that transcripts involved in stress, cell death, detoxification, and the MAPK/JNK pathways were affected by the venom of *N. vitripennis* [29].

Other interesting reporters involved in inflammation and the immune response showed significant alterations in transcriptional activity after co-treatment of HEK293T cells with TNFα and *N. vitripennis* venom. The Interferon Regulatory Factor 1 (IRF1) reporter for instance, is a member of the interferon regulatory TF family and serves as a transcriptional activator of interferon alpha, beta, and gamma. IRF1 is known to regulate host defense against pathogens, tumor prevention, and development of the immune system. Accordingly, in the natural host, the venom is known to allow or even stimulate certain antimicrobial defenses of the host next to the expected interference of the venom with host melanization and coagulation responses [30].

The reporter arrays revealed that the venom from *N. vitripennis* has a wide-spread impact on multiple mammalian signaling cascades. However, it would be interesting to investigate what the effect would be on transcriptional activity when separate venom compounds would be used to induce the cells. Therefore, not only proteins or peptides, but also small chemical compounds like amines or steroids that are possibly present in the venom from *N. vitripennis*, should be isolated from the complete venom mixture and further investigated for their effects on the cellular signaling cascades that were found to be targeted by the complete venom.

2.2. Effect of Venom on NF-κB Signaling Targets

Since previous results on L929sA cells showed clear anti-inflammatory activity of the venom [8], we decided to dig deeper in the NF-κB regulating effects of the *N. vitripennis* venom. Therefore, a PCR array experiment was set up to evaluate the expression of NF-κB target genes under different conditions. We decided to use Raw264.7 macrophage cells that show activation of the NF-κB signaling cascade when stimulated with LPS. We treated these cells with *N. vitripennis* venom, together with or without LPS induction. Significant relative expression of 84 genes was evaluated by RT-qPCR as presented in Table S1. The genes that show a significant differential expression of more than 2-fold higher or lower in three different comparisons are summarized in Table 2 in which the NF-κB target genes are classified according to their properties and/or functions.

LPS stimulation increased transcription of 36 NF-κB target genes in the macrophage cells and down-regulated the expression level of one target gene. The highest up-regulated genes can be found in the groups of the cytokines/chemokines, stress response genes and growth factors. Several TFs, including TFs that take part in the NF-κB cascade, show an elevated transcription level. Previous results showed that RT-qPCR on Raw264.7 cells stimulated with *N. vitripennis* venom resulted in no significant effect in mRNA levels of NF-κB inhibitor alpha (IκBα) and A20, that are both early response genes [8]. The NF-κB Signaling Targets PCR Array resulted in one NF-κB target gene that showed down-regulated expression, and nine that were up-regulated after venom treatment. Remarkable is the up-regulation of Relb expression by venom treatment, which could possibly lead to further elevation in expression of other NF-κB target genes. On the other hand, the reporter array performed after venom treatment on HEK293T cells resulted in the induction of several TFs, like ATF2/3/4 and MEF2 (see Table 1) potentially attributing to the up-regulated expression levels shown here.

Table 2. Effect of *N. vitripennis* venom in Raw264.7 cells, either induced with lipopolysaccharide (LPS) or not, on NF-κB signaling targets. Fold regulation of all tested NF-κB signaling target genes are presented for 3 different comparisons. When $p > 0.05$, insignificant values are between brackets; when $|FR| > 2$, values are in bold. (Abb = abbreviation; FR = fold regulation).

NF-κB signaling target genes	Abb	FR LPS-treated *versus* untreated	FR venom-treated *versus* untreated	FR LPS- and venom-treated *versus* LPS-treated
Cytokines/chemokines and their modulators				
Chemokine (C–C motif) ligand 22	Ccl22	**917.635**	(1.032)	(−1.532)
Chemokine (C–C motif) ligand 5	Ccl5	**1254.881**	(2.107)	(1.750)
Chemokine (C–C motif) receptor 5	Ccr5	1.625	(1.875)	(2.346)
Chemokine (C–X–C motif) ligand 10	Cxcl10	**483.835**	(2.254)	(−1.025)
Chemokine (C–X–C motif) ligand 3	Cxcl3	**111.806**	(3.484)	(34.595)
Interleukin 15	Il15	(3.325)	(2.527)	**13.408**
Interleukin 1α	Il1a	**1273.082**	(2.414)	(−1.242)
Interleukin 1β	Il1b	**15,647.327**	**5.885**	**−4.392**
Interleukin 1 receptor antagonist	Il1rn	**26.052**	(3.074)	(−1.145)
Interleukin 6	Il6	**525.452**	(2.419)	−1.472
Lymphotoxin A	Lta	**25.056**	(3.037)	(1.094)
Tumor necrosis factor	Tnf	**36.399**	(−1.244)	−1.193
Immunoreceptors				
CD40 antigen	Cd40	**68.505**	**9.933**	(4.098)
CD80 antigen	Cd80	**3.113**	(2.621)	(2.723)
CD83 antigen	Cd83	(3.545)	(2.949)	**29.395**
Tumor necrosis factor receptor superfamily, member 1b	Tnfrsf1b	**22.445**	(1.469)	(1.708)
Proteins involved in antigen presentation				
Complement component 3	C3	**3.651**	(2.035)	(1.175)
Complement factor B	Cfb	**14.677**	−(1.007)	**−4.199**
Cell adhesion molecules				
Intercellular adhesion molecule 1	Icam1	(1.405)	(1.212)	**17.631**
Vascular cell adhesion molecule 1	Vcam1	(−1.148)	(1.387)	**3.481**
Acute phase proteins				
Coagulation factor III	F3	**41.407**	(4.466)	(−1.979)
Stress response genes				
NAD(P)H dehydrogenase, quinone 1	Nqo1	(−1.247)	**24.637**	(3.014)
Prostaglandin-endoperoxide synthase 2	Ptgs2	**698.790**	(1.715)	(−1.605)
Superoxide dismutase 2, mitochondrial	Sod2	**4.919**	−1.052	(1.301)
Regulators of apoptosis				
B-cell leukemia/lymphoma 2 related protein A1a	Bcl2a1a	**33.896**	**5.027**	(3.713)
Bcl2-like 1	Bcl2l1	**2.677**	(1.255)	(1.275)
Baculoviral IAP repeat-containing 2	Birc2	(−1.428)	(1.513)	**6.288**
Baculoviral IAP repeat-containing 3	Birc3	1.874	(1.558)	**3.599**
Fas (TNF receptor superfamily member 6)	Fas	**11.621**	**3.261**	**6.857**
Tnf receptor-associated factor 2	Traf2	(−1.056)	1.983	**5.107**

Table 2. *Cont.*

NF-κB signaling target genes	Abb	FR LPS-treated *versus* untreated	FR venom-treated *versus* untreated	FR LPS- and venom-treated *versus* LPS-treated
Growth factors, ligands and their modulators				
Colony stimulating factor 1 (macrophage)	Csf1	**35.675**	**55.854**	**117.792**
Colony stimulating factor 2 (granulocyte-macrophage)	Csf2	**530**	(1.548)	(−1.436)
Colony stimulating factor 3 (granulocyte)	Csf3	**15,821.676**	**5.486**	**−31.724**
Platelet derived growth factor, B polypeptide	Pdgfb	**3.395**	(1.479)	(1.092)
Transcription factors and regulators				
Interferon regulatory factor 1	Irf1	**2.945**	(2.039)	**13.953**
Microphthalmia-associated transcription factor	Mitf	(−1.549)	(1.434)	**7.247**
Myelocytomatosis oncogene	Myc	**25.814**	**17.898**	(8.083)
Nuclear factor of kappa light polypeptide gene enhancer in B-cells 1, p105	Nfkb1	**4.526**	(1.599)	**3.399**
Nuclear factor of kappa light polypeptide gene enhancer in B-cells 2, p49/p100	Nfkb2	(1.368)	(1.001)	**10.255**
Nuclear factor of kappa light polypeptide gene enhancer in B-cells inhibitor, alpha	Nfkbia	**7.963**	(1.102)	(1.508)
Reticuloendotheliosis oncogene	Rel	**2.664**	(1.121)	**3.864**
Avian reticuloendotheliosis viral (v-rel) oncogene related B	Relb	(1.334)	**3.591**	**8.675**
Signal transducer and activator of transcription 1	Stat1	**2.855**	−1.257	(1.111)
Signal transducer and activator of transcription 3	Stat3	1.517	−1.104	−1.444
Miscellaneous				
Cyclin D1	Ccnd1	**−5.051**	**−2.667**	**−7.439**
Growth arrest and DNA-damage-inducible 45 beta	Gadd45b	**8.330**	(2.325)	**10.985**
Matrix metallopeptidase 9	Mmp9	**17.819**	(1.635)	**−5.805**

Before cells are stimulated with LPS, NF-κB subunits like p50 and p52 can already be bound to NF-κB binding sites in the promoters of a number of genes, exhibiting a certain range of expression levels [31]. After cellular stimulation with LPS, other NF-κB family members enter the nucleus and bind to those genes and to other genes, leading to enhanced gene transcription. *N. vitripennis* venom could influence this process before cells are stimulated with LPS resulting in the down-regulated expression level of NF-κB target gene, cyclin D1.

Transcription of the stress response gene, NAD(P)H dehydrogenase quinone 1 (Nqo1), is nearly 25 times up-regulated after venom treatment. This gene has been demonstrated to play an important role in protecting cells against oxidative stress [32]. Venoms of several animals, including the parasitoid wasp *Aphidius ervi*, are known to cause oxidative stress in their host organism [33,34]. The elevation of the Nqo1 gene expression after venom treatment in macrophage cells could therefore suggest a protective function. Remarkably, the venom induced transcription of two growth factors, colony stimulating factor 1 and 3 (Csf1 and Csf3) that stimulate the bone marrow progenitor cells to differentiate into macrophages or granulocytes, respectively. These proteins have an important role in innate immunity and inflammation [35]. Especially transcription of Csf1 is highly up-regulated (nearly 56 times), which even doubles when the cells are immune challenged with LPS.

However, most differentially expressed genes could be observed when venom was added to the macrophage cells, combined with an immune challenge of LPS. Transcription of two of the NF-κB target genes, tested by the NF-κB Signaling Target PCR Array, were previously tested by RT-qPCR: IκBα and interleukin 6 (IL-6) [8]. Although the suppression of IκBα transcription could not be validated in this PCR Array experiment, the 2-fold inhibition of IL-6 transcription on the other hand could be confirmed by a 1.5-fold down-regulation in the PCR Array. Transcription of the cytokine IL-1β, produced by activated macrophages and an important mediator of inflammatory response, shows a nearly 6-fold up-regulation after venom-treatment, while being 4.3 times down-regulated after co-treatment of venom and LPS. Apparently, the cells need to be immune challenged, in this case by LPS, in order for the venom to be able to suppress the inflammatory response.

Previous experiments focused on the inhibitory effect of venom on the canonical NF-κB pathway [8], in which IκB kinase beta (IKKβ) phosphorylates IκBs at *N*-terminal sites to trigger their ubiquitin-dependent degradation and induce nuclear entry of RelA:p50 dimers [36]. The remark has to be made that aside from this classical NF-κB signaling pathway, an alternative non-canonical signaling cascade has been identified that activates NF-κB signaling based on processing of the NFκB2:p100 precursor protein by IKKα [37]. In this alternative NF-κB pathway, the stimulation by LPS (among others) results in the nuclear translocation of the dimer RelB:p52 [38]. While many target genes are shared between the canonical and the non-canonical pathways, some promoters of NF-κB target genes are only recognized by RelB:p52 dimers and not by RelA:p50 dimers [31,39]. Several NF-κB target genes can bind multiple members of the NF-κB family, suggesting that they can be activated by both the canonical and the non-canonical pathways. For instance, Icam1 and Gadd45b can bind all five members of the NF-κB family (p52, p50, Rela, Relb and c-Rel) in U937 cells after LPS stimulation [31]. When Raw264.7 cells were treated with *N. vitripennis* venom, transcription of Icam1 and Gadd45b was up-regulated 17.63 and 10.99 times, respectively. Therefore, it can be suggested that *N. vitripennis* venom potentially has an inhibitory effect on the canonical pathway, but a stimulatory effect on the alternative NF-κB signaling pathway explaining the up-regulation of some inflammatory genes. This hypothesis however needs to be further investigated. Interestingly, *N. vitripennis* venom up-regulated transcription of four different NF-κB subunits when cells were co-treated with LPS, of which Nfkb2 (p52) and Relb, that take part in the non-canonical NF-κB pathway, show the highest elevations.

Interesting is the number of up-regulated NF-κB target genes involved in apoptosis regulation after venom and LPS co-treatment. Formesyn and colleagues have previously proven that serine proteases and metalloproteases in *N. vitripennis* venom cause apoptosis in non-host insect cells and suggested their possible role in immune related processes [40]. In this PCR array on murine macrophages, pro- as well as anti-apoptotic mediators were differentially expressed: Fas activation induces apoptosis [41], while Birc2, Birc3 and Traf2 are known for their anti-apoptotic effects [42,43].

An acute stimulation, for instance by LPS, is known to create two distinct waves of NF-κB recruitment to target promoters: a fast recruitment to immediately accessible promoters and a late recruitment to promoters requiring stimulus-dependent modifications in chromatin structure to make NF-κB sites accessible [44]. A relatively long (6 h) LPS-treatment was chosen in the PCR Array experiment. Unfortunately, possible transient effects caused by *N. vitripennis* venom, due to the short mRNA half-life of many of the early response genes, could not be observed by this experiment and time kinetics would need to be performed in order to have a glimpse on the complete venom profile. Next to possible time course effects, the mechanistic complexity of the venom can be considerable, since different cell lines can show large differences in their responses to specific compounds. Future experiments should therefore also incorporate different cell lines together with time kinetics in response to separate venom compounds.

Next to an NF-κB binding site, the promoter of the genes tested by this PCR array may also contain binding sites for other TFs. Instead of an up-regulated transcription by NF-κB binding to the promoter, expression of these genes can also be targeted by other TFs. When performing a TFSEARCH, the promoter sequence of a gene of interest can be screened for possible TF binding sites by correlating this sequence against the TRANSFAC MATRIX database. The upstream gene sequence 2000 nucleotides before and 50 nucleotides after the start of the gene (where the promoter sequence is assumed to be located) is inserted into the TFSEARCH program and an 85.0 threshold is used.

This was performed for five genes that showed the highest significant up-regulated fold regulation (FR) after venom-and LPS-treatment, and for four genes that had a negative FR after venom- and LPS-treatment. All nine genes obviously contained the NF-κB binding site, as the genes included in the array were previously demonstrated to be NF-κB target genes. The TF binding sites that were predicted by the program to be present in all five up-regulated genes were listed in Table 3. This means that the up-regulated transcription of these genes, next to NF-κB binding to their promoter, could also be the result of binding of these other TFs to their promoters. When looking at these TF binding sites in the promoters of the four down-regulated genes, cAMP response element-binding protein (CRE-BP) can be a possible candidate for causing the synthesis of several of the tested genes while at the same time not having an increasing effect in transcription of Csf3, Mmp9 and Ccnd1, since these genes do not contain a binding site for CRE-BP. On the other hand, a role of CCAAT-enhancer-binding proteins (C/EBPs) in the activation of NF-κB target genes can also be proposed. C/EBP binding sites can be found in promoter regions of several tested genes with an up-regulated transcription, while being absent in the promoters of some of the down-regulated genes (see Table 3). Additionally, C/EBPs can interact with p50 homodimers activating NF-κB target genes, even in the absence of canonical NF-κB activation [45]. Interestingly, the reporter array performed on HEK293T cells treated with *N. vitripennis* venom, also showed significant up-regulation of the C/EBP reporter (Table 1).

Table 3. TF binding sites that are commonly present in the promoters of 5 up-regulated genes tested by the NF-κB Signaling Targets PCR Array, predicted by TFSEARCH (threshold 85.0). The presence of these TF binding sites is shown for 4 down-regulated genes when cells were treated with venom and LPS. "x" represents the presence of the respective TF site in the promoter region of that particular gene. At the bottom, the fold regulations are presented for the selected genes.

NF-κB signaling target genes	Cd83	Csf1	IL15	Irf1	Icam1	Il1b	Csf3	Mmp9	Ccnd1
				TF binding site					
NF-Kap	x	x	x	x	x	x	x	x	x
C/EBP	x	x	x	x	x	x	x	x	-
C/EBPa	x	x	x	-	x	-	x	x	-
AML-1a	x	x	x	x	x	x	x	x	x
CdxA	x	x	x	x	x	x	x	x	x
CRE-BP	x	x	x	x	x	x	-	-	-
deltaE	x	x	x	x	x	x	x	x	x
GATA-1	x	x	x	x	x	x	x	x	x
GATA-2	x	x	x	x	x	x	x	x	x
GATA-3	x	x	x	x	x	-	x	x	x
GATA-X	x	x	x	x	x	x	x	-	x
HSF2	x	x	x	x	x	x	-	x	-
MZF1	x	x	x	x	x	x	x	x	x
Nkx-2.	x	x	x	x	x	x	x	x	x
Oct-1	x	x	x	x	x	x	x	x	-
SRY	x	x	x	x	x	x	x	x	x
TATA	x	x	x	x	x	x	-	x	x
				Fold regulation					
venom- and LPS- *vs.* LPS-treated	29.4	117.8	13.4	14.0	17.6	−4.4	−31.7	−5.8	−7.4
venom-treated *vs.* untreated	3.0	55.9	2.5	2.0	1.2	5.9	5.5	1.6	−2.7

The presence (or absence) of many other TF binding sites in specific gene promoters could possibly be an aid to explaining the apparent contradiction between the up-regulation of several inflammatory genes by *N. vitripennis* venom, but at the same time the anti-inflammatory activity of the venom on the other hand [8]. Other mechanisms can play a role in the change of gene expression: microRNAs can post-transcriptionally bind to the 3′-UTR (untranslated region) of their target mRNAs and repress protein production [46], or epigenetic phenomena like DNA methylation or histone modification can cause differential gene expression [47]. Until now, none of these processes have been investigated for the gene alterations caused by parasitoid venoms.

It needs to be commented that venom from *N. vitripennis* is a complex mixture of at least 80 different proteins. Maybe even more other venom compounds like peptides or other bio-molecules could be present, all having their own effects on gene regulation. Regarding to venomous effects on the complex apoptosis process, it would be interesting to see how the separate venom compounds, especially the serine proteases and metalloproteases as tested in insect cells [40], would affect the mammalian cell death process. However, it would also be interesting to examine what the effects of individual venom compounds would be on mammalian cells, concerning gene regulation, but also with regard to translational regulation of the NF-κB pathway.

Many of the target genes of NF-κB signaling are involved in multiple inflammatory or autoimmune diseases. Some of them are promising targets in certain therapies [48,49], while others can themselves be used as agents in processes involved in human diseases [50,51]. Interestingly, *N. vitripennis* venom significantly altered the expression of some of these possible drug targets, presenting the venom with an exciting potential as therapeutic in several diseases (Table 4). However, keep in mind that the physiological processes affected by the complex *Nasonia* venom only hint at possible interesting therapies for human disease conditions. Since until now specific mechanistic insights are lacking, further studies focusing on the affected NF-κB targets need to be performed, ideally with the responsible venom compounds.

Table 4. NF-κB target genes that were differentially transcribed and can be a possible drug target of *N. vitripennis* venom. Three different comparisons are presented: venom-treated *versus* untreated, venom- and LPS-treated *versus* LPS-treated cells and LPS-treated *versus* untreated. When $p > 0.05$, insignificant values are between brackets; when |FR| >2, values are in bold. (Abb = abbreviation; FR = fold regulation; Ref = reference).

Possible drug targets of venom	Abb	FR venom *versus* untreated	FR venom- and LPS-treated *versus* LPS-treated	FR LPS-treated *versus* untreated	Potential targeted diseases	Reference
NAD(P)H dehydrogenase, quinone 1	Nqo1	**24.64**	(3.014)	(−1.247)	acute leukemia	[52]
Cyclin D1	Ccnd1	**−2.67**	**−7.44**	**−5.05**	breast cancer	[53]
Interferon regulatory factor 1	Irf1	(2.039)	**13.95**	**2.95**	breast cancer	[51]
Matrix metallopeptidase 9	Mmp9	(1.635)	**−5.80**	**17.82**	cancer	[54]
Colony stimulating factor 3 (granulocyte)	Csf3	**5.49**	**−31.72**	**15,821.68**	inflammatory arthritis	[49]
Interleukin 1 beta	Il1b	**5.88**	**−4.39**	**15,647.33**	autoinflammatory diseases	[55]
Complement factor B	Cfb	(−1.007)	**−4.20**	**14.68**	complement mediated inflammatory diseases	[56]

The second highest up-regulated NF-κB target gene transcription tested after the sole addition of venom on the cells, was NAD(P)H:quinone oxidoreductase 1 (Nqo1). This enzyme detoxifies quinones and reduces oxidative stress. Low activity of this enzyme is associated with increased risk of acute leukemia in adults [52]. Inducing this detoxification enzyme in certain mammalian cell types could potentially benefit patients suffering from acute leukemia or could prevent people from developing this disease.

For the other venom targets with potential therapeutic application, the cells needed to be immune challenged with LPS, offering a possible use in diseases characterized by constitutive activity of NF-κB, like autoimmune diseases [57] or many cancers [58]. With regard to anti-inflammatory drugs, colony stimulating factor 3 (Csf3) with a nearly 32-fold decrease in transcription after venom and LPS incubation compared to LPS induction, seems to be the most interesting interaction to further investigate. This regulator of granulopoiesis has a critical role in driving joint inflammatory diseases, like rheumatoid arthritis (RA), and its antagonists may be of therapeutic value [49]. Anti-IL-1β is a commonly used drug with the name canakinumab utilized in several auto inflammatory diseases [55]. Previously mentioned inhibition of IL-1β transcription by venom- and LPS-treatment may therefore hint at a therapeutic role of *N. vitripennis* venom.

Intriguingly, the possible anti-cancer role of *N. vitripennis* venom is displayed by the regulation of different NF-κB targets. Transcription of Cyclin D1, involved in the G1-S phase transition of cells, is significantly suppressed by *N. vitripennis* venom, with and without LPS-treatment. Since pharmacological inhibition of cyclin D1/CDK4 complexes is suggested to be a useful strategy to inhibit the growth of tumors, the potential of *N. vitripennis* venom in cancer treatments may be valuable to further investigate. In contrast, the expression of Irf1 is significantly up-regulated after the co-treatment of venom and LPS, which was confirmed in the reporter array showing up-regulated transcriptional activity of the IRF1 reporter after co-treatment of venom and TNFα in HEK293T cells. A functional role of Irf1 was established in the growth suppression of breast cancer cells and it was implicated in acting as a tumor suppressor gene in breast cancer by controlling apoptosis [51]. The potential of *N. vitripennis* venom as therapeutic agent in several oncogenetic diseases is therefore interesting to look at. The possible drug targets in Table 4 show the exciting potential of *Nasonia* venom, but need to be interpreted with the necessary precaution. Future studies also require incorporate alterations at the protein level, modulations of diverse immune pathways or biological signals and determination of the exact effect of the individual responsible venom components.

3. Experimental Section

3.1. Isolation of Crude Wasp Venom

N. vitripennis wasps were reared on pupae of the flesh fly, *S. crassipalpis*, and maintained at 25 °C with a daily 16:8 light:dark cycle. Female wasps were allowed to host feed on flesh fly pupae for 24 h. Venom gland reservoirs were dissected into insect saline buffer (ISB) (150 mM NaCl, 10 mM KCl, 4 mM CaCl$_2$, 2 mM MgCl$_2$, 10 mM Hepes) [59] and centrifuged at 12,000× g for 10 min at 4 °C. The supernatant containing the venom was transferred to a clean microcentrifuge tube and stored frozen at −70 °C. Total protein in crude venom was determined colorimetrically at 595 nm using a Coomassie Protein Assay Reagent (No. 23200, Thermo Fisher Scientific, Rockford, IL, USA).

3.2. Cell Culture and Treatments

For the reporter array, human embryonic kidney cells 293T (HEK293T, kind gift from Prof. Kathleen Van Craenenbroeck, Ghent University, Ghent, Belgium) were cultured as adherent monolayers in 25 cm^2 flasks in Opti-MEM reduced serum medium without phenol red (Thermo Fisher Scientific, Rockford, IL, USA) supplemented with 5% FBS (International Medical Products, Brussels, 1160, Belgium) and 100 U/mL penicillin (Thermo Fisher Scientific, Rockford, IL, USA) and 0.1 mg/mL streptomycin (Thermo Fisher Scientific, Rockford, IL, USA). Cells were grown in an atmosphere with 5% CO$_2$ at 37 °C. Prior to the assays, cells were grown in 75 cm^2 flasks and cultured in such a way that they are sub confluent prior to collection. The Cignal Finder™ 45-Pathway Reporter Arrays (Qiagen, SABiosciences corp., Frederick, MD, USA) consisting of 45 dual luciferase reporter assays, were used according to the manufacturer's instructions. One hour before transfection, fresh Opti-MEM medium supplemented with 2% FBS was added to the cells and cells were placed back in the incubator. Meanwhile, DNA reporter constructs were first dissolved in 50 μL/mL Opti-MEM and incubated for 5 minutes at room temperature. Subsequently, PEI transfection reagent (Thermo Fisher Scientific, Rockford, IL, USA) was diluted in Opti-MEM without serum or antibiotics, and 50 μL/well was dispensed into 96-well white tissue-culture plates. For reverse transfection, freshly grown cells were counted and adjusted to 1.4 million cells/mL in Opti-MEM containing 2% FBS and 1% non-essential amino acids (NEAA, Thermo Fisher Scientific, Rockford, IL, USA), without antibiotics. Cells (50 µL; 7×104 cells/well) were added to each well and incubated overnight at 37 °C with 5% CO_2. Transfection media were removed 6–8 h after transfection and replaced with 100 µL fresh Opti-MEM, supplemented with 5% FBS and 1% NEAA, and incubated again overnight at 37 °C. One hour prior to venom induction, media were removed and replaced with fresh Opti-MEM supplemented with 0.5% FBS and 1% NEAA. Cells (two array plates) were then induced with *N. vitripennis* venom at a concentration of 2.5 µg/mL for 8 h, the other 2 plates were treated with ISB and served as controls. Each plate contains a duplicate of the reporter.

For the PCR array, mouse macrophage-like Raw264.7 cells (kind gift from Prof. Kathleen Van Craenenbroeck, Ghent University, Ghent, Belgium) were maintained in RPMI 1640 medium (Thermo Fisher Scientific, Rockford, IL, USA) at 37 °C in 5% CO_2 humidified air. Medium was supplemented with 10% FBS, 100 U/mL penicillin and 0.1 mg/mL streptomycin. Twenty-four hours before induction, cells were seeded in multiwell dishes so that they were confluent at the time of the experiment. Raw264.7 cells were submitted to the following different treatments: untreated; 6-h LPS induction (1 µg/mL); 6 h and 15 min incubation with 10 µg/mL of *N. vitripennis* venom; 6 h LPS induction and 6 h and 15 min venom incubation. The 4 treatments were performed in duplicate.

3.3. Reporter Array Analysis

Dual-luciferase reporter activity was determined using a dual-luciferase reporter assay system (Promega, Madison, WI, USA), following the manufacturer's instructions using a Victor 3TM 1420 Multilabel Counter plate reader (PerkinElmer, Waltham, MA, USA). Induced TFs were reported as luminescence ratios by dividing the Firefly signal by the Renilla signal. Subsequently, normalization was performed using the ratio of control wells on every plate (negative and positive controls). Data was evaluated by Student's *t*-tests and Mann-Whitney *U* tests. A value of $p < 0.05$ was considered statistically significant. All statistical analysis were performed with Prism 5.0 (GraphPad Software, Inc., La Jolla, CA, USA, 2011).

3.4. Total RNA Extraction and Reverse Transcription

RNA was isolated by using TRIzol Reagent (Thermo Fisher Scientific, Rockford, IL, USA) as described previously [60]. An on-column DNase digestion was performed and the concentration of RNA in all 8 samples was measured. Equal amounts of RNA were reverse-transcribed into cDNA using the RT^2 First Strand Kit (Qiagen, Frederick, MD, USA) following the manufacturer's instructions.

3.5. Real-Time PCR-Based Array Analysis

The relative expression of NF-κB signaling target genes was determined in each of the 8 samples by RT-qPCR (RT^2 Profiler Mouse NF-κB Signaling Targets PCR Array; SABiosciences Corp., Frederick, MD, USA) using the qPCR master mix (RT^2 SYBR Green; SABiosciences Corp., Frederick, MD, USA) according to the supplier's directions. Mixes were pipetted into 384-well PCR array plates to evaluate the expression of 84 NF-κB signaling target genes. RT-qPCR was performed in technical duplicates (Roche LightCycler 480, 384-well block, Mannheim, Germany). Raw data from the real-time PCR were uploaded using a PCR array data analysis template available at RT^2 Profiler PCR Array Data Analysis version 3.5 (http://www.sabiosciences.com/pcr/arrayanalysis.php). Quality controls included within the array plates confirmed the lack of DNA contamination and successfully tested for RNA quality and PCR performance. The integrated Web-based software package for the PCR array system automatically performed all comparative threshold cycle (ΔΔCt)-based fold-change calculations from the uploaded data. For these calculations, the average expression of 3 housekeeping genes (β-actin, glyceraldehydes-3-phosphate dehydrogenase and β2-microglobulin) was used for normalization of the data. After normalization, the relative expression of each gene was averaged for the 2 samples in each condition. Fold regulations in average gene expression were expressed as the difference in expression of venom-treated compared with untreated cells, or of venom- and LPS-treated compared with only LPS-treated cells. A fold change ≥ 2.5 with $p \leq 0.05$ was considered significant.

4. Conclusions

By using high-throughput screening tools such as the reporter and PCR arrays performed here, the multi-facetted effects of venom can become visible, pinpointing interesting pathways and targets for further investigation. The effect of *N. vitripennis* venom on 45 intracellular signaling cascades was analyzed with a commercially available reporter array in HEK293T cells either immune challenged with TNFα or not. Interesting pathways were affected, of which several are related with an early stress response and others that need to be activated by steroid compounds. Whether steroids, possibly present in the venom, induced these reporters or venom stimulation contributes to the release of steroids in cells, still needs to be further investigated. Since most of the affected pathways are involved in multiple biological processes, more detailed research needs to be performed on both transcript and protein level in order to unravel how the venom affects these processes. In addition, the possible protein interactions with other TFs might be investigated, because the composition of the TF complexes influences their promoter specificity and hence their target pathways. Furthermore, other induction time points or venom concentrations may also provide useful information. The NF-κB Signaling Target PCR Array performed on Raw264.7 macrophages treated with *N. vitripennis* venom and/or LPS uncovered several new ideas on how the venom exerts its complex effects. Interestingly, the proinflammatory cytokine IL-1β was significantly suppressed after the venom and LPS co-treatment, indicating the anti-inflammatory action of the *Nasonia* venom. Previous data describing the inhibition of the canonical NF-κB pathway by the venom [8], is not that obvious when looking at the expression of a large number of NF-κB target genes. Several aspects encourage us to be cautious in interpreting the results. The complexity of the venom mixture applied on the cells creates multiple crosstalk effects that could be circumvented when working with separate venom compounds. The fact that one time-point was used, only gives a glimpse of the complete picture. Additionally, changes in transcription expression do not always translate into the same changes at protein level, since alterations in translation efficiency and post-translational modifications can lead to a different end-result. Possible effects on the non-canonical NF-κB pathway next to the canonical pathway, in addition to the presence of multiple TF sites in the promoter of NF-κB target genes, make the puzzle more difficult to solve. However, keeping all these remarks in mind, still several interesting hints for future research could be noted, which was the original intent of this experiment. Some NF-κB target genes that were differentially expressed by the addition of *N. vitripennis* venom, with or without LPS-treatment, are drug targets of major diseases, hinting to possible future biomedical application as therapeutic agent in several human diseases. Performing time kinetics and dose-responses by the separated venom compounds seems the logical next step.

Supplementary Materials: Supplementary materials can be accessed at: http://www.mdpi.com/2072-6651/7/6/2051/s1.

Acknowledgments: The authors would like to thank Karen Heyninck for laboratory assistance and help with the design of the experiments.

Author Contributions: Experiments were designed and conducted by Ellen L. Danneels, Ellen M. Formesyn and Dirk C. de Graaf. Data was analyzed by Ellen L. Danneels and Ellen M. Formesyn. All authors, Ellen L. Danneels, Ellen M. Formesyn and Dirk C. de Graaf, contributed to the writing of the manuscript. No external funding was obtained.

Conflicts of Interest: The authors declare no conflict of interest.

References

1. De Graaf, D.C.; Aerts, M.; Brunain, M.; Desjardins, C.A.; Jacobs, F.J.; Werren, J.H.; Devreese, B. Insights into the venom composition of the ectoparasitoid wasp *Nasonia vitripennis* from bioinformatic and proteomic studies. *Insect Mol. Biol.* **2010**, *19*, 11–26. [CrossRef] [PubMed]

2. Ye, J.L.; Zhao, H.W.; Wang, H.J.; Bian, J.M.; Zheng, R.Q. A defensin antimicrobial peptide from the venoms of *Nasonia vitripennis*. *Toxicon* **2010**, *56*, 101–106. [CrossRef] [PubMed]

3. Paterson, I.; Anderson, E.A. The renaissance of natural products as drug candidates. *Science* **2005**, *310*, 451–453. [CrossRef] [PubMed]

4. Mishra, B.B.; Tiwari, V.K. Natural products: An evolving role in future drug discovery. *Eur. J. Med. Chem.* **2011**, *46*, 4769–4807. [CrossRef] [PubMed]

5. Fry, B.G.; Roelants, K.; Champagne, D.E.; Scheib, H.; Tyndall, J.D.A.; King, G.F.; Nevalainen, T.J.; Norman, J.A.; Lewis, R.J.; Norton, R.S.; *et al.* The toxicogenomic multiverse: Convergent recruitment of proteins into animal venoms. *Annu. Rev. Genomics Hum. Genet.* **2009**, *10*, 483–511. [CrossRef] [PubMed]

6. Vyas, V.K.; Brahmbhatt, K.; Bhatt, H.; Parmar, U.; Patidar, R. Therapeutic potential of snake venom in cancer therapy: Current perspectives. *Asian Pac. J. Trop. Biomed.* **2013**, *3*, 156–162. [CrossRef]

7. Son, D.J.; Lee, J.W.; Lee, Y.H.; Song, H.S.; Lee, C.K.; Hong, J.T. Therapeutic application of anti-arthritis, pain-releasing, and anti-cancer effects of bee venom and its constituent compounds. *Pharmacol. Ther.* **2007**, *115*, 246–270. [CrossRef] [PubMed]

8. Danneels, E.L.; Gerlo, S.; Heyninck, K.; van Craenenbroeck, C.K.; de Bosscher, B.K.; Haegeman, G.; de Graaf, D.C. How the venom from the ectoparasitoid wasp *Nasonia vitripennis* exhibits anti-inflammatory properties on mammalian cell lines. *PLoS ONE* **2014**, *9*, e96825. [CrossRef] [PubMed]

9. Zhang, G.M.; Schmidt, O.; Asgari, S. A calreticulin-like protein from endoparasitoid venom fluid is involved in host hemocyte inactivation. *Dev. Comp. Immunol.* **2006**, *30*, 756–764. [CrossRef] [PubMed]

10. Mortimer, N.T.; Goecks, J.; Kacsoh, B.Z.; Mobley, J.A.; Bowersock, G.J.; Taylor, J.; Schlenke, T.A. Parasitoid wasp venom SERCA regulates *Drosophila* calcium levels and inhibits cellular immunity. *Proc. Natl. Acad. Sci. USA* **2013**, *110*, 9427–9432. [CrossRef] [PubMed]

11. Cherniack, E.P. Drugs from bugs, Part 1: The "new" alternative medicine for the 21st century? *Altern. Med. Ref.* **2010**, *15*, 124–135.

12. Vallabhapurapu, S.; Karin, M. Regulation and function of NF-κB transcription factors in the immune system. *Annu. Rev. Immunol.* **2009**, *27*, 693–733. [CrossRef] [PubMed]

13. D'Acquisto, F.; May, M.J.; Ghosh, S. Inhibition of nuclear factor κB (NF-B): An emerging theme in anti-inflammatory therapies. *Mol. Interv.* **2002**, *2*, 22–35. [CrossRef] [PubMed]

14. Gasparini, C.; Celeghini, C.; Monasta, L.; Zauli, G. NF-κB pathways in hematological malignancies. *Cell. Mol. Life Sci.* **2014**, *71*, 2083–2102. [CrossRef] [PubMed]

15. Pahl, H.L. Activators and target genes of Rel/NF-κB transcription factors. *Oncogene* **1999**, *18*, 6853–6866. [CrossRef] [PubMed]

16. NF-κB Transcription Factors. Available online: http://www.bu.edu/nf-kb/gene-resources/target-genes/ (accessed on 1 June 2015).

17. Wu, Z.; Miyamoto, S. Many faces of NF-κB signaling induced by genotoxic stress. *J. Mol. Med.* **2007**, *85*, 1187–1202. [CrossRef] [PubMed]

18. Yan, R.; Li, Y.; Zhang, L.; Xia, N.; Liu, Q.; Sun, H.; Guo, H. Augmenter of liver regeneration attenuates inflammation of renal ischemia/reperfusion injury throuth the NF-κB pathway in rats. *Int. Urol. Nephrol.* **2015**, *47*, 861–868. [CrossRef] [PubMed]

19. Maraslioglu, M.; Weber, R.; Korff, S.; Blattner, C.; Nauck, C.; Henrich, D.; Jobin, C.; Marzi, I.; Lehnert, M. Activation of NF-κB after chronic ethanol intake and haemorrhagic shock/resuscitation in mice. *Br. J. Pharmacol.* **2013**, *170*, 506–518. [CrossRef] [PubMed]

20. Halle, M.; Hall, P.; Tornvall, P. Cardiovascular disease associated with radiotherapy: Activation of nuclear factor kappa-B. *J. Intern. Med.* **2011**, *269*, 469–477. [CrossRef] [PubMed]

21. Samuel, T.; Fadlalla, K.; Gales, D.N.; Putcha, B.D.K.; Manne, U. Variable NF-κB pathway responses in colon cancer cells treated with chemotherapeutic drugs. *BMC Cancer* **2014**, *14*. [CrossRef] [PubMed]

22. Kaltschmidt, B.; Kaltschmidt, C.; Hofmann, T.G.; Hehner, S.P.; Droge, W.; Schmitz, M.L. The pro- or anti-apoptotic function of NF-kappa B is determined by the nature of the apoptotic stimulus. *Eur. J. Biochem.* **2000**, *267*, 3828–3835. [CrossRef] [PubMed]

23. Bednarski, B.K.; Baldwin, A.S.; Kim, H.J. Addressing reported pro-apoptotic functions of NF-κB: Targeted inhibition of canonical NF-κB enhances the apoptotic effects of doxorubicin. *PLoS ONE* **2009**, *4*, e6992. [CrossRef] [PubMed]

24. Kilberg, M.S.; Pan, Y.X.; Chen, H.; Leung-Pineda, V. Nutritional control of gene expression: How mammalian cells respond to amino acid limitation. *Annu. Rev. Nutr.* **2005**, *25*, 59–85. [CrossRef] [PubMed]

25. Chaveroux, C.; Jousse, C.; Cherasse, Y.; Maurin, A.C.; Parry, L.; Carraro, V.; Derijard, B.; Bruhat, A.; Fafournoux, P. Identification of a novel amino acid response pathway triggering ATF2 phosphorylation in mammals. *Mol. Cell. Biol.* **2009**, *29*, 6515–6526. [CrossRef] [PubMed]

26. Sun, X.; Lin, Y.; Huang, Q.; Shi, J.; Qiu, L.; Kang, M.; Chen, Y.; Fang, C.; Ye, T.; Dong, S.; *et al.* Di(2-ethylhexyl) phthalate-induced apoptosis in rat INS-1 cells is dependent on activation of endoplasmic reticulum stress and suppression of antioxidant protection. *J. Cell. Mol. Med.* **2015**, *19*, 581–594. [CrossRef] [PubMed]

27. Mrinalini; Siebert, A.; Wright, J.; Martinson, E.; Wheeler, D.; Werren, J. Parasitoid venom induces metabolic cascades in fly hosts. *Metabolomics* **2015**, *11*, 350–366. [CrossRef]

28. Ronci, M.; Catanzaro, G.; Pieroni, L.; Po, A.; Besherat, Z.M.; Greco, V.; Mortera, S.L.; Screpanti, I.; Ferretti, E.; Urbani, A.; *et al.* Proteomic analysis of human sonic hedgehog (SHH) medulloblastoma stem-like cells. *Mol. BioSyst.* **2015**. [CrossRef] [PubMed]

29. Danneels, E.L.; Formesyn, E.M.; Hahn, D.A.; Denlinger, D.L.; Cardoen, D.; Wenseleers, T.; Schoofs, L.; de Graaf, D.C. Early changes in the pupal transcriptome of the flesh fly *Sarcophaga crassipalpis* to parasitization by the ectoparasitic wasp, *Nasonia vitripennis*. *Insect Biochem. Mol. Biol.* **2013**, *43*, 1189–1200. [CrossRef] [PubMed]

30. Danneels, E.L.; Rivers, D.B.; de Graaf, D.C. Venom proteins of the parasitoid wasp *Nasonia vitripennis*: Recent discovery of an untapped pharmacopee. *Toxins* **2010**, *2*, 494–516. [CrossRef] [PubMed]

31. Schreiber, J.; Jenner, R.G.; Murray, H.L.; Gerber, G.K.; Gifford, D.K.; Young, R.A. Coordinated binding of NF-kappaB family members in the response of human cells to lipopolysaccharide. *Proc. Natl. Acad. Sci. USA* **2006**, *103*, 5899–5904. [CrossRef] [PubMed]

32. Dinkova-Kostova, A.T.; Talalay, P. Persuasive evidence that quinone reductase type 1 (DT diaphorase) protects cells against the toxicity of electrophiles and reactive forms of oxygen. *Free Radic. Biol. Med.* **2000**, *29*, 231–240. [CrossRef]

33. Falabella, P.; Riviello, L.; Caccialupi, P.; Rossodivita, T.; Teresa, V.M.; de Luisa, S.M.; Tranfaglia, A.; Varricchio, P.; Gigliotti, S.; Graziani, F.; *et al.* A gamma-glutamyl transpeptidase of *Aphidius ervi* venom induces apoptosis in the ovaries of host aphids. *Insect Biochem. Mol. Biol.* **2007**, *37*, 453–465. [CrossRef] [PubMed]

34. Katkar, G.D.; Sundaram, M.S.; Hemshekhar, M.; Sharma, D.R.; Santhosh, M.S.; Sunitha, K.; Rangappa, K.S.; Girish, K.S.; Kemparaju, K. Melatonin alleviates *Echis carinatus* venom-induced toxicities by modulating inflammatory mediators and oxidative stress. *J. Pineal Res.* **2014**, *56*, 295–312. [CrossRef] [PubMed]

35. Takahashi, K.; Naito, M.; Takeya, M. Development and heterogeneity of macrophages and their related cells through their differentiation pathways. *Pathol. Int.* **1996**, *46*, 473–485. [CrossRef] [PubMed]

36. Karin, M.; Ben-Neriah, Y. Phosphorylation meets ubiquitination: The control of NF-κB activity. *Annu. Rev. Immunol.* **2000**, *18*, 621–623. [CrossRef] [PubMed]

37. Xiao, G.; Harhaj, E.W.; Sun, S.C. NF-κB-inducing kinase regulates the processing of NF-κB2 p100. *Mol. Cell* **2001**, *7*, 401–409. [CrossRef]

38. Mordmuller, B.; Krappmann, D.; Esen, M.; Wegener, E.; Scheidereit, C. Lymphotoxin and lipopolysaccharide induce NF-kappaB-p52 generation by a co-translational mechanism. *EMBO Rep.* **2003**, *4*, 82–87. [CrossRef] [PubMed]

39. Bonizzi, G.; Bebien, M.; Otero, D.C.; Johnson-Vroom, K.E.; Cao, Y.X.; Vu, D.; Jegga, A.G.; Aronow, B.J.; Ghosh, G.; Rickert, R.C.; *et al.* Activation of IKK alpha target genes depends on recognition of specific kappa B binding sites by RelB:p52 dimers. *EMBO J.* **2004**, *23*, 4202–4210. [CrossRef] [PubMed]

40. Formesyn, E.M.; Heyninck, K.; de Graaf, D.C. The role of serine- and metalloproteases in *Nasonia vitripennis* venom in cell death related processes towards a *Spodoptera frugiperda* Sf21 cell line. *J. Insect Physiol.* **2013**, *59*, 795–803. [CrossRef] [PubMed]

41. Huang, D.C.; Hahne, M.; Schroeter, M.; Frei, K.; Fontana, A.; Villunger, A.; Newton, K.; Tschopp, J.; Strasser, A. Activation of Fas by FasL induces apoptosis by a mechanism that cannot be blocked by Bcl-2 or Bcl-x(L). *Proc. Natl. Acad. Sci. USA* **1999**, *96*, 14871–14876. [CrossRef] [PubMed]

42. Wang, Y.; Tang, X.; Yu, B.; Gu, Y.; Yuan, Y.; Yao, D.; Ding, F.; Gu, X. Gene network revealed involvements of Birc2, Birc3 and Tnfrsf1a in anti-apoptosis of injured peripheral nerves. *PLoS ONE* **2012**, *7*, e43436. [CrossRef] [PubMed]

43. Lin, Y.; Ryan, J.; Lewis, J.; Wani, M.A.; Lingrel, J.B.; Liu, Z.G. TRAF2 exerts its antiapoptotic effect by regulating the expression of Kruppel-like factor LKLF. *Mol. Cell. Biol.* **2003**, *23*, 5849–5856. [CrossRef] [PubMed]

44. Saccani, S.; Pantano, S.; Natoli, G. Two waves of nuclear factor kappa B recruitment to target promoters. *J. Exp. Med.* **2001**, *193*, 1351–1359. [CrossRef] [PubMed]

45. Dooher, J.E.; Paz-Priel, I.; Houng, S.; Baldwin, A.S., Jr.; Friedman, A.D. C/EBPalpha, C/EBPalpha oncoproteins, or C/EBPbeta preferentially bind NF-κB p50 compared with p65, focusing therapeutic targeting on the C/EBP:p50 interaction. *Mol. Cancer Res.* **2011**, *9*, 1395–1405. [CrossRef] [PubMed]

46. Cannell, I.G.; Kong, Y.W.; Bushell, M. How do microRNAs regulate gene expression? *Biochem. Soc. Trans.* **2008**, *36*, 1224–1231. [CrossRef] [PubMed]

47. Jaenisch, R.; Bird, A. Epigenetic regulation of gene expression: How the genome integrates intrinsic and environmental signals. *Nat. Genet.* **2003**, *33*, S245–S254. [CrossRef] [PubMed]

48. Dziadziuszko, R.; Jassem, J. Epidermal growth factor receptor (EGFR) inhibitors and derived treatments. *Ann. Oncol.* **2012**, *23*, 193–196. [CrossRef] [PubMed]

49. Lawlor, K.E.; Campbell, I.K.; Metcalf, D.; O'Donnell, K.; van Nieuwenhuijze, A.; Roberts, A.W.; Wicks, I.P. Critical role for granulocyte colony-stimulating factor in inflammatory arthritis. *Proc. Natl. Acad. Sci. USA* **2004**, *101*, 11398–11403. [CrossRef] [PubMed]

50. Murray, H.W. Interferon-gamma and host antimicrobial defense—Current and future clinical-applications. *Am. J. Med.* **1994**, *97*, 459–467. [CrossRef]

51. Bouker, K.B.; Skaar, T.C.; Riggins, R.B.; Harburger, D.S.; Fernandez, D.R.; Zwart, A.; Wang, A.; Clarke, R. Interferon regulatory factor-1 (IRF-1) exhibits tumor suppressor activities in breast cancer associated with caspase activation and induction of apoptosis. *Carcinogenesis* **2005**, *26*, 1527–1535. [CrossRef] [PubMed]

52. Smith, M.T.; Wang, Y.X.; Kane, E.; Rollinson, S.; Wiemels, J.L.; Roman, E.; Roddam, P.; Cartwright, R.; Morgan, G. Low NAD(P)H: Quinone oxidoreductase 1 activity is associated with increased risk of acute leukemia in adults. *Blood* **2001**, *97*, 1422–1426. [CrossRef] [PubMed]

53. Grillo, M.; Bott, M.J.; Khandke, N.; McGinnis, J.P.; Miranda, M.; Meyyappan, M.; Rosfjord, E.C.; Rabindran, S.K. Validation of cyclin D1/CDK4 as an anticancer drug target in MCF-7 breast cancer cells: Effect of regulated overexpression of cyclin D1 and siRNA-mediated inhibition of endogenous cyclin D1 and CDK4 expression. *Breast Cancer Res. Treat.* **2006**, *95*, 185–194. [CrossRef] [PubMed]

54. Hua, H.; Li, M.J.; Luo, T.; Yin, Y.C.; Jiang, Y.F. Matrix metalloproteinases in tumorigenesis: An evolving paradigm. *Cell. Mol. Life Sci.* **2011**, *68*, 3853–3868. [CrossRef] [PubMed]

55. Dinarello, C.A. Anti-inflammatory agents: Present and future. *Cell* **2010**, *140*, 935–950. [CrossRef] [PubMed]

56. Ruiz-Gomez, G.; Lim, J.; Halili, M.A.; Le, G.T.; Madala, P.K.; Abbenante, G.; Fairlie, D.P. Structure-activity relationships for substrate-based inhibitors of human complement factor B. *J. Med. Chem.* **2009**, *52*, 6042–6052. [CrossRef] [PubMed]

57. Ellrichmann, G.; Thone, J.; Lee, D.H.; Rupec, R.A.; Gold, R.; Linker, R.A. Constitutive activity of NF-κB in myeloid cells drives pathogenicity of monocytes and macrophages during autoimmune neuroinflammation. *J. Neuroinflamm.* **2012**, *9*. [CrossRef]

58. Voboril, R.; Weberova-Voborilova, J. Constitutive NF-kappaB activity in colorectal cancer cells: Impact on radiation-induced NF-κB activity, radiosensitivity, and apoptosis. *Neoplasma* **2006**, *53*, 518–523. [PubMed]

59. Formesyn, E.M.; Danneels, E.L.; de Graaf, D.C. Proteomics of the venom of the parasitoid *Nasonia vitripennis*. In *Parasitoid Viruses: Symbionts and Pathogens*, 19th ed.; Beckage, N.E., Drezen, J., Eds.; Academic Press, Elsevier: London, UK, 2013; pp. 233–246.

60. De Bosscher, K.; Vanden Berghe, W.; Beck, I.M.E.; van Molle, W.; Hennuyer, N.; Hapgood, J.; Libert, C.; Staels, B.; Louw, A.; Haegeman, G.; *et al.* A fully dissociated compound of plant origin for inflammatory gene repression. *Proc. Natl. Acad. Sci. USA* **2005**, *102*, 15827–15832. [CrossRef] [PubMed]

© 2015 by the authors; licensee MDPI, Basel, Switzerland. This article is an open access article distributed under the terms and conditions of the Creative Commons Attribution (CC-BY) license (http://creativecommons.org/licenses/by/4.0/).

toxins

Article

Anti-Fibrotic Effect of Natural Toxin Bee Venom on Animal Model of Unilateral Ureteral Obstruction

Hyun Jin An [1], Kyung Hyun Kim [1], Woo Ram Lee [1], Jung Yeon Kim [1], Sun Jae Lee [1], Sok Cheon Pak [2], Sang Mi Han [3] and Kwan Kyu Park [1,*]

[1]   Department of Pathology, College of Medicine, Catholic University of Daegu, 3056-6, Daemyung-4-Dong, Nam-gu, Daegu 705-718, Korea; ahj119@cu.ac.kr (H.J.A.); khkim1@cu.ac.kr (K.H.K.); wooramee@cu.ac.kr (W.R.L.); kjy1118@cu.ac.kr (J.Y.K.); patho.dr.lee@gmail.com (S.J.L.)

[2]   School of Biomedical Sciences, Charles Sturt University, Panorama Avenue, Bathurst, NSW 2795, Australia; spak@csu.edu.au

[3]   Department of Agricultural Biology, National Academy of Agricultural Science, RDA, 300, Nongsaengmyeong-ro, Wansan-gu, Jeonju-si, Jeollabuk-do 560-500, Korea; sangmih@korea.kr

*   Correspondence: kkpark@cu.ac.kr; Tel.: +82-53-650-4149; Fax: +82-53-650-4843

Academic Editor: Glenn F. King

Received: 22 December 2014; Accepted: 1 May 2015; Published: 29 May 2015

Abstract: Progressive renal fibrosis is the final common pathway for all kidney diseases leading to chronic renal failure. Bee venom (BV) has been widely used as a traditional medicine for various diseases. However, the precise mechanism of BV in ameliorating the renal fibrosis is not fully understood. To investigate the therapeutic effects of BV against unilateral ureteral obstruction (UUO)-induced renal fibrosis, BV was given intraperitoneally after ureteral ligation. At seven days after UUO surgery, the kidney tissues were collected for protein analysis and histologic examination. Histological observation revealed that UUO induced a considerable increase in the number of infiltrated inflammatory cells. However, BV treatment markedly reduced these reactions compared with untreated UUO mice. The expression levels of TNF-α and IL-1β were significantly reduced in BV treated mice compared with UUO mice. In addition, treatment with BV significantly inhibited TGF-β1 and fibronectin expression in UUO mice. Moreover, the expression of α-SMA was markedly withdrawn after treatment with BV. These findings suggest that BV attenuates renal fibrosis and reduces inflammatory responses by suppression of multiple growth factor-mediated pro-fibrotic genes. In conclusion, BV may be a useful therapeutic agent for the prevention of fibrosis that characterizes progression of chronic kidney disease.

Keywords: bee venom; renal fibrosis; inflammation; UUO

1. Introduction

Chronic kidney disease involves renal inflammation, interstitial fibrosis, and tubular atrophy [1]. Progressive renal fibrosis is the final common pathway for all kidney diseases leading to chronic renal failure [2]. Histologically, renal fibrosis is characterized by interstitial infiltration of mononuclear cells, accumulation of myofibroblasts, proliferation of interstitial fibroblasts, accumulation of extracellular matrix (ECM) proteins, and tubular atrophy [3]. Inflammation is involved in the initiation and maintenance of renal damage, and a decreased inflammatory response results in the loss of renal fibrosis [4]. The classic view on the connection between inflammation and fibrosis is that they are mediated in a paracrine fashion; in which inflammatory cells secrete pro-fibrotic cytokines that act on resident fibroblasts and tubular cells to promote fibrogenesis [5]. The development and progression of renal fibrosis primarily involves differentiation of renal fibroblasts into myofibroblasts and infiltration of inflammatory cells, including dendritic cells, lymphocytes, macrophages, and mast cells [6]. These cells are regulated by numerous cytokines and growth factors, such as transforming growth factor-β1

(TGF-β1), tumor necrosis factor-α (TNF-α), interleukin-1β (IL-1β), IL-6, fibroblast growth factor (FGF), and platelet-derived growth factor (PDGF). Thus, a therapeutic intervention that blocks the activation of these cytokine and growth factor receptors could improve antifibrotic effects to slow progression of renal fibrosis [7].

Because a large variety of pathophysiologically distinct diseases converge finally into renal fibrosis, this makes it a unique target for treatment. Unfortunately, there are no effective therapies in most other types of organ fibrosis [8]. Thus, more systematic and safer agents are required. Purified bee venom (BV) is a mixture of natural toxins produced by honeybees (*Apis mellifera*), and has been widely used as a traditional medicine for various diseases, including arthritis, rheumatism, pain, cancerous tumors, and skin diseases [9,10]. However, the anti-fibrotic effects of BV on renal fibrosis have not been reported.

Therefore, this study investigated the anti-fibrotic effect of BV on the expression of pro-inflammatory cytokines and on the activation of growth factors related with the development of progressive renal fibrosis in an animal model of unilateral ureteral obstruction (UUO).

2. Results and Discussion

2.1. Histological Examination of the UUO Mice with or without Treatment with Bee Venom

The morphological changes in the kidney tissue caused by UUO were visualized in sections stained by hematoxylin and eosin (Figure 1A). Tubular dilatation with flattening of epithelial cells was visualized in UUO kidneys. However, BV treatment significantly reduced these changes when compared to the UUO group. BV attenuated renal histologic damage in UUO mice. The extent of collagen deposition was viewed using Masson's trichrome staining of renal tissue (Figure 1B) and renal fibrosis was calculated using a well described semiquantitative score derived from the percentage of the positive staining per grid field (Figure 1C). Seven days after UUO surgery, the UUO group demonstrated significant interstitial fibrosis compared with the NC group. However, there was a significant reduction in the number of collagen fibers in the BV treated mice.

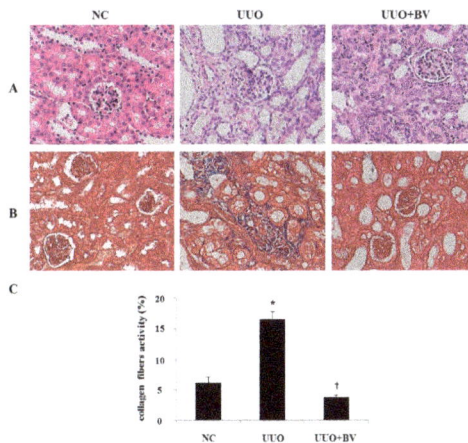

Figure 1. BV inhibits renal fibrosis in obstructed kidney. (**A**) Histological sections of mouse kidney stained with H&E at seven days after UUO surgery. (**B**) Kidney sections are stained with Masson's trichrome, which accentuates interstitial fibrosis by staining collagen blue. (**C**) Masson's trichrome staining was used to evaluate the extent of renal fibrosis which was subsequently quantified. NC, normal control; UUO, kidney injury induced by UUO; UUO+BV, UUO treated with 0.01 mg/kg of BV. Representative images from each study group. Magnification 400×. Results are expressed as means ± SE of three independent determinations. * $p < 0.05$ *vs.* NC group. † $p < 0.05$ *vs.* UUO group.

2.2. Bee Venom Suppresses Pro-Inflammatory Cytokines in the Kidneys of UUO

During UUO, obstruction of the ureter is followed by inflammatory cell infiltration, and by secretion of pro-inflammatory cytokines including TNF-α and IL-1β [11]. To investigate the inflammatory changes in UUO, the expression of TNF-α and IL-1β was determined by immunohistochemical staining, Western blotting and RT-PCR. Immunohistochemistry results showed that UUO kidneys had a marked increase in TNF-α and IL-1β positive cells compared with NC kidneys (Figure 2A). Western blotting and RT-PCR results also demonstrated that the expression of TNF-α and IL-1β was increased in the UUO group (Figure 2D,E). However, the BV treatment group showed significantly reduced expression of pro-inflammatory cytokines compared with the UUO group. There were no obvious expression changes in the kidney of both the NC and BV alone-treated group (Figure not shown). These observations indicate that BV effectively inhibits the expression levels of pro-inflammatory cytokines in UUO mice.

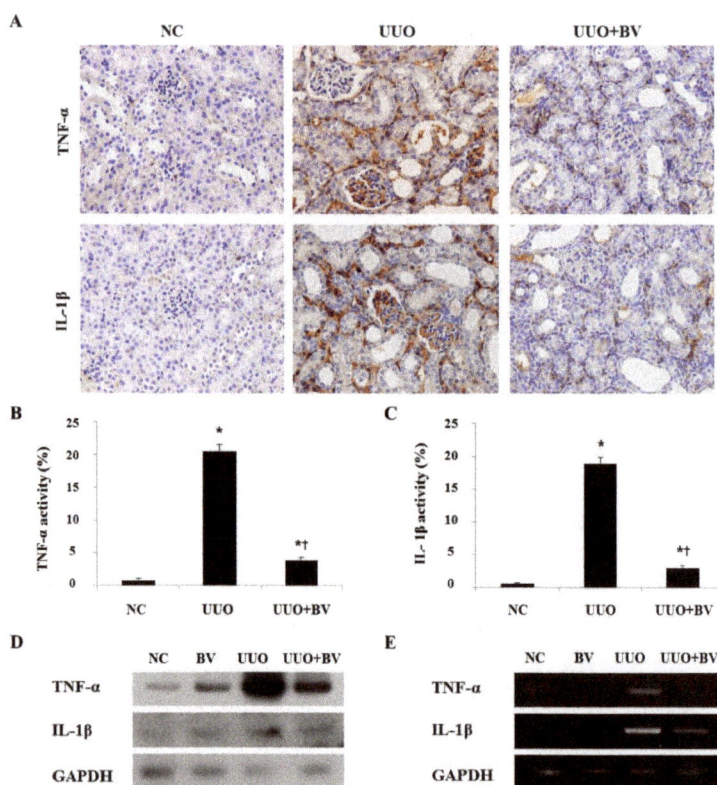

Figure 2. BV attenuates the expression of pro-inflammatory cytokine in obstructed kidneys. (A) Representative macrographs show immunohistochemical staining for TNF-α and IL-1β in the kidneys at seven days after UUO surgery. (B,C) Immunohistochemical staining was used to evaluate the extent of pro-inflammatory cytokines, which was subsequently quantified. (D) Western blot analysis shows that BV suppresses the protein expression of TNF-α and IL-1β in UUO kidneys. (E) RT-PCR results show that BV suppresses the mRNA expression of TNF-α and IL-1β in UUO kidneys. GAPDH levels were analyzed as an internal control. NC, normal control; UUO, kidney injury induced by UUO; UUO+BV, UUO treated with 0.01 mg/kg of BV. Representative images from each study group. Magnification 400×. Results are expressed as means $\pm$ SE of three independent determinations. * $p < 0.05$ *vs.* NC group. † $p < 0.05$ *vs.* UUO group.

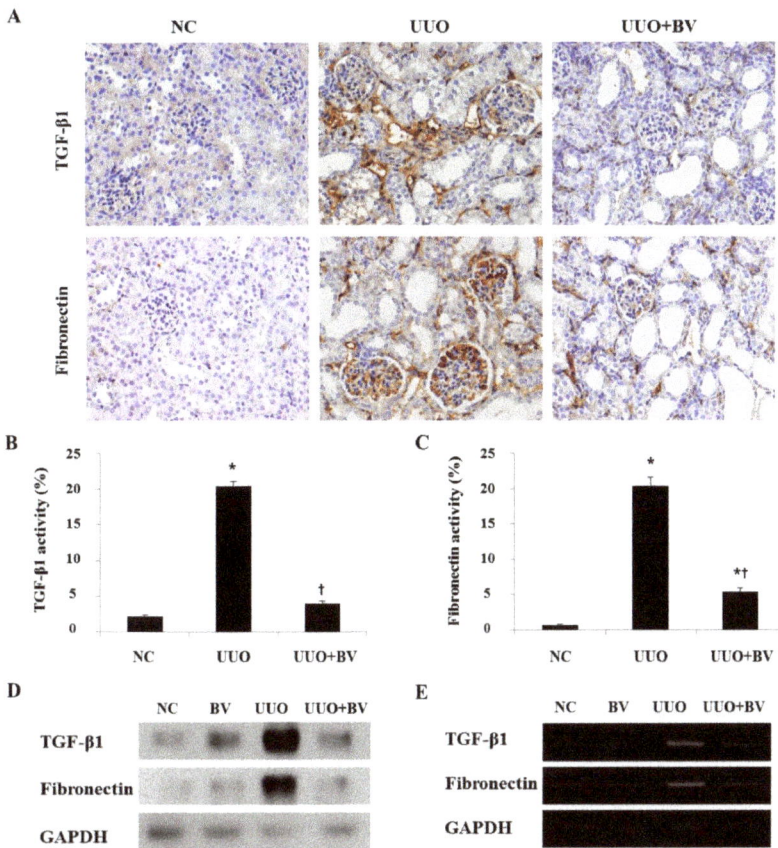

Figure 3. BV attenuates the expression of TGF-β1 and fibronectin in obstructed kidneys. (**A**) Immunohistochemical staining for TGF-β1 and fibronectin in the kidneys at seven days after UUO surgery. (**B,C**) Immunohistochemical staining was used to evaluate the extent of fibrotic genes, which was subsequently quantified. (**D**) Western blot analysis shows that BV suppresses the protein expression of TGF-β1 and fibronectin in UUO kidneys. (**E**) RT-PCR results show that BV suppresses the mRNA expression of TGF-β1 and fibronectin in UUO kidneys. GAPDH levels were analyzed as an internal control. Representative images from each study group. Magnification 400×. Results are expressed as means ± SE of three independent determinations. * $p < 0.05$ *vs.* NC group. ✝ $p < 0.05$ *vs.* UUO group.

2.3. Bee Venom Inhibits the Fibrotic Gene Expression in an Animal Model of UUO

UUO initially produces inflammation, which gradually progresses to fibrosis in the kidney with increased expression of cytokines, such as TGF-β1 [12]. During the development of fibrosis, TGF-β1 expression is upregulated and is known to promote fibrosis under a variety of circumstances, including ECM remodeling [13]. The ECM protein fibronectin is focally deposited in renal fibrosis, where it contributes to inflammatory signaling [14]. As shown in Figure 3A, TGF-β1 and fibronectin positive cells are limited to the tubular basement membranes of the non-obstructed kidney, whereas large amounts of those cells are present in the interstitial space of the obstructed kidney. The cells positive for TGF-β1 and fibronectin were increased in UUO mice, but they were decreased by BV treatment. These observations were confirmed through Western blotting and RT-PCR. The expression of TGF-β1 and fibronectin was increased in UUO, however this increase was abolished by BV treatment (Figure 3D,E). These results suggest that BV effectively blocks fibrotic changes and suppresses the accumulation of ECM in obstructive kidneys.

2.4. UUO-Induced Renal Myofibroblast Activation is Suppressed in Bee Venom Treated Mice

To investigate the ability of BV to suppress myofibroblast activation, this study examined the expression of α-SMA, a representative marker of activated myofibroblasts, by immunofluorescence staining. In normal kidneys, α-SMA positive cells are found only in the blood vessel wall. However, α-SMA positive cells are scattered in the interstitial space of obstructed kidneys, whereas this population of cells was reduced significantly by BV treatment (Figure 4). This data shows clearly that BV plays a critical role in the inactivation of renal fibroblasts after obstructive injury.

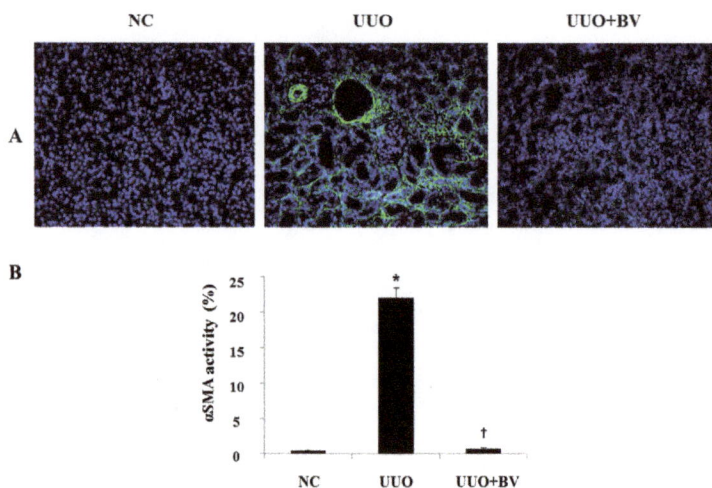

Figure 4. BV abolishes the expression of α-SMA in obstructed kidneys. (**A**) Immunofluorescence staining shows that BV treatment reduces α-SMA positive cells in the kidneys at seven days after UUO surgery. Visible green color indicates α-SMA. Representative images from each study group. (**B**) Immunofluorescence staining was used to evaluate the extent of α-SMA, which was subsequently quantified. Magnification 200×. Results are expressed as means ± SE of three independent determinations. * $p < 0.05$ *vs.* NC group. † $p < 0.05$ *vs.* UUO group.

2.5. Discussion

Natural toxin BV contains a variety of peptides, including adolapin, apamin, melittin, and mast cell degranulating peptide along with enzymes, biological amines, and other nonpeptide components [10]. In our previous study, we demonstrated that BV can reduce hepatic fibrosis via anti-fibrogenic mechanism [15]. Our another study has reported anti-inflammatory effects of BV against *Propionibacterium* acnes-induced inflammatory skin disease in an animal model [16]. However, the effects of BV during renal fibrosis have not been reported. Thus, this study examined the therapeutic effects of BV on the progression of renal fibrosis using the UUO model.

Recent reports have shown that obstruction-mediated renal injuries were involved in the mechanisms of inflammatory cytokines, chemokines, and fibrosis-related gene expressions [1,17,18]. TNF-α and IL-1β as key pro-inflammatory cytokines are considered to play important roles in renal fibrosis, and are produced by several types of inflammatory cells [19]. Renal fibrosis is indicated by increasing TGF-β1 activity and collagen deposition, which is stimulated by TNF-α [20,21]. IL-1β is secreted by macrophages in the fibrotic lesions of the kidney [22]. On the basis of this information, this study investigated whether BV could have an effect on these pro-inflammatory cytokines in renal fibrosis. In UUO mice, the numbers of positive cells for TNF-α and IL-1β were increased, but they were decreased by BV treatment. These results demonstrate that BV is an effective blocker of inflammatory cytokine expression.

TGF-β1 is widely recognized as a strong inducer of fibrosis in renal structures during UUO [23]. During renal fibrosis, TGF-β1 directly regulates the expression of ECM proteins, activates resident fibroblasts and myofibroblasts, and down regulates ECM degradation [13]. Inflammatory reaction in the obstructed kidney stimulates the expression of TGF-β1 [24]. In the present study, the kidneys that received UUO surgery showed increased expression of TGF-β1 and production of fibronectin, a major ECM protein, compared with normal kidneys as demonstrated by immunohistochemistry, Western blot and RT-PCR analyses. However, this increase was abolished by BV treatment in UUO kidneys.

As a consequence of interstitial inflammation, interstitial myofibroblasts were increased and resident interstitial fibroblasts were activated. Interstitial myofibroblasts are the major source of tubulointerstitial ECM and are the best prognostic indicators of disease progression in both human and animal glomerulonephritis [25–27]. To investigate the ability of BV to suppress myofibroblast activation *in vivo*, this study examined the effect of BV on the expression of α-SMA, a hallmark of myofibroblasts, in UUO mice. The expression of α-SMA was increased in UUO mice, while this effect was significantly decreased with BV treatment. These results suggest that renal fibrosis is a complex result of various factors and the present study demonstrated that BV can effectively prevent renal fibrosis.

In summary, these findings suggest that BV attenuates renal fibrosis and reduces inflammatory responses by suppression of multiple growth factor-mediated pro-fibrotic genes. Therefore, BV may be a useful therapeutic agent for the prevention of fibrosis that characterizes progression of chronic kidney disease.

3. Experimental Section

3.1. Collection of Bee Venom

Colonies of natural honeybees (*Apis mellifera* L.) used in this study were maintained at the National Academy of Agricultural Science, Korea. BV was collected by the collecting device (Chung Jin Biotech Co., Ltd., Ansan, Korea) in a sterile manner under strict laboratory conditions. In brief, the BV collector was placed on the hive, and the bees were given enough electric shocks to cause them to sting a glass plate, from which dried bee venom was later scraped off. The collected venom diluted in cold sterile water and then centrifuged at 10,000 g for 5 min at 4 °C to discard residues from the supernatant. BV was lyophilized by freeze dryer and refrigerated at 4 °C for later use. BV used in the experiment was confirmed with size exclusion gel chromatography (AKTA Explorer, GE Healthcare, Pittsburgh, PA, USA) by dissolving in 0.02 M phosphate buffer with 0.25M NaCl adjusted to pH 7.2 using a Superdex Peptide column (Amersham Biosciences, GE Healthcare, Pittsburgh, PA, USA).

3.2. Animal Model

The animal model was established using male Balb/c mice (20–25 g) that were individually housed in polycarbonate cages and maintained under constant temperature (22 ± 2 °C) and humidity (55%). Mice had free access to food, water and were subjected to an artificial light-dark cycle of 12:12 hours. All surgical and experimental procedures used in current study were approved by the IRB committee at Catholic University of Daegu Medical Center (protocol number 2013-1125-CU-AEC-16-Y). In UUO operation, the abdominal cavity was exposed by a midline incision, and the left ureter was isolated and ligated with 5-0 silk at two points. Balb/c mice were randomly divided into three groups: (1) non-treated mice (Normal Control, NC); (2) UUO mice (UUO); and (3) UUO mice were injected with BV (UUO+BV) (*n* = 6, each group). Intraperitoneal injection of BV at a concentration of 0.01 mg/kg was given immediately after ureteral ligation. Then, BV was given intraperitoneal injection 2 days after UUO operation. The kidneys were collected for mRNA and protein analysis including histologic examination at day 7 post UUO surgery.

3.3. Western Blot Analysis

Tissues were lysed in a lysis buffer (50 mM Tris pH 8.0, 150 mM NaCl, 5 mM EDTA, 0.5% NP-40, 100 mM PMSF, 1 M DTT, 10 mg/mL leupeptin and aprotinin; all from Sigma-Aldrich, St. Louis, MO, USA). After incubation for 30 min on ice, samples were centrifuged at 8000 g for 30 min at 4 °C. Then, supernatant was collected. The protein concentration was determined with the Bradford assay (Bio-Rad Laboratories, Hercules, CA, USA). Total protein (10–50 µg) was separated on 8% to 12% SDS-polyacrylamide gels and transferred to PVDF membrane (Millipore Corporation, Bedford, MA, USA) using standard SDS-PAGE gel electrophoresis procedure. Membranes were blocked in 5% skim milk in TBS-T (10 mM Tris, 150 mM NaCl and 0.1% Tween-20) for 2 h at room temperature. Then, membrane was probed with primary antibody for 4 hours and a horseradish peroxidase (HRPO)-conjugated secondary antibody (anti-mouse, anti-rabbit and anti-goat) was used for detection. Signals were detected using an enhanced chemiluminescence detection system (Amersham, Piscataway, NJ, USA). Primary antibodies used in this study were the following: anti-TNF-α, anti-fibronectin, and anti-α-smooth muscle actin (α-SMA, Abcam, MA, USA), anti-TGF-β1 (R&D Systems, Minneapolis, MN, USA), and anti-IL-1β, anti-collagen type I, and anti-glyceraldehyde-3-phosphate-dehydrogenase (GAPDH) from Santa Cruz (Dallas, TX, USA). All primary antibodies were diluted at 1:1000. Signal intensity was quantified by image analyzer (Las 3000, Fuji, Japan).

3.4. Histological Analysis

All tissue specimens were fixed in 10% formalin for at least 24 h at room temperature. After fixation, perpendicular sections to the anterior–posterior axis of the kidney were dehydrated in graded ethanol, cleared in xylene, and embedded in paraffin. Thin sections (3 µm) were mounted on glass slides, dewaxed, rehydrated to distilled water, and stained with hematoxylin and eosin (H & E). As part of the histological evaluation, all slides were examined by a pathologist, without knowledge of the previous treatment, under a light microscope.

3.5. Immunohistochemical Staining

Paraffin-embedded tissue sections at 5 µm thickness were deparaffinized with xylene, dehydrated in gradually decreasing concentrations of ethanol, and then treated with 3% hydrogen peroxidase in methanol for 10 min to block endogenous peroxidase activity. Tissue sections were immersed in 10 mM sodium citrate buffer (pH 6.0) for 5 min at 95 °C. The last step was repeated using fresh 10 mM sodium citrate solution (pH 6.0). Sections were allowed to remain in the same solution while cooling for 20 min and rinsed in PBS. Sections were incubated with primary antibody (1:100 dilution) for 1 h at 37 °C. Primary antibodies were following: anti-TNF-α and anti-fibronectin (Abcam), anti-IL-1β (Santa Cruz), anti-TGF-β1 (R&D Systems). Signal was visualized using an Envision system (DAKO, CA, USA) for 30 min at 37 °C. DAB (3,3'-diaminobenzidine tetrahydrochloride) was used as the coloring reagent and hematoxylin was used as counter stain.

3.6. Immunofluorescent Staining

Paraffin-embedded tissue sections were deparaffinized with xylene and dehydrated in gradually decreasing concentrations of ethanol. Tissue sections were then placed in blocking serum (5% bovine serum albumin in PBS) at room temperature for 1 h. Primary antibody (1:500 dilution) was incubated at room temperature for 2 h, and secondary antibody incubation (1:200 dilution) was performed at room temperature for 2 h. Antibodies were following: α-SMA (Abcam), and goat anti-mouse IgG secondary antibody conjugated with FITC (Invitrogen, Carlsbad, CA, USA). Slides were mounted using VECTASHIELD Mounting Medium (VECTOR Laboratories, Burlingame, CA, USA). Specimens were examined and photographed using a fluorescence microscope (Nikon, Tokyo, Japan).

3.7. Reverse Transcription-Polymerase Chain Reaction (RT-PCR)

Total RNA was extracted from the frozen kidney with TRIzol Reagent (Gibco, Grand Island, NY, USA) according to the manufacturer's recommendations. Purity and quantity of RNA preparation were measured at optical densities of 260 nm and 280 nm. First stand cDNA was synthesized with oligo-d(T) primer and M-MLV reverse transcriptase (Promega, Madison, WI, USA). Aliqout of cDNA was used for PCR using primer sets specific to mouse TNF-α, IL-1β, TGF-β1, fibronectin, and GAPDH. Primer sequences are following: TNF-α forward primer, 5'-AGT GGT GCC AGC CGA TGG GTT GT-3'; TNF-α backward primer, 5'-GCT GAG TTG GTC CCC CTT CTC CAG-3'; IL-1β forward primer, 5'-CAT GAG CAC CTT CTT TTC CT-3'; IL-1β backward primer, 5'-TGT ACC AGT TGG GGA ACT CT-3'; TGF-β1 forward primer, 5'-CCT GCT GCT TTC TCC CTC AAC C-3'; TGF-β1 backward primer 5'-CTG GCA CTG CTT CCC GAA TGT C-3'; fibronectin forward primer, 5'-TGT GAC AAC TGC CGT AGA CC-3'; fibronectin backward primer, 5'-GAC CAA CTG TCA CCA TTG AGG-3'; GAPDH forward primer, 5'-GTG GAC ATT GTT GCC ATC AAC G-3'; GAPDH backward primer, 5'-GAG GGA GTT GTC ATA TTT CTC G-3'. PCR products were visualized by 1.5% agarose gel electrophoresis with ethidium bromide staining.

3.8. Statistical Analysis

Data are presented as means $\pm$ SE. Student's *t*-test was used to assess the significance of independent experiments. The criterion $p < 0.05$ was used to determine statistical significance.

4. Conclusions

This study demonstrated that natural toxin BV inhibits the development and progression of renal fibrosis in an animal model of UUO. Anti-fibrotic effects of BV are associated with inactivation of multiple cytokine and growth factors as well as inhibition of inflammatory responses. These results suggest that BV could have therapeutic potential for the treatment of renal fibrosis due to its interaction with multiple growth factor-mediated pro-fibrotic genes.

Acknowledgments: This work was carried out with the support of "Cooperative Research Program for Agriculture Science & Technology Development (Project No. PJ01132501)" Rural Development Administration, Korea.

Author Contributions: Hyun Jin An and Kwan Kyu Park designed the study and prepared the manuscript. Hyun Jin An, Kyung Hyun Kim, Woo Ram Lee, Jung Yeon Kim, Sun Jae Lee and Sang Mi Han performed overall experiments. Sok Cheon Pak discussed the study. All authors have read and approved the final version of this manuscript.

Conflicts of Interest: The authors declare no conflict of interest.

References

1. Anders, H.J.; Ryu, M. Renal microenvironments and macrophage phenotypes determine progression or resolution of renal inflammation and fibrosis. *Kidney Int.* **2011**, *80*, 915–925. [PubMed]
2. Harris, R.C.; Neilson, E.G. Toward a unified theory of renal progression. *Annu. Rev. Med.* **2006**, *57*, 365–380. [CrossRef] [PubMed]
3. Eddy, A.A. Molecular insights into renal interstitial fibrosis. *J. Am. Soc. Nephrol.* **1996**, *7*, 2495–2508. [PubMed]
4. Liu, N.; Guo, J.K.; Pang, M.; Tolbert, E.; Ponnusamy, M.; Gong, R.; Bayliss, G.; Dworkin, L.D.; Yan, H.; Zhuang, S. Genetic or pharmacologic blockade of EGFR inhibits renal fibrosis. *J. Am. Soc. Nephrol.* **2012**, *23*, 854–867. [CrossRef] [PubMed]
5. Liu, Y. Cellular and molecular mechanisms of renal fibrosis. *Nat. Rev. Nephrol.* **2011**, *7*, 684–696. [CrossRef] [PubMed]
6. Pradere, J.P.; Gonzalez, J.; Klein, J.; Valet, P.; Gres, S.; Salant, D.; Bascands, J.L.; Saulnier-Blache, J.S.; Schanstra, J.P. Lysophosphatidic acid and renal fibrosis. *Biochim. Biophys. Acta* **2008**, *1781*, 582–587. [CrossRef] [PubMed]
7. Liu, N.; He, S.; Tolbert, E.; Gong, R.; Bayliss, G.; Zhuang, S. Suramin alleviates glomerular injury and inflammation in the remnant kidney. *PLoS ONE* **2012**, *7*, e36194. [CrossRef] [PubMed]

8. Boor, P.; Sebekova, K.; Ostendorf, T.; Floege, J. Treatment targets in renal fibrosis. *Nephrol. Dial. Transplant.* **2007**, *22*, 3391–3407. [CrossRef] [PubMed]

9. Billingham, M.E.; Morley, J.; Hanson, J.M.; Shipolini, R.A.; Vernon, C.A. Letter: An anti-inflammatory peptide from bee venom. *Nature* **1973**, *245*, 163–164. [CrossRef] [PubMed]

10. Son, D.J.; Lee, J.W.; Lee, Y.H.; Song, H.S.; Lee, C.K.; Hong, J.T. Therapeutic application of anti-arthritis, pain-releasing, and anti-cancer effects of bee venom and its constituent compounds. *Pharmacol. Ther.* **2007**, *115*, 246–270. [CrossRef] [PubMed]

11. Bascands, J.L.; Schanstra, J.P. Obstructive nephropathy: Insights from genetically engineered animals. *Kidney Int.* **2005**, *68*, 925–937. [CrossRef] [PubMed]

12. Meng, X.M.; Nikolic-Paterson, D.J.; Lan, H.Y. Inflammatory processes in renal fibrosis. *Nat. Rev. Nephrol.* **2014**, *10*, 493–503. [CrossRef] [PubMed]

13. Wynn, T.A. Cellular and molecular mechanisms of fibrosis. *J. Pathol.* **2008**, *214*, 199–210. [CrossRef] [PubMed]

14. Chen, H.Y.; Huang, X.R.; Wang, W.; Li, J.H.; Heuchel, R.L.; Chung, A.C.; Lan, H.Y. The protective role of Smad7 in diabetic kidney disease: Mechanism and therapeutic potential. *Diabetes* **2011**, *60*, 590–601. [CrossRef] [PubMed]

15. Kim, S.J.; Park, J.H.; Kim, K.H.; Lee, W.R.; Chang, Y.C.; Park, K.K.; Lee, K.G.; Han, S.M.; Yeo, J.H.; Pak, S.C. Bee venom inhibits hepatic fibrosis through suppression of pro-fibrogenic cytokine expression. *Am. J. Chin. Med.* **2010**, *38*, 921–935. [CrossRef] [PubMed]

16. An, H.J.; Lee, W.R.; Kim, K.H.; Kim, J.Y.; Lee, S.J.; Han, S.M.; Lee, K.G.; Lee, C.K.; Park, K.K. Inhibitory effects of bee venom on Propionibacterium acnes-induced inflammatory skin disease in an animal model. *Int. J. Mol. Med.* **2014**, *34*, 1341–1348. [CrossRef] [PubMed]

17. Misseri, R.; Meldrum, K.K. Mediators of fibrosis and apoptosis in obstructive uropathies. *Curr. Urol. Rep.* **2005**, *6*, 140–145. [CrossRef] [PubMed]

18. Lloyd, C.M.; Dorf, M.E.; Proudfoot, A.; Salant, D.J.; Gutierrez-Ramos, J.C. Role of MCP-1 and RANTES in inflammation and progression to fibrosis during murine crescentic nephritis. *J. Leukoc. Biol.* **1997**, *62*, 676–680. [PubMed]

19. Misseri, R.; Rink, R.C.; Meldrum, D.R.; Meldrum, K.K. Inflammatory mediators and growth factors in obstructive renal injury. *J. Surg. Res.* **2004**, *119*, 149–159. [CrossRef] [PubMed]

20. Guo, G.; Morrissey, J.; McCracken, R.; Tolley, T.; Liapis, H.; Klahr, S. Contributions of angiotensin II and tumor necrosis factor-alpha to the development of renal fibrosis. *Am. J. Physiol. Ren. Physiol.* **2001**, *280*, F777–F785.

21. Meldrum, K.K.; Misseri, R.; Metcalfe, P.; Dinarello, C.A.; Hile, K.L.; Meldrum, D.R. TNF-alpha neutralization ameliorates obstruction-induced renal fibrosis and dysfunction. *Am. J. Physiol. Regul. Integr. Comp. Physiol.* **2007**, *292*, R1456–R1464. [CrossRef] [PubMed]

22. Nikolic-Paterson, D.J.; Main, I.W.; Tesch, G.H.; Lan, H.Y.; Atkins, R.C. Interleukin-1 in renal fibrosis. *Kidney Int. Suppl.* **1996**, *54*, S88–S90. [PubMed]

23. Garcia-Sanchez, O.; Lopez-Hernandez, F.J.; Lopez-Novoa, J.M. An integrative view on the role of TGF-beta in the progressive tubular deletion associated with chronic kidney disease. *Kidney Int.* **2010**, *77*, 950–955. [CrossRef] [PubMed]

24. Lan, H.Y. Diverse roles of TGF-beta/Smads in renal fibrosis and inflammation. *Int. J. Biol. Sci.* **2011**, *7*, 1056–1067. [CrossRef] [PubMed]

25. Roberts, I.S.; Burrows, C.; Shanks, J.H.; Venning, M.; McWilliam, L.J. Interstitial myofibroblasts: Predictors of progression in membranous nephropathy. *J. Clin. Pathol.* **1997**, *50*, 123–127. [CrossRef] [PubMed]

26. Goumenos, D.S.; Brown, C.B.; Shortland, J.; el Nahas, A.M. Myofibroblasts, predictors of progression of mesangial IgA nephropathy? *Nephrol. Dial. Transplant.* **1994**, *9*, 1418–1425. [PubMed]

27. Hewitson, T.D.; Wu, H.L.; Becker, G.J. Interstitial myofibroblasts in experimental renal infection and scarring. *Am. J. Nephrol.* **1995**, *15*, 411–417. [PubMed]

© 2015 by the authors; licensee MDPI, Basel, Switzerland. This article is an open access article distributed under the terms and conditions of the Creative Commons Attribution (CC-BY) license (http://creativecommons.org/licenses/by/4.0/).

toxins

MDPI

Article

Influence of Honeybee Sting on Peptidome Profile in Human Serum

Jan Matysiak [1,*], **Agata Światły** [1], **Joanna Hajduk** [1], **Joanna Matysiak** [2] **and Zenon J. Kokot** [1]

[1] Department of Inorganic & Analytical Chemistry, Poznan University of Medical Sciences;
6 Grunwaldzka Street, Poznań 60-780, Poland; agata_swiatly@wp.pl (A.S.); jo.hajduk@gmail.com (J.H.);
zkokot@ump.edu.pl (Z.J.K.)

[2] Ward of Paediatric Diseases, L. Perzyna Regional Unified Hospital in Kalisz 79 Poznańska Street,
Kalisz 62-800, Poland; jkamatysiak@gmail.com

* Correspondence: jmatysiak@ump.edu.pl; Tel.: +48-61-854-66-11; Fax: +48-61-854-66-09

Academic Editor: Sokcheon Pak

Received: 31 March 2015; Accepted: 15 May 2015; Published: 22 May 2015

Abstract: The aim of this study was to explore the serum peptide profiles from honeybee stung and non-stung individuals. Two groups of serum samples obtained from 27 beekeepers were included in our study. The first group of samples was collected within 3 h after a bee sting (stung beekeepers), and the samples were collected from the same person a second time after at least six weeks after the last bee sting (non-stung beekeepers). Peptide profile spectra were determined using MALDI-TOF mass spectrometry combined with Omix, ZipTips and magnetic beads based on weak-cation exchange (MB-WCX) enrichment strategies in the mass range of 1–10 kDa. The samples were classified, and discriminative models were established by using the quick classifier, genetic algorithm and supervised neural network algorithms. All of the statistical algorithms used in this study allow distinguishing analyzed groups with high statistical significance, which confirms the influence of honeybee sting on the serum peptidome profile. The results of this study may broaden the understanding of the human organism's response to honeybee venom. Due to the fact that our pilot study was carried out on relatively small datasets, it is necessary to conduct further proteomic research of the response to honeybee sting on a larger group of samples.

Keywords: honeybee venom; peptidome profiling; MALDI; sting

1. Introduction

The diagnostic algorithm for *Hymenoptera* venom hypersensitivity (insect sting allergy) is a serious issue in allergological practice. Measurement of specific IgE-antibodies' (sIgE) concentration and skin tests represent the routinely used methods to demonstrate the response of the organism to the honeybee sting [1,2]. However, these tests are not sufficient for a proper diagnosis, because of their non-specificity. A number of prospective studies analyzing the biochemistry response of the organism to the venom have been published in the last decade [1–6]. Additionally, the reaction after a bee sting at the metabolomic level was studied by our group [7]. However, there is still a lack of work analyzing the human organism's response to *Hymenoptera* sting at the proteomic or peptidomic level.

Proteomics is one of the most promising approaches for the identification of potential biomarkers and for assessing differences between individuals of different health status. Searching for characteristic indicators of physiological or pathophysiological state related to allergic diseases, like asthma or chronic obstructive pulmonary disease, can lead to the implementation of specific and accurate diagnostic methods, noninvasive monitoring of the condition and new drug development [8,9]. The clinical research employed two proteomics strategies in the field of biomarker searching: the classical approach and peptide profiling. In the classical approach, proteins/peptides are first separated by applying two-dimensional electrophoresis (2DE), then isolated and identified using the MS/MS technique [10], which is both sample and time consuming and becomes very tedious, particularly when the analyses are repeated several times. The major aspect of the second approach is not the identification of particular proteins/peptides, but the development of peptide-protein patterns of a whole sample, called proteome pattern analysis or proteome profiling. The goal of this methodology is the recognition of changes in the proteome/peptidome, which distinguish the studied groups [11]. The description of correlations within the panels of multiple peptides representing markers of a given morbid unit provides grounds for designing rapid, specific and sensitive diagnostic tests. The multi-component peptide profiling manifests higher specificity than individual un-correlated markers do [11]. There are many studies using this approach for the diagnosis of cancer [12–14] and other diseases, like endometriosis, adenomyosis [15] and acute hepatitis E [16]. MALDI-TOF MS (matrix-assisted laser desorption/ionization time-of-flight mass spectrometry) is widely used for searching biomarkers in biological samples [12,16], because of its high sensitivity, high speed of analysis, low consumption of analyte and the prevalence of singly-charged ions [17]. However, the complexity of serum samples makes proteomic characterization difficult. This is caused by the presence of high abundant proteins and peptides, which might mask low abundant components with predictive, prognostic or diagnostic potential [18]. This limitation on the dynamic range may be partially overcome by using depletion methods that yield a defined subset of the proteome [19,20]. The aim of this study was to explore the serum peptide profiles from honeybee stung and non-stung individuals. The advanced chemometric analysis was applied to evaluate the organism response to honeybee sting at the peptidomic level. Since the discriminating pattern of dozens of peptides formed by a subset of m/z (mass-to-charge ratio) signals has been taken into account, no identification of single compounds was required. We decided to use three serum sample preparation procedures, including different enrichment strategies: magnetic beads based on weak-cation exchange (MB-WCX) and two types of micropipette tips with prepacked C18 reverse phase (Omix and ZipTips). Using this techniques allows the preconcentration and purification of serum samples and led to generating complementary peptide profiles. This is the first report in the available literature on clinical studies that is focused on the characterization of serum peptidomic profiles after a sting.

2. Results and Discussion

Two groups of serum samples obtained from 27 beekeepers (24 male, three female) were included in our study. The first group of samples was collected within 3 h after a bee sting (stung beekeepers). The samples were collected from the same person a second time after at least six weeks after the last bee sting, a minimum of six weeks from the end of the beekeeping season (non-stung beekeepers). MALDI-TOF MS combined with Omix, ZipTips and MB-WCX enrichment strategies were used in the study to detect LMW protein/peptide profile spectra in the mass range of 1–10 kDa. The mechanism of the magnetic beads used relies on weak-cation exchange. The MB-WCX kit is based on super-paramagnetic micro-particles, which have negatively-charged functional groups at the surface of the beads [21]. In our study, we used two C18 SPE micropipette tips made by different manufacturers: Omix (Agilent Technologies, Great Britain) and ZipTip (Millipore, NH, USA), which are based on reversed-phase chromatography. The compounds are separated due to their chemical and physical properties, which determine their separation between a mobile liquid phase and a solid stationary phase. The compounds, which are not bound, are washed away. Finally, the bound molecules are eluted from the solid phase by the elution solvent [22,23]. After normalization and alignment of the all processed spectra, with a signal-to-noise threshold equal to or greater than five on the average spectrum, 149 unique peaks were detected in the Omix dataset, 153 unique peaks in the ZipTips dataset and 127 unique peaks in the MB-WCX dataset. MALDI-TOF-MS averaged peptidome profiles obtained using Omix, ZipTips and MB-WCX are shown in Figures 1–3, respectively. The samples were classified, and discriminative models were established by using the quick classifier (QC), genetic algorithm (GA) and supervised neural network (SNN) algorithms in ClinProTools software to analyze all of the detected peaks. In order to assess the influence of time series, *in vitro* diagnostic tests for allergy to bee venom were performed in our earlier studies [24]. The diagnostic tests were carried out directly after a bee sting and after at least six weeks after the sting and showed no significant differences in the levels of total IgE antibodies, honeybee venom-specific IgE antibodies, phospholipase A_2 (the main allergen of honeybee venom)-specific IgE antibodies, serum tryptase and honeybee venom-specific IgG4 antibodies.

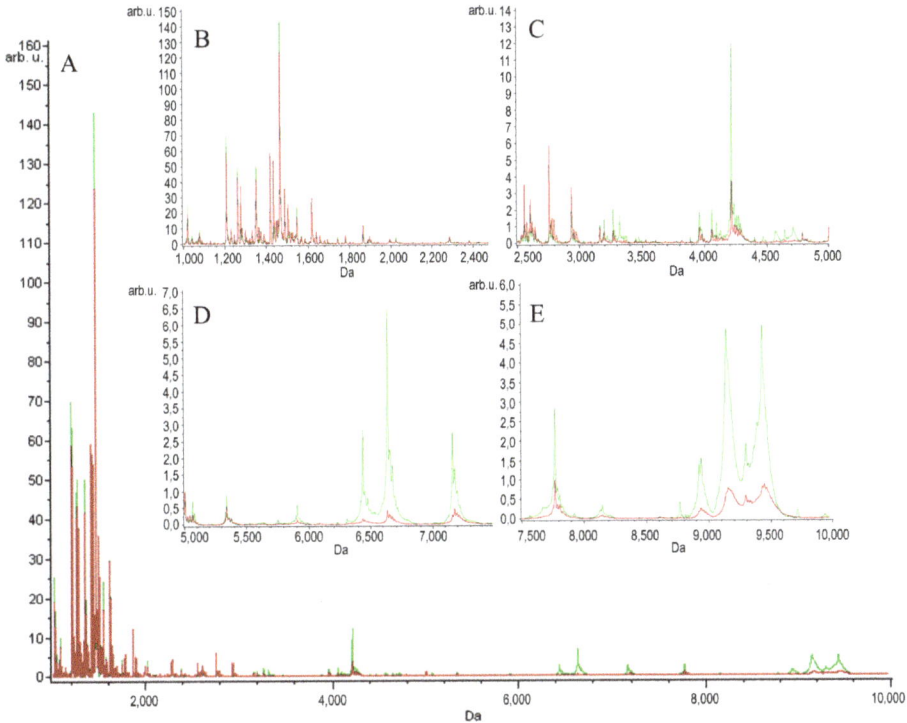

Figure 1. Average MALDI-TOF spectra of serum samples of stung (red) and non-stung (green) individuals pretreated with Omix over the full scan range of m/z 1–10 kDa (**A**); zoomed spectra over the m/z 1–2.5 kDa range (**B**); zoomed spectra over the m/z 2.5–5.0 kDa range (**C**); zoomed spectra over the m/z 5–7.5 kDa range (**D**); zoomed spectra over the m/z 7.5–10 kDa range (**E**).

To identify the discriminatory power of all detected peaks, the QC algorithm was used. QC is a univariate sorting algorithm. For proper classification of the peak, averages of the peak areas are stored in the model together with p-values obtained from a t-test. The peak areas are sorted per peak. Then, a weighted average over all peaks is calculated [25]. The best detection value of this model was obtained for the samples pretreated with the Omix strategy (cross-validation: 94.81%; recognition capability: 94.65%). However, for the samples pretreated with ZipTips and MB, applying this algorithm also led to high detection values (Table 1). Using multivariate analysis (GA and SNN) also enabled distinguishing the groups of the stung and non-stung individuals. The genetic algorithm is used to select a combination of peaks that are most relevant for the separation. During the selection of the peaks, the best capable peak combinations are taken into account. This is done by optimizing a function, which aims at optimal class separation with high variance between classes [25,26]. By applying GA for MS data analysis, it was shown that the spectra obtained after sample enrichment using Omix allowed building a model of higher average cross-validation (97.01%) and average recognition capability (100%) compared to the spectra obtained using the magnetic beads and ZipTips strategy (Table 1). SNN is the algorithm based on the classification of the prototype. It identifies some characteristic spectra of each class. Spectra are called prototypes and can be considered as prototypical samples of that class [25]. The SNN algorithm allowed efficiently discriminating stung and non-stung individuals. Comparing all of the enrichment strategies used, the highest cross-validation (97.31%) and recognition capability (100%) was obtained for Omix.

Feng Qiu *et al.* chose MB-WCX as the best magnetic bead for pre-extraction samples for proteomic analysis in breast cancer research [20]. This method of sample purification is being increasingly used as a relatively simple technique [26]. However, there have been reports criticizing the reproducibility and robustness of magnetic beads [27,28]. Ali Tiss *et al.* proved that the spectra obtained for samples analyzed using ZipTips showed lower background noise and better signal-to-noise ratios compared to the four types of magnetic beads; while the achieved MS profiles are very comparable between samples treated with pipette tips pre-packed with solid phase material, ZipTips and Omix [29], which is in agreement with our findings.

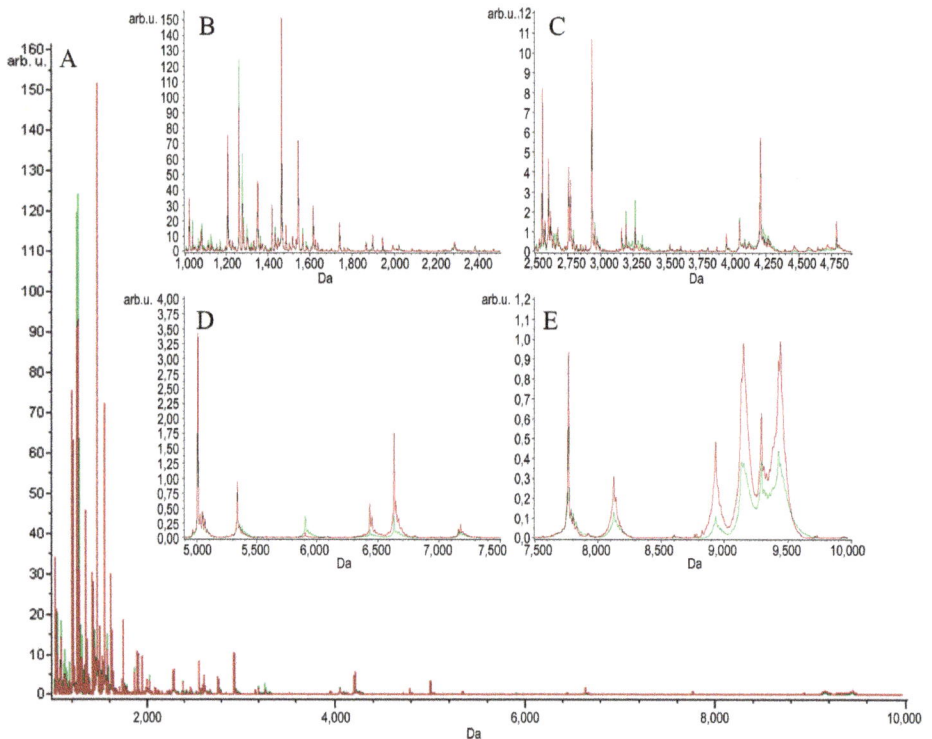

Figure 2. Average MALDI-TOF spectra of serum samples of stung (red) and non-stung (green) individuals pretreated with ZipTips over the full scan range of m/z 1–10 kDa (**A**); zoomed spectra over the m/z 1–2.5 kDa range (**B**); zoomed spectra over the m/z 2.5–5.0 kDa range (**C**); zoomed spectra over the m/z 5–7.5 kDa range (**D**); zoomed spectra over the m/z 7.5–10 kDa range (**E**).

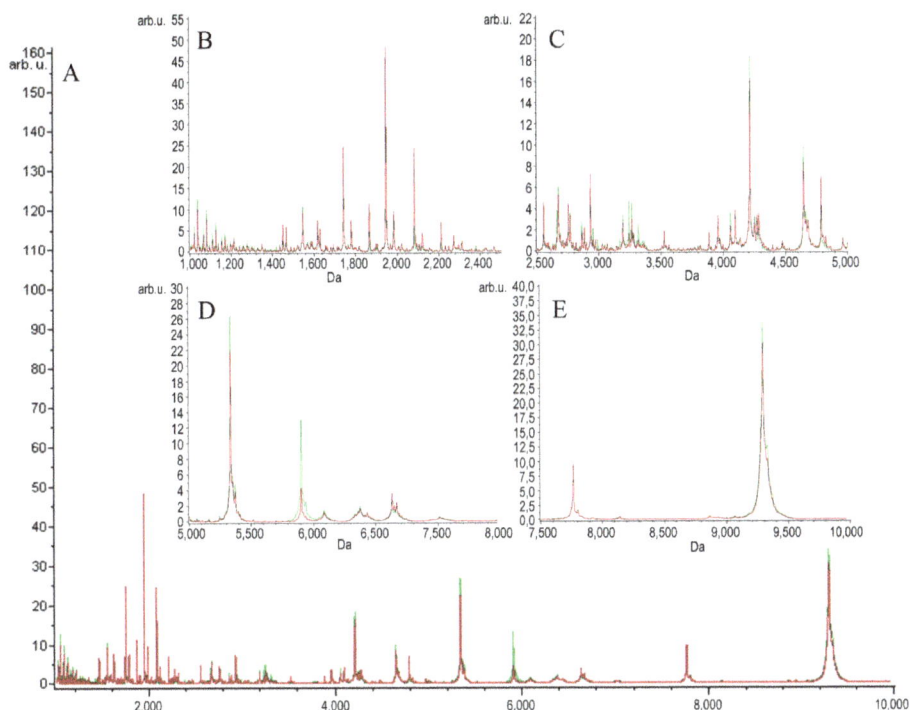

Figure 3. Average MALDI-TOF spectra of serum samples of stung (red) and non-stung (green) individuals pretreated with magnetic beads based on weak-cation exchange (MB-WCX) over the full scan range of m/z 1–10 kDa (**A**); zoomed spectra over the m/z 1–2.5 kDa range (**B**); zoomed spectra over the m/z 2.5–5.0 kDa range (**C**); zoomed spectra over the m/z 5–7.5 kDa range (**D**); zoomed spectra over the m/z 7.5–10 kDa range (**E**).

Table 1. Statistical analysis of MALDI-TOF spectra of serum samples of stung and non-stung individuals pretreated with Omix, ZipTips and MB-WCX. QC, quick classifier; GA, genetic algorithm; SNN, supervised neural network.

Sample enrichment strategy	Algorithm	Cross validation (%)	Recognition capability (%)	Number of rejected spectra (%)
	QC	94.81	94.65	
Omix	GA	97.01	100.00	12.96
	SNN	97.31	100.00	
	QC	84.97	87.20	
ZipTips	GA	87.08	97.52	0.00
	SNN	53.57	50.57	
	QC	80.41	85.64	
MB-WCX	GA	90.54	99.38	0.00
	SNN	85.16	98.13	

It is noteworthy that the amount of excluded spectra is also a critical element in the evaluation of the different methods used in peptide profiling. The least number of excluded spectra (0%) was obtained for samples analyzed using MB and ZipTips. For samples pretreated with Omix, the average number of rejected spectra was 12.96%. From earlier studies, it is known that the number of failures (shown by the number of rejected spectra) is lower for samples purified with ZipTips compared to

MB [29]. ROC curve analysis was used both for the classification and significant feature selection. This diagnostic performance displays the relation between sensitivity and specificity at different thresholds. The area under the curve (AUC) indicates the diagnostic accuracy of the test methods, the greater the area (value close to one), the more precise the method [30]. The ROC curve indicated that using the Omix enrichment strategy allowed obtaining as many as 10 peptides that demonstrated an AUC value of 1.0, which yields the best possible prediction method characterized by 100% sensitivity and 100% specificity. Their m/z values were: 1277.60 Da, 1436.01 Da, 5656.11 Da, 6432.84 Da, 6472.15 Da, 6528.35 Da, 6631.08 Da, 6432.89 Da, 6631.19 Da and 6632.70 Da. For serum samples pretreated with magnetic beads, the highest discrimination quality (AUC: 0.87) was shown by a peptide with a mass of 2120.29 Da, whereas using ZipTips allowed reaching an AUC of 0.9 for a peptide mass of 1277.49 Da. The sensitivities and specificities of the discrimination quality of peaks obtained in our study were satisfactory in all sample enrichment strategies used according to the data presented in the available literature [20,31,32].

Every single peptide indicator has an inherent specificity and sensitivity that cannot be improved. However, multiple indicators can be combined to achieve improvement in research parameters. In our study, different diagnostic models generated by the QC, GA and SNN algorithms analysis comprised at least several indicators. The m/z values of discriminating peaks from three models (QC, GA, SNN) are shown in Table 2. Analyzing samples pretreated with Omix, it was shown that there is one peak with an m/z value of 1299.62 Da, which is by all three models. Applying MB as the enrichment strategy also allowed obtaining one mutual discriminating peak by all three models with an m/z value of 3240.99 Da. In the analysis using each of the depletion methods, there are several mutual peaks between at least two models. The various m/z values of discriminating peaks are caused by different peak classification in the algorithms used. Applying these established models, serum samples derived from stung individuals could be distinguished from non-stung controls.

Table 2. Discriminating peaks between stung and non-stung individuals for serum samples pretreated with Omix, ZipTips and MB-WCX.

Enrichment Strategy	Omix			Ziptips			MB-WCX		
Algorithm	QC	GA	SNN	QC	GA	SNN	QC	GA	SNN
	1277.63	1299.62	1656.15	1037.36	1993.78	8933.73	1450.64	4661.58	3240.99
	1299.62	1458.18	1261.58	1078.03	1037.36		1628.00	1779.41	1866.30
	1436.04	1332.67	6432.85	1125.57	2379.82		1779.41	3240.99	1628.00
	1505.18	1217.39	2755.95	1207.37	1299.90		1866.30	3934.87	2883.66
	1656.15	8766.23	1299.62	1277.47	1217.26		2082.49	1153.78	2120.27
	4210.32		1277.63	1299.90			2120.27		2082.49
	4467.44		1436.04	1364.67			2210.55		1450.64
	4568.85		1505.18	1466.78			2554.59		1617.58
	4711.11		7157.01	1584.84			2660.80		3955.05
	6432.85		1217.39	1897.73			2754.17		4053.49
m/z value of peaks	6450.18		2973.00	4466.21			3240.99		2092.96
used for	6472.33		1904.97	8933.73			4053.49		5001.79
classification (Da)	6528.33		7766.34				5940.75		9062.91
	6631.12		6631.12						2688.19
	6648.76								6509.33
	6670.30								6527.47
	7672.67								8863.22
	8919.94								4963.28
	9136.06								
	9290.45								
	9310.43								
	9333.33								
	9384.58								
	9425.00								

Since the first publication of Lancet in 2002, SELDI (Surface enhanced laser desorption ionization)has been explored for cancer diagnosis [33]. After this article, several papers have been published that confirmed both the MALDI and SELDI methods to be promising diagnostic tools in different diseases [34–37]. At the same time, it also suffered from criticism about reproducibility from different labs. However, our findings showed that MALDI-TOF-MS is a useful method for investigating the organism's response to physiological or pathophysiological stimuli at the proteomic or peptidomic level. All of the statistical algorithms used in this study allow distinguishing the analyzed groups with high statistical significance, which confirms the influence of honeybee sting on the serum peptidome profile.

3. Experimental Section

3.1. Study Participants and Serum Samples

In the study, the participating volunteers were recruited at the beekeepers' meetings. Two groups of serum samples obtained from 27 beekeepers (24 male, 3 female) were included in our study. The first group of samples was collected within 3 h after a bee sting (stung beekeepers). The samples were collected from the same person a second time after at least 6 weeks after the last bee sting, a minimum of 6 weeks from the end of the beekeeping season (non-stung beekeepers). The blood sampling was carried out after overnight fasting to reduce the influence of food components on the peptide profile. The volunteers' age range was from 20–80 years. The storage temperature of the samples was $-80\,^{\circ}\mathrm{C}$ until the analysis. The study was conducted with the approval of the Bioethics Committee of the Poznan University of Medical Sciences, Poland (Resolution No. 324/11) and fulfilled the requirements of the Helsinki declaration. Consent to participate in the study was written by all beekeepers.

3.2. Chemicals and Reagents

Trifluoroacetic acid (TFA), ultrapure water and α-cyano-4-hydroxycinnamic acid (HCCA) were supplied by Sigma Aldrich (St. Louis, MO, USA). Ethanol, isopropanol and acetonitrile (ACN) were supplied by J.T. Baker (Center Valley, PA, USA). All used reagents were of analytical grade or better.

3.3. MALDI-TOF-MS Analysis

Directly before MALDI-TOF-MS analysis, large molecular weight proteins were removed from serum samples. Depletion was performed using micropipette tips C18 Omix (Agilent Technologies, Waldbronn, Germany) [22], ZipTip (Millipore, Bedford, MA, USA) [23] and magnetic beads WCX (Bruker, Bremen, Germany) [21], according to the manufacturer's instruction. Then, samples were mixed with 0.5 µL of one of the matrix solutions, α-cyano-4-hyroxycinnamic acid (HCCA), and put directly onto the MALDI plate (AnchorChip, Bruker Daltonics, Bremen, Germany). Each sample was spotted in triplicate on the plate. Samples were analyzed using an UltrafleXtreme MALDI-TOF mass spectrometer (Bruker Daltonics, Bremen, Germay). The analyzer worked in the linear mode, and positive ions were recorded in a mass range of m/z between 1 kDa and 10 kDa. Typical instrument settings were: Ion Source 1, 25.09 kV; Ion Source 2, 23.80 kV; lens, 6.40 kV; pulsed ion extraction, 260 ns; matrix suppression mass cut off, m/z 700 Da for a mass range 1–10 kDa. Two thousand spectra (laser shots) were picked up for one analysis. For external calibration of the mass spectrometer, the ClinProt standards (1:5 mixture v/v of Peptide Calibration Standard and Protein Calibration Standard I) were analyzed. The average mass deviation was better than 100 ppm. Each sample was analyzed three times. Inter-day and intra-day reproducibility of MS results on three representative serum samples have been analyzed in triplicate in three consecutive days and three times within a day. The data have shown that the coefficient of variance (CV) for five selected m/z peaks with the highest amplitude was less than 10%. The data collection was obtained with FlexControl 3.4 software (Bruker Daltonics, Bremen, Germany, 2011), and the spectra were saved automatically in FlexAnalysis 3.4 software (Bruker Daltonics, Bremen Germany, 2011).

3.4. Statistical Analysis

In order to determine the optimal model allowing the discrimination of the analyzed samples, chemometric software for biomarker detection ClinProTools 3.0 (Bruker Daltonics, Bremen, Germany, 2011) was used. The baseline was set by the "Top Hat Baseline" algorithm. Smoothing of the spectra was obtained by Savitzky–Golay. For the samples' classification, the following algorithms were used: GA, QC and SNN. For these algorithms' cross-validation, recognition capability and the number of rejected spectra were calculated. These indicators of the model's performance are useful predictors of the model's capability to distinguish between two studied groups. The cross-validation values reflect the model's ability to handle variability among test spectra. It can be used to predict how the model will behave in the future. In our research, random cross-validation was set. A random subset of data points is taken over all classes and left out of the model generation procedure. For model generation, the remaining points are used. The absent data points are classified against the model. The obtained classification results are stored for the model. This formula is repeated for a defined number of iterations. The following settings have been used: random mode, 20% to leave out, 10 iterations. The results are averaged and returned as the prediction capability. The specificity and sensitivity of a test and evaluation of the discrimination quality of the peak were calculated by the statistical method receiver operating characteristic (ROC) curve. The generic principle behind this diagnostic performance is to give a graphical overview between sensitivity and specificity at different thresholds. The best prediction method would obtain 100% sensitivity and 100% specificity. The precise way to characterize the ROC curve is the area under the curve (AUC). It indicates the diagnostic accuracy of the test methods: the greater the area (value close to 1), the more precise the method. In ClinProTools, the ROC curve is used as a test separating two groups. The threshold is represented by the peak area or the intensity of the peak. On the diagram on the x-axis, the specificity is given (false positives) and on the y-axis the sensitivity (true positives). Statistical significance was assumed when the *p*-value was <0.001.

4. Conclusions

It can be concluded that the implementation of different sample enrichment strategies (Omix, ZipTips and MB-WCX) linked with MALDI TOF MS and different chemometric algorithms allowed obtaining high differentiation between stung and non-stung individuals. Furthermore, it has been shown that *Hymenoptera* sting changes the serum peptidomic profile. Mass spectrometry-based serum peptidomic profiling is a technique that may broaden the understanding of the human body response to honeybee venom. Due to the fact that our pilot study was carried out on relatively small datasets, it is necessary to conduct further proteomic research of the response to honeybee sting on a larger group of samples.

Acknowledgments: The project was supported by the Polish National Science Centre (2012/05/B/NZ7/02535).

Author Contributions: Jan Matysiak, Joanna Matysiak and Zenon J. Kokot conceived of and designed the experiments. Jan Matysiak, Agata Świątły and Joanna Hajduk performed the experiments. Jan Matysiak, Agata Świątły and Joanna Hajduk analyzed the data. Jan Matysiak, Agata Świątły and Joanna Hajduk contributed reagents/materials/analysis tools. Jan Matysiak, Agata Świątły, Joanna Hajduk, Joanna Matysiak and Zenon J. Kokot wrote the paper.

Conflicts of Interest: The authors declare no conflict of interest.

References

1. Rueff, F.; Jappe, U.; Przybilla, B. Standards and pitfalls of *in vitro* diagnostics of hymenoptera venom allergy. *Hautarzt* **2010**, *61*, 938–945. [CrossRef] [PubMed]
2. Rieger-Ziegler, V.; Rieger, E.; Kranke, B.; Aberer, W. Hymenoptera venom allergy: Time course of specific IgE concentrations during the first weeks after a sting. *Int. Arch. Allergy Immunol.* **1999**, *120*, 166–168. [CrossRef] [PubMed]

3. Kemeny, D.M.; Harries, M.G.; Youlten, L.J.; Mackenzie-Mills, M.; Lessof, M.H. Antibodies to purified bee venom proteins and peptides. I. Development of a highly specific RAST for bee venom antigens and its application to bee sting allergy. *J. Allergy Clin. Immunol.* **1983**, *71*, 505–514. [CrossRef] [PubMed]

4. Matysiak, J.; Matysiak, J.; Bręborowicz, A.; Kokot, Z.J. Diagnosis of hymenoptera venom allergy—With special emphasis on honeybee (Apis mellifera) venom allergy. *Ann. Agric. Environ. Med.* **2013**, *20*, 875–879. [PubMed]

5. Müller, U.R. New developments in the diagnosis and treatment of hymenoptera venom allergy. *Int. Arch. Allergy Immunol.* **2001**, *124*, 447–453. [CrossRef] [PubMed]

6. Simons, F.E.; Frew, A.J.; Ansotegui, I.J.; Bochner, B.S.; Golden, D.B.; Finkelman, F.D. Risk assessment in anaphylaxis: Current and future approaches. *J. Allergy Clin. Immunol.* **2007**, *120*, 2–24. [CrossRef]

7. Matysiak, J.; Dereziński, P.; Klupczyńska, A.; Matysiak, J.; Kaczmarek, E.; Kokot, Z.J. Effects of a honeybee sting on the serum free amino acid profile in humans. *PLoS One* **2014**, *9*, 1–12. [CrossRef]

8. Szefler, S.J.; Wenzel, S.; Brown, R.; Erzurum, S.C.; Fahy, J.V.; Hamilton, R.G.; Hunt, J.F.; Kita, H.; Liu, A.H.; Panettieri, R.A.; *et al.* Asthma outcomes: Biomarkers. *J. Allergy Clin. Immunol.* **2012**, *129*, 9–23. [CrossRef]

9. Taylor, R.D. Using biomarkers in the assessment of airways disease. *J. Allergy Clin. Immunol.* **2011**, *128*, 927–934. [CrossRef] [PubMed]

10. Cristea, I.M.; Gaskell, S.J.; Whetton, A.D. Proteomics techniques and their application to hematology. *Blood* **2004**, *103*, 3624–3634. [CrossRef] [PubMed]

11. Li, L.; Tang, H.; Wu, Z.; Gong, J.; Gruidl, M.; Zou, J.; Tockman, M.; Clark, R.A. Data mining techniques for cancer detection using serum proteomic profiling. *Artif. Intell. Med.* **2004**, *32*, 71–83. [CrossRef] [PubMed]

12. Pietrowska, M.; Polańska, J.; Suwiński, R.; Wideł, M.; Rutkowski, T.; Marczyk, M.; Domińczyk, I.; Ponge, L.; Marczak, Ł.; Polański, A.; *et al.* Comparison of peptide cancer signatures identified by mass spectrometry in serum of patients with head and neck, lung and colorectal cancers: Association with tumor progression. *Int. J. Oncol.* **2012**, *40*, 148–156. [PubMed]

13. Pietrowska, M.; Polanska, J.; Marczak, L.; Behrendt, K.; Nowicka, E.; Stobiecki, M.; Polanski, A.; Tarnawski, R.; Widlak, P. Mass spectrometry-based analysis of therapy-related changes in serum proteome patterns of patients with early-stage breast cancer. *J. Trans. Med.* **2010**, *8*. [CrossRef]

14. He, J.; Zeng, Z.; Xiang, Z.; Yang, P. Mass spectrometry-based serum peptide profiling in hepatocellular carcinoma with bone metastasis. *World J. Gastroenterol.* **2014**, *20*, 3025–3032. [CrossRef] [PubMed]

15. Long, X.; Jiang, P.; Zhou, L.; Zhang, W. Evaluation of novel serum biomarkers and the proteomic differences of endometriosis and adenomyosis using MALDI-TOF–MS. *Arch. Gynecol. Obstet.* **2013**, *288*, 201–205. [CrossRef] [PubMed]

16. Taneja, S.; Ahmad, I.; Sen, S.; Kumar, S.; Arora, R.; Gupta, V.K.; Aggarwal, R.; Narayanasamy, K.; Reddy, V.S.; Jameel, S. Plasma peptidome profiling of acute hepatitis E patients by MALDI-TOF/TOF. *Proteome Sci.* **2011**, *9*. [CrossRef] [PubMed]

17. Pusch, W.; Kostrzewa, M. Application of MALDI-TOF mass spectrometry in screening and diagnostic research. *Curr. Pharm. Design* **2005**, *11*, 2577–2591. [CrossRef]

18. Šalplachta, J.; Řehulka, P.; Chmelík, J. Identification of proteins by combination of size-exclusion chromatography with matrix-assisted laser desorption/ionization time-of-flight mass spectrometry and comparison of some desalting procedures for both intact proteins and their tryptic digests. *J. Mass Spectrom.* **2004**, *39*, 1395–1401. [CrossRef] [PubMed]

19. Palmblad, M.; Vogel, J.S. Quantitation of binding, recovery and desalting efficiency of peptides and proteins in solid phase extraction micropipette tips. *J. Chromatogr. B* **2005**, *814*, 309–313. [CrossRef]

20. Velstra, B.; van der Burgt, Y.E.M.; Mertens, B.J.; Mesker, W.E.; Deelder, A.M.; Tollenaar, R.A. Improved classification of breast cancer peptide and protein profiles by combining two serum workup procedures. *J. Cancer Res. Clin. Oncol.* **2012**, *138*, 1983–1992. [CrossRef] [PubMed]

21. Instructions for Use. Available online: https://www.bruker.com/fileadmin/user_upload/8-PDF-Docs/Separations_MassSpectrometry/InstructionForUse/IFU_223983_223987_MB-WCX_Rev1.pdf (accessed on 27 March 2015).

22. Agilent Bond Elut OMIX Pipette Tips for Micro Extractions. Available online: http://www.chem.agilent.com/Library/primers/Public/5990--9049EN-Omix-Sept11-lo.pdf (accessed on 27 March 2015).

23. User Guide for Reversed-Phase ZipTip Pipette Tips for Sample Preparation. Available online: http://personal.rhul.ac.uk/upba/211/Zip-tip.pdf (accessed on 27 March 2015).

24. Matysiak, J. Assessment the Risk Factors of Allergic Reactions after a Bee Sting in the Beekeepers and Their Family Members. Ph.D. Thesis, Poznan University of Medical Sciences, Poznań, Poland, May 2012.

25. Basics on data preparation, model generation and spectra classification, Clinprotools software for biomarker detection and evaluation. In *ClinProTools 3.0 User Manual*; Bruker Daltonic GmbH: Bremen, Germany, 2011; pp. 51–96.

26. Villanueva, J.; Lawlor, K.; Toled-Crow, R.; Tempst, P. Automated serum peptide profiling. *Nat. Protoc.* **2006**, *1*, 880–891. [CrossRef] [PubMed]

27. Villanueva, J.; Philip, J.; Chaparro, C.A.; Li, Y. Correcting common errors in identifying cancer-specific serum peptide signatures. *J. Proteome Res.* **2005**, *4*, 1060–1072. [CrossRef] [PubMed]

28. Gustafsson, M.; Hirschberg, D.; Palmberg, C.; Jornvall, H.; Bergman, T. Integrated sample preparation and MALDI mass spectrometry on a microfluid compact disk. *Anal. Chem.* **2004**, *76*, 345–350. [CrossRef] [PubMed]

29. Tiss, A.; Smith, C.; Camuzeau, S.; Kabir, M.; Gayther, S.; Menon, U.; Waterfield, M.; Timms, J.; Jacobs, I.; Cramer, R. Serum peptide profiling using MALDI mass spectrometry avoiding the pitfalls of coated magnetic beads using well-established ZipTip technology. *Pract. Proteomics* **2007**, *7*, 77–89. [CrossRef]

30. Lin, J.; Bruni, F.M.; Fu, Z.; Maloney, J.; Bardina, L.; Boner, A.L.; Gimenez, G.; Sampson, H.A. A bioinformatics approach to identify patients with symptomatic peanut allergy using peptide microarray immunoassay. *J. Allergy Clin. Immunol.* **2012**, *129*, 1321–1328. [CrossRef] [PubMed]

31. Sandanayake, N.S.; Camuzeaux, S.; Sinclair, J.; Blyuss, O.; Andreola, F.; Chapman, M.H.; Webster, G.J.; Smith, R.C.; Timms, J.F.; Pereira, S.P. Identification of potential serum peptide biomarkers of biliary tract cancer using MALDI MS profiling. *BMC Clin. Pathol.* **2014**, *14*, 7. [CrossRef] [PubMed]

32. Kentsis, A.; Lin, Y.; Kurek, K.; Calicchio, M.; Wang, Y.; Monigatti, F.; Campagne, F.; Lee, R.; Horwitz, B.; Steen, H.; Bachur, R. Discovery and validation of urine markers of acute pediatric appendicitis using high-accuracy mass spectrometry. *Ann. Emerg. Med.* **2010**, *55*, 62–70. [CrossRef] [PubMed]

33. Petricoin, E.F.; Ardekani, A.M.; Hitt, B.A.; Levine, P.J.; Fusaro, V.A.; Steinberg, S.M.; Mills, G.B.; Simone, C.; Fishman, A.; Kohn, E.C.; *et al.* Use of proteomic patterns in serum to identify ovarian cancer. *Lancet* **2002**, *359*, 572–577. [CrossRef] [PubMed]

34. Xu, W.; Hu, Y.; He, X.; Li, J.; Pan, T.; Liu, H.; Wu, X.; He, H.; Ge, W.; Yu, J.; *et al.* Serum profiling by mass spectrometry combined with bioinformatics for the biomarkers discovery in diffuse large B-cell lymphoma. *Tumour Biol.* **2014**, *36*, 2193–2199. [CrossRef] [PubMed]

35. Wu, S.P.; Lin, Y.W.; Lai, H.C.; Chu, T.Y.; Kuo, Y.L.; Liu, H.S. SELDI-TOF MS profiling of plasma proteins in ovarian cancer. *Taiwan J. Obstet. Gynecol.* **2006**, *45*, 26–32. [CrossRef] [PubMed]

36. Bruegel, M.; Planert, M.; Baumann, S.; Focke, A.; Bergh, F.T.; Leichtle, A.; Ceglarek, U.; Thiery, J.; Fiedler, G.M. Standardized peptidome profiling of human cerebrospinal fluid by magnetic bead separation and matrix-assisted laser desorption/ionization time-of-flight mass spectrometry. *J. Proteomics* **2009**, *72*, 608–615. [CrossRef] [PubMed]

37. Cheng, A.J.; Chen, L.C.; Chien, K.Y.; Chen, Y.J.; Chang, J.T.; Wang, H.M.; Liao, C.T.; Chen, I.H. Oral cancer plasma tumor marker identified with bead-based affinity-fractionated proteomic technology. *Clin. Chem.* **2005**, *51*, 2236–2244. [CrossRef] [PubMed]

© 2015 by the authors; licensee MDPI, Basel, Switzerland. This article is an open access article distributed under the terms and conditions of the Creative Commons Attribution (CC-BY) license (http://creativecommons.org/licenses/by/4.0/).

toxins

MDPI

Article

Honeybee (*Apis mellifera*) Venom Reinforces Viral Clearance during the Early Stage of Infection with Porcine Reproductive and Respiratory Syndrome Virus through the Up-Regulation of Th1-Specific Immune Responses

Jin-A Lee [1], Yun-Mi Kim [1], Pung-Mi Hyun [2], Jong-Woon Jeon [2], Jin-Kyu Park [2], Guk-Hyun Suh [3], Bock-Gie Jung [4,*] and Bong-Joo Lee [1,*]

[1] Department of Veterinary Infectious Diseases, College of Veterinary Medicine, Chonnam National University, Gwangju 500-757, Korea; dvmjina@gmail.com (J.-AL.); kyli75@hanmail.net (Y.-M.K.)
[2] Wissen Co., Ltd., #410 Bio Venture Town, 461-8, Daejeon 305-811, Korea; muzekokomi@hanmail.net (P.-M.H.); confessor@hanmail.net (J.-W.J.); jkypark@live.co.kr (J.-K.P.)
[3] Department of Veterinary Internal Medicine, College of Veterinary Medicine, Chonnam National University, Gwangju 500-757, Korea; ghsuh@chonnam.ac.kr
[4] Department of Pulmonary Immunology, University of Texas Health Science Center at Tyler, 11937 US Hwy 271, Tyler, TX 75708-3154, USA
* Authors to whom correspondence should be addressed; skulljung@naver.com (B.-G.J.); bjlee@chonnam.ac.kr (B.-J.L.); Tel.: +1-800-428-7432 (B.-G.J.); +82-62-530-2850 (B.-J.L.); Fax: +1-903-877-7989 (B.-G.J.); +82-62-530-2857 (B.-J.L.).

Academic Editor: Sokcheon Pak
Received: 9 February 2015; Accepted: 18 May 2015; Published: 22 May 2015

Abstract: Porcine reproductive and respiratory syndrome (PRRS) is a chronic and immunosuppressive viral disease that is responsible for substantial economic losses for the swine industry. Honeybee venom (HBV) is known to possess several beneficial biological properties, particularly, immunomodulatory effects. Therefore, this study aimed at evaluating the effects of HBV on the immune response and viral clearance during the early stage of infection with porcine reproductive and respiratory syndrome virus (PRRSV) in pigs. HBV was administered via three routes of nasal, neck, and rectal and then the pigs were inoculated with PRRSV intranasally. The CD4$^+$/CD8$^+$ cell ratio and levels of interferon (IFN)-γ and interleukin (IL)-12 were significantly increased in the HBV-administered healthy pigs via nasal and rectal administration. In experimentally PRRSV-challenged pigs with virus, the viral genome load in the serum, lung, bronchial lymph nodes and tonsil was significantly decreased, as was the severity of interstitial pneumonia, in the nasal and rectal administration group. Furthermore, the levels of Th1 cytokines (IFN-γ and IL-12) were significantly increased, along with up-regulation of pro-inflammatory cytokines (TNF-α and IL-1β) with HBV administration. Thus, HBV administration—especially via the nasal or rectal route—could be a suitable strategy for immune enhancement and prevention of PRRSV infection in pigs.

Keywords: honeybee venom (HBV); porcine reproductive and respiratory syndrome virus (PRRSV); immune enhancing effect; viral clearance effect

1. Introduction

Honeybee (*Apis mellifera*) venom (HBV) has long been used as a therapeutic agent in alternative medicine for alleviation of pain, inflammation, and some immune system-related diseases such as rheumatoid arthritis and multiple sclerosis [1]. Moreover, accumulating evidence indicates that HBV exerts immunomodulatory effects on Th1 responses. Nam *et al.* [2] reported that whole HBV administration induces development of a Th1 lineage from the CD4+ T lymphocyte population and enhances the expression of interferon-gamma (IFN-γ) in a mouse model. In addition, Perrin-Cocon *et al.* [3] and Ramoner *et al.* [4] reported that a phospholipase in HBV induces the maturation of dendritic cells and activates dendritic-cell associated immune responses. Whole HBV contains at least 18 active ingredients, which include enzymes, biogenic amines, a protease inhibitor, and several biologically active peptides such as melittin, apamine, and adolapine [1]. Melittin, the principal component that is extracted from the water-soluble phase of HBV, has been reported to show multiple pharmacological effects, such as anti-microbial [5,6], anti-viral [7], and anti-cancer [8] properties. For this reason, many commercial products containing HBV components have been developed and manufactured with the emphasis on the melittin content as either the sole ingredient or a mixture of water-soluble components. On the other hand, the lipid-soluble portion of HBV has received less attention and has been usually neglected by the manufacturers of HBV. Some lipid-soluble components of HBV such as chrysin and pinocembrin-known as "flavonoids"-have been reported to show antiviral and antimicrobial effects [9,10]. Therefore, we developed a new HBV product that combines the lipid-soluble ingredients (chrysin and pinocembrin) with melittin to enhance the immunostimulatory and viral clearance effects of HBV. At present, it is uncertain whether immunostimulatory effect of HBV exists in pigs and whether HBV can be applied as an immunostimulatory agent for prevention of viral diseases in the swine industry. Nevertheless, there is growing evidence supporting the concept that HBV mainly enhances the CD4+/CD8+ T lymphocyte subset ratio, relative mRNA expression levels of IL-18 and IFN-γ and serum lysozyme activity, as shown in our previous study [11].

Porcine reproductive and respiratory syndrome (PRRS) is an economically important disease affecting the swine-producing industry in many countries [12]. This syndrome is characterized by reproductive failure in sows (such as abortions and stillbirths), increased pre-weaning mortality, induced respiratory disorders, and poor growth performance in growing pigs [13]. Infected pigs also have weak and delayed adaptive immune responses, and consequently, the susceptibility of such pigs to other viral infections and secondary bacterial infections is increased [14].

Thus, on the basis of our previous report, this study aimed at addressing several objectives: (a) evaluating the immunostimulatory efficacy of HBV in the pig immune system; (b) assessing the viral clearance activity of an HBV product containing melittin, chrysin and pinocembrin in pigs experimentally infected with PRRS virus (PRRSV; as an initial step towards the prevention of the viral diseases) and to elucidate the host immune responses, especially the pro-inflammatory response and changes in Th1-related cytokines, including IFN-γ, IL-12, TNF-α, and IL-1β (which are associated with protection from PRRSV); and (c) determining whether nasal, neck, and rectal administration of HBV induce comparable immune responses and viral clearance effects.

2. Results

2.1. Characteristics of HBV

The HBV components pinocembrin, chrysin and melittin were detected by high-performance liquid chromatography (HPLC). The retention times of pinocembrin, chrysin and melittin were 37, 38, and 42.5 min, respectively (Figure 1).

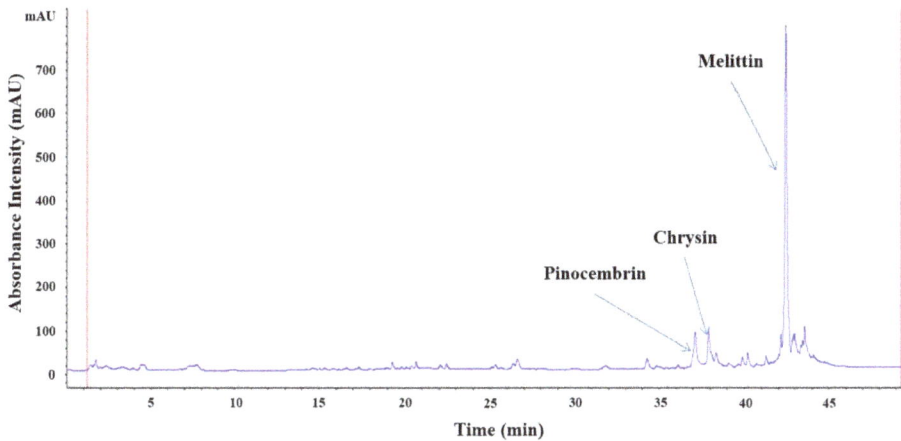

Figure 1. The high-performance liquid chromatography (HPLC) chemical fingerprint of honeybee venom (HBV) at 270 nm. Melittin, pinocembrin and chrysin were detected as components (mAU: arbitrary units).

2.2. Effects of HBV on T Lymphocyte CD4⁺/CD8⁺ Subsets in Healthy Pigs

The CD4⁺/CD8⁺ cell ratio in all HBV-administration groups tended to increase during the entire experimental period in comparison with the mock group (Figure 2). In particular, the difference in the CD4⁺/CD8⁺ cell ratio was significant between the mock group and groups "nasal injection" (NSI) and "rectal injection" (RI) at 4 ($p < 0.05$ for group RI and $p < 0.01$ for group NSI) and seven days post HBV-administration (DPA; $p < 0.01$).

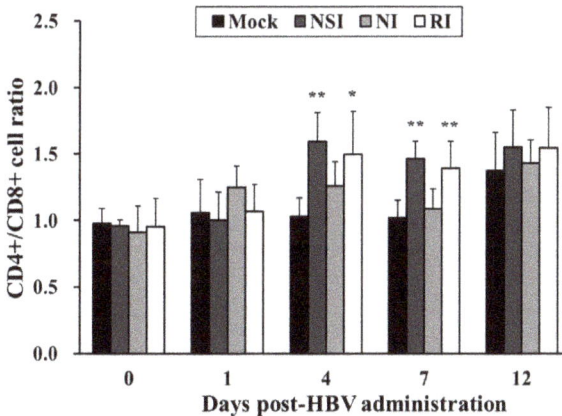

Figure 2. Effects of honeybee venom (HBV) on T lymphocyte CD4⁺/CD8⁺ subsets in the peripheral blood. After administration of HBV to pigs via three routes—nasal (NSI group), neck (NI group), and rectal (RI group)—the lymphocytes were isolated one day before HBV administration (0) and at 1, 4, 7 and 12 days post-HBV administration (DPA), and were analyzed for T lymphocyte CD4⁺/CD8⁺ subsets using flow cytometry. A higher CD4⁺/CD8⁺ T lymphocyte ratio was evident in the NSI and RI groups compared to the mock group at 4 and 7 DPA. The data are presented as mean ± SD of five pigs in each group. Significant differences with the mock group are presented as * $p < 0.05$ and ** $p < 0.01$.

2.3. Effects of HBV on Cytokine Expression Levels in Healthy Pigs

The level of IL-12 significantly increased in HBV-administration groups at 1 DPA ($p < 0.01$ compared to the mock group) and at 4 DPA ($p < 0.05$ compared to the mock group; Figure 3A). Furthermore, the level of IFN-γ was significantly higher in group NSI at 4 and 7 DPA ($p < 0.01$ and $p < 0.05$, respectively) and group RI at 4 DPA ($p < 0.05$) in comparison with the mock group (Figure 3B). The relative expression levels of TNF-α and IL-1β (considered pro-inflammatory cytokines) did not show any statistically significant differences between the mock group and the HBV-administration groups (Figure 3C,D).

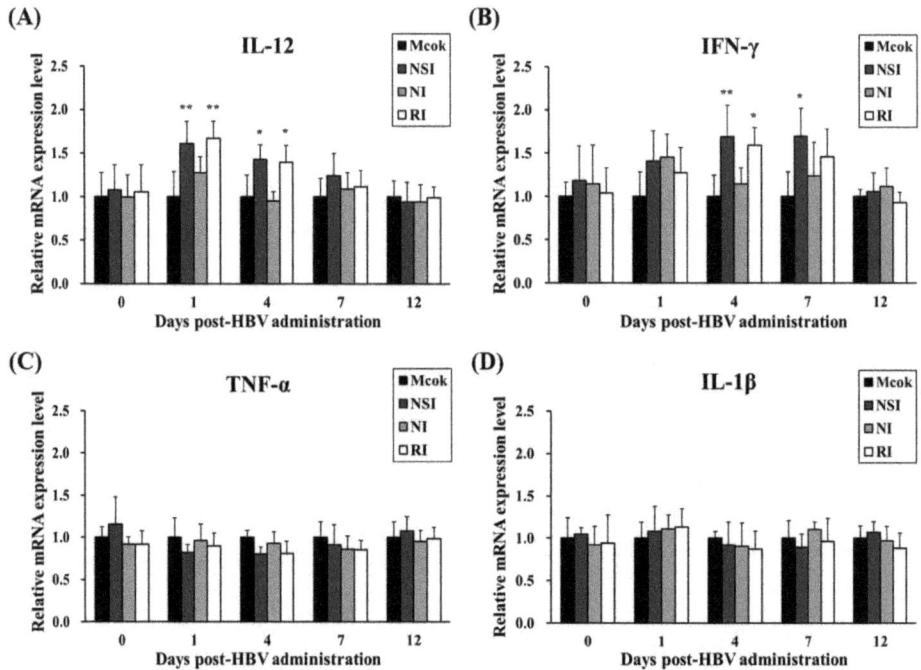

Figure 3. Effects of honeybee venom (HBV) on the mRNA expression levels of Th1 cytokines and pro-inflammatory cytokines. The expression levels of (**A**) IL-12, (**B**) IFN-γ, (**C**) TNF-α, and (**D**) IL-1β in peripheral blood mononuclear cells (PBMCs) were measured by quantitative real-time PCR and are presented after normalization to β-actin levels. The data are presented as mean $\pm$ SD of five pigs in each group. Significant differences with the mock group are presented as * $p < 0.05$ and ** $p < 0.01$. (NSI; nasal injection group, NI; neck injection group, and RI; rectal injection group).

2.4. Effects of HBV on Viral Clearance in the PRRSV Infected Pigs

The load of the PRRSV genome in serum was measured using real-time quantitative PCR and groups NSI and RI group showed a marked reduction in the load of the PRRSV genome in serum at 4, 7, and 12 days post-inoculation (DPI; $p < 0.05$ at 4 and 12 DPI, $p < 0.01$ at 7 DPI) compared to the mock group (Figure 4A). At the end of the experiment, lung, bronchial lymph nodes (BLNs) and tonsils were collected and analyzed for the load of the PRRSV genome. The viral genome load in lung tissue of groups NSI and RI showed a significant reduction ($p < 0.05$ for group RI and $p < 0.01$ for group NSI compared to the mock group). Similarly, the PRRSV load of BLNs and tonsil tissues was significantly decreased ($p < 0.05$) in groups NSI and RI compared to the mock group (Figure 4B).

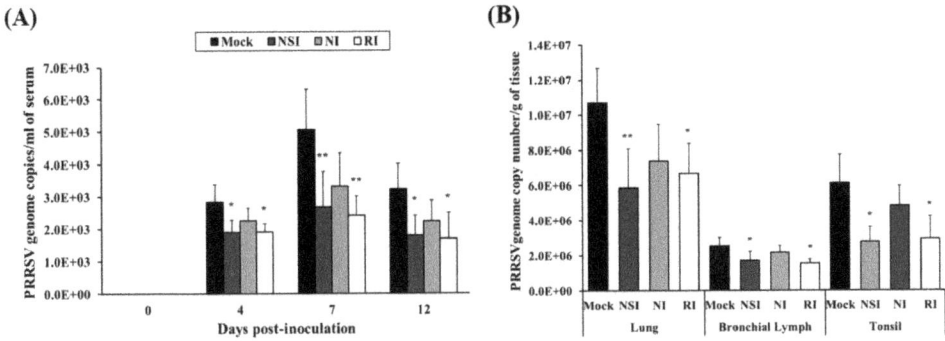

Figure 4. Effects of honeybee venom (HBV) on viral clearance of porcine reproductive and respiratory syndrome virus (PRRSV) in experimentally infected pigs. HBV was administered to pigs via three routes—nasal, neck, and rectal—followed by experimental PRRSV inoculation through both nostril. For quantification of the PRRSV genome, (**A**) serum was collected at 0, 4, 7, and 12 days post-inoculation (DPI), and (**B**) tissue samples (lung, bronchial lymph nodes and tonsils) were collected at post-mortem examination. The viral genome load in the serum samples was significantly decreased in groups NSI and RI compared to the mock group at 4, 7, and 12 DPI. In the tissue samples, groups NSI and RI showed a significant reduction in the viral load in the lung, bronchial lymph nodes (BLNs), and tonsil tissues, in comparison with the mock group. The data are expressed as mean ± SD of five pigs in each group. Significant differences with the mock group are presented as * $p < 0.05$, and ** $p < 0.01$. (NSI; nasal injection group, NI; neck injection group, and RI; rectal injection group).

2.5. Evaluation of Body Temperature and Histopathological Examination of the PRRSV Infected Pigs

After experimental infection with PRRSV, rectal temperature of all pigs rapidly increased to over 39.5 °C, and groups NSI and RI showed significant decrease compared to the mock group at 7 and 12 DPI (Figure 5A). The tissue lesions in the HBV-administration groups were milder compared to those of the mock group (Figure 5C–F). Especially, the differences were evident in the NSI and RI group compared with that of the mock group ($p < 0.05$; Figure 5B).

2.6. Effects of HBV on T Lymphocyte CD4+/CD8+ Subsets in the PRRSV Infected Pigs

The CD4+/CD8+ cell ratio in groups NSI and RI tended to increase during the entire study period in comparison with the mock group. In particular, the difference in the CD4+/CD8+ cell ratio was significant between the mock group and groups NSI and RI at 4 and 7 DPI ($p < 0.05$; Figure 6).

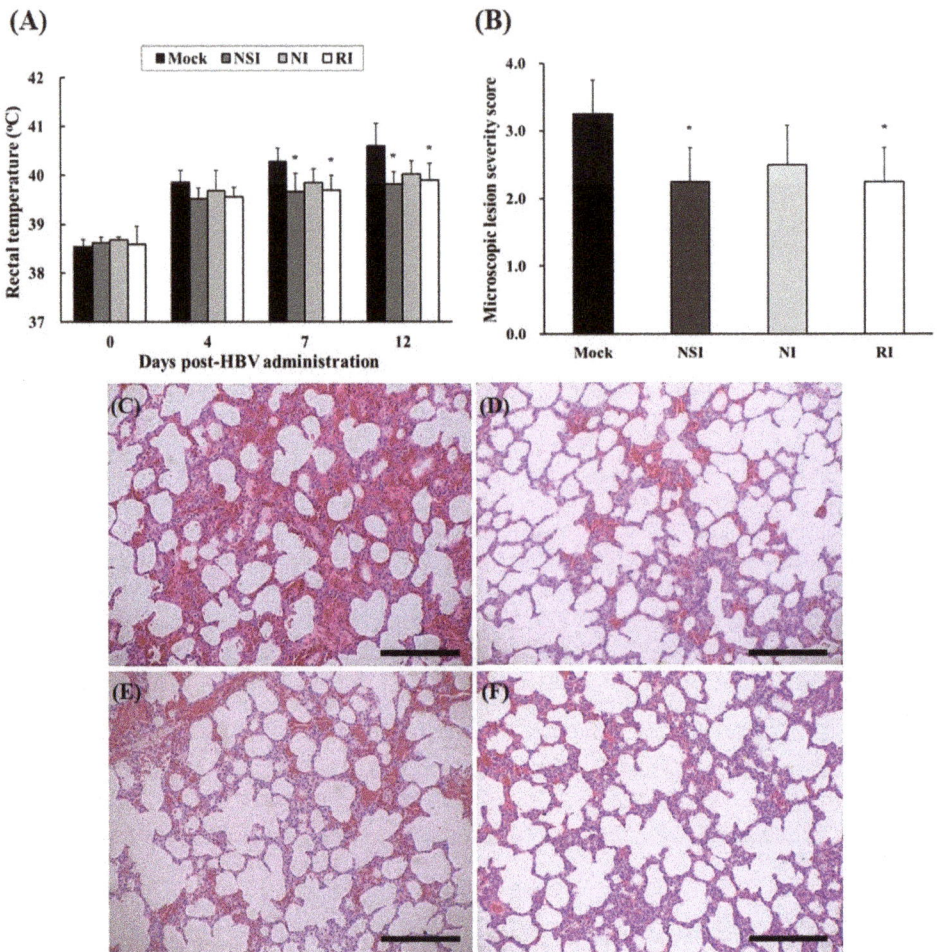

Figure 5. Changes of rectal temperature and histopathological evaluation of lung tissue in the pigs experimentally infected with porcine reproductive and respiratory syndrome virus (PRRSV). Rectal temperature was measured at 4, 7, and 12 DPI (**A**) and the lung tissue samples were analyzed by means of hematoxylin and eosin (H&E) staining in (**C**) the mock group, (**D**) group "nasal injection" (NSI), (**E**) group "neck injection" (NI), and (**F**) group "rectal injection" (RI) for evaluation of the severity of interstitial pneumonia. All pigs showed increased rectal temperature over 39.5 °C, and the groups NSI and RI showed a significant decrease in rectal temperature at 7 and 12 DPI in comparison with the mock group. Mild interstitial pneumonia was evident in the HBV-administration groups, and the lung microscopic lesion score (**B**) was significantly decreased in groups NSI and RI compared to the mock group (* $p < 0.05$). The data are expressed as mean ± SD of five pigs in each group. In all panels, magnification is at 200× and the scale bars are 100 μm.

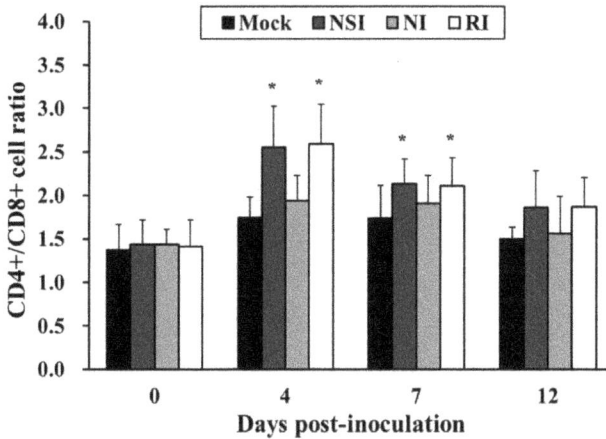

Figure 6. Effects of honeybee venom (HBV) on T lymphocyte $CD4^+/CD8^+$ subsets in the peripheral blood in pigs experimentally infected with porcine reproductive and respiratory syndrome (PRRSV). The PBMCs were isolated at 0, 4, 7, and 12 days post-inoculation (DPI) and were analyzed for the $CD4^+/CD8^+$ cell ratio by using flow cytometry. The $CD4^+/CD8^+$ T lymphocyte subset ratio was significantly increased in groups "nasal injection" (NSI) and "rectal injection" (RI) as compared to the mock group at 4 and 7 DPI. The data are presented as mean ± SD of five pigs in each group. Significant differences with the mock group are presented as $* p < 0.05$.

2.7. Effects of HBV on Cytokine Expression Profiles in the PRRSV Infected Pigs

The relative mRNA expression levels of Th1 cytokines (IFN-γ and IL-12) and pro-inflammatory cytokines (TNF-α and IL-1β) were measured in PBMCs at the early stage of PRRSV infection (4, 7, and 12 DPI). All cytokines investigated showed similar increasing kinetic patterns (Figure 7). Especially, the level of IL-12 was significantly increased in groups NSI and RI at 4 DPI ($p < 0.001$ for group NSI and $p < 0.01$ for group RI compared to the mock group) and 7 DPI ($p < 0.05$ for group NSI compared to the mock group; Figure 7A). Similarly, IFN-γ levels were markedly increased in groups NSI and RI at 4 DPI ($p < 0.001$ for groups NSI and RI compared to the mock group), at 7 DPI ($p < 0.05$ for groups NSI and RI compared to the mock group), and at 12 DPI ($p < 0.05$ for group NSI compared to the mock group; Figure 7B). The pro-inflammatory cytokines also showed statistically significant differences between the mock group and groups NSI and RI. Notably, the level of TNF-α significantly increased at 4 DPI ($p < 0.01$ for groups NSI and RI compared to the mock group) and at 7 DPI ($p < 0.001$ for group NSI and $p < 0.01$ for group RI compared to the mock group; Figure 7C). The level of IL-1β also markedly increased at 4 DPI ($p < 0.05$ for groups NSI and RI compared to the mock group) and at 7 DPI ($p < 0.05$ for group NSI and $p < 0.01$ for group RI compared to the mock group; Figure 7D).

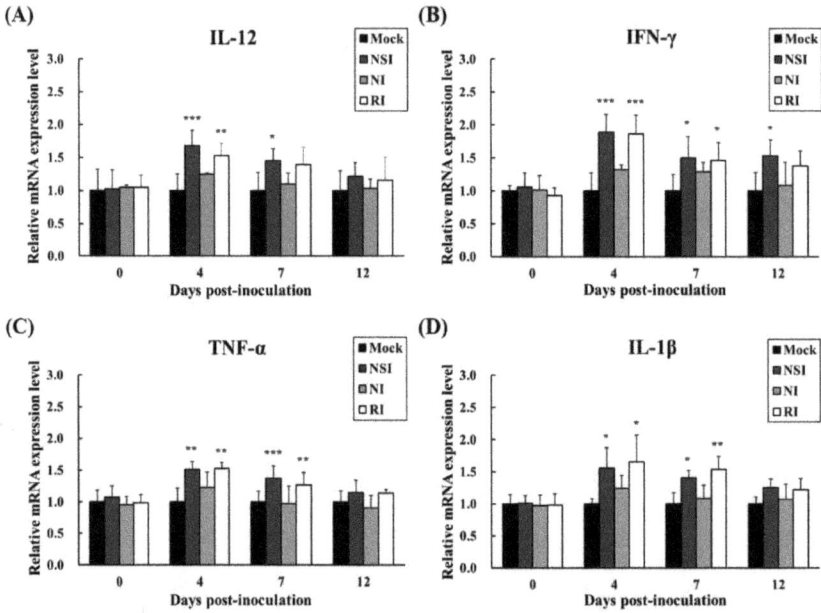

Figure 7. Effects of honeybee venom (HBV) on expression levels of Th1 and pro-inflammatory cytokines in experimentally PRRSV-infected pigs. The expression levels of (**A**) IL-12, (**B**) IFN-γ, (**C**) TNF-α, and (**D**) IL-1β were measured by means of quantitative real-time PCR and are presented after normalization to β-actin followed by the ΔΔCt method. The data are presented as mean ± SD of five pigs in each group. Significant differences with the mock group are presented as * $p < 0.05$, ** $p < 0.01$, and *** $p < 0.001$. (NSI; nasal injection group, NI; neck injection group, and RI; rectal injection group).

3. Discussion

In the present study, HBV was administered to pigs through three routes—nasal, neck, and rectal—and then the CD4$^+$/CD8$^+$ cell ratio and the expression levels of Th1 and pro-inflammatory cytokines were evaluated. The neck region is a common route for administration of classical vaccines in pigs. The nasal and rectal regions are newly highlighted sites for vaccine administration, and this approach can induce both mucosal and systemic immunity. The mucosal regions, which are classified into ocular, nasal, oral, pulmonary, and vaginal or rectal, comprise ~80% of all immune cells in the body [15]. Particulate antigens (such as microspheres, liposomes, and bacterial ghosts) that are delivered via a mucosal route are recognized by microfold cells (M cells) located in mucosa-associated lymphoid tissues and are then presented to professional antigen-presenting cells (APCs) such as dendritic cells and macrophages. Consequently, APCs forcefully stimulate differentiation of T cells and effectively induce specific adaptive immune responses [16]. In particular, optimal stimulation of mucosal immunity is effective for protection from many infectious diseases because it can stimulate both mucosal and systemic immunity, leading to inhibition of entry of pathogens into the body [15]. According to the present study, the CD4$^+$/CD8$^+$ cell ratio in all HBV-administration groups tends to increase during the entire experimental period. Notably, the change in the CD4$^+$/CD8$^+$ cell ratio is statistically significant in group RI at 4 and 7 DPA and in group NSI at 7 DPA. The number and ratio of two main T lymphocyte subsets, CD4$^+$ cells (or T helpers) and CD8$^+$ cells (or T cytotoxic) were used as the most meaningful parameters for evaluation of immune homeostasis and response of the intrinsic immune system [17]. Furthermore, low CD4$^+$/CD8$^+$ cell ratio are usually related to acute viral diseases and hemophilia, whereas high ratios are observed individually with stimulation of immuno-functional ability [11]. Simultaneously with the CD4$^+$/CD8$^+$ cell ratio, the level of Th1 cytokines (IL-12 and IFN-γ) is significantly increased in groups NSI and RI at 1, 4, and 7 DPA. These findings indicate that administration of HBV effectively induces Th1-related immune responses up to 7 DPA, and this phenomenon may correlate with the increasing CD4$^+$/CD8$^+$ cell ratio through stimulation of the CD4$^+$ T cell population when HBV is administered via the nasal or rectal route. These results are consistent with other reports showing that HBV administration to broiler chicks (using a spray method) significantly increases INF-γ expression levels and the CD4$^+$/CD8$^+$ cell ratio [11].

PRRSV is a unique viral pathogen because this virus can evade host immune surveillance or effectively control both innate and adaptive immune regulators [18,19]. PRRSV-infected pigs typically show insufficient anti-PRRSV immunity because of low levels of Th1 cytokine (IFN-γ and IL-12) production, which are typically involved in cellular immunity [20]. IFN-γ is known to possess potent antiviral properties against PRRSV infection through the inhibition of viral replication and to be an important mediator of cellular immunity [21]. Furthermore, IL-12 plays a critical role in host immunity against viral infections through the promotion of IFN-γ production and Th1 cell type differentiation [19]. Therefore, the PRRSV-induced suppression of Th1-related cytokines may result in prolonged immunosuppression, followed by weaker antibody responses, increased levels of viremia, and greater severity of the disease [22]. Simultaneously with the repression of Th1 responses, PRRSV acts via unique mechanisms whereby the production of pro-inflammatory cytokines, such as TNF-α and IL-1β, is limited during the PRRSV infection, in contrast to other viral respiratory diseases of pigs [23]. TNF-α performs an important function in the inflammatory response and in the protection of the cells from viral infection or in promotion of selective elimination of virus-infected cells via IFN-independent mechanism [24]. The PRRSV-induced suppressive action on TNF-α and IL-1β production, as a strategy for modulation of host immune responses, thus contributes to the unique clinical features; the absence of a fever correlates with the development of more severe interstitial pneumonia after PRRSV infection [22]. Accordingly, there are reports about weak expression of pro-inflammatory cytokines (TNF-α and IL-1β) in pigs infected with PRRSV-1 and PRRSV-2 [14]. In the present study, levels of Th1 cytokines (IFN-γ and IL-12) and pro-inflammatory cytokines (TNF-α and IL-1 β) were significantly increased in groups NSI and RI, especially 4 and 7 DPI compared to the mock group. In addition, the PRRSV genome loads in serum and PRRSV-infected tissues (lung, BLNs and tonsils) were considerably reduced in HBV-administration groups, especially in groups NSI and RI. Similarly, the viral load of PRRSV-infected tissues is markedly reduced in groups NSI and RI. Additionally, histopathological analysis of the lung tissue revealed that groups NSI and RI undergo less pathological changes in comparison with the mock group. Pigs that are experimentally infected with PRRSV typically show mild respiratory and systemic clinical signs, where the absence of a fever correlates with more severe interstitial pneumonia [14]. Therefore, these findings suggest that HBV administration, especially via the nasal or rectal route, enhances clearance of PRRSV in experimentally infected pigs, and these viral clearance effect of HBV may be related to the up-regulation of Th1 and partial pro-inflammatory immune responses. These findings are similar to those by Dwivedi *et al.* [19], who found that vaccines consisting of the PLGA nanoparticle-entrapped killed PRRS virus exhibit high protective efficacy in pigs against a PRRSV challenge (in comparison with classically vaccinated pigs); specifically, there is significant up-regulation of Th1 cytokines (IFN-γ and IL-12) and several types of immune cells, including CD3$^+$CD8$^+$, CD4$^+$CD8$^+$, and γδ T cells.

In summary, the data from our results suggest that administration of HBV via a nasal or rectal route can effectively enhance the Th1-specific systemic immunity in healthy pigs and reduce the viral load and severity of interstitial pneumonia in PRRSV-infected pigs. These immunomodulatory effects and PRRSV clearance that are induced by the administration of HBV may be related to promotion of Th1-specific responses and inhibition of PRRSV-specific mechanisms that involve down-regulation of pro-inflammatory cytokines. Thus, our findings—in conjunction with further research on HBV in a field study (for confirmation of the protective efficacy against a naturally occurring PRRSV infection)—may lead to new immunostimulatory and preventive strategies for PRRSV infection in the swine industry.

4. Materials and Methods

4.1. Preparation and Characterization of HBV

Crude HBV was obtained using a Large Quantity Bee-Venom Collector (P10-1003672, Wissen, Daejeon, Korea). We extracted 2.5g of crude HBV with 250 mL of ultra-filtered water and ethyl acetate three times at room temperature (RT) and passed the extract through a filter containing a nylon membrane (0.45-µm pore diameter; Millipore, Billerica, MA, USA) under vacuum driven filtration system. The two filtrates were mixed and then concentrated in vacuum at 40 °C. The final HBV concentrates (10 mg) were dissolved in 1 mL of ultra-filtered water and analyzed by HPLC. The HBV samples that we used for HPLC analysis were passed through a 0.45-µm filter (Millipore) before injection into a UPLC® BEH C18 column (1.7 µm, 2.1 × 100 mm; Waters Corporation, Milford, MA, USA). The gradient ratios of mobile phases A (0.1% trifluoroacetic acid in methanol) and B (0.1% trifluoroacetic acid in distilled water) were set at 0:100 for 0–10 min, with a flow rate of 0.2 mL/min, and then were set at 50:50 for 20–30 min until the last step, which were set at 100:0 at 50 min, with a flow rate of 0.3 mL/min. The wavelengths for detection was 270 nm. For the experimental use, the final HBV fine powder was dissolved in a solvent consisting of 95.7% distilled water, 3.5% ethanol, and 0.8% propylene glycol, by volume. The HBV concentration was 2.1 mg/mL, which was the optimal concentration according to our preliminary experiments.

4.2. PRRS Virus

The PRRSV type II LMY strain (GenBank accession number DQ473474), which was originally isolated in South Korea from a pig with naturally occurring PRRS. LMY strain was kindly provided by Prof. Dr. Kyoung-oh Cho (Chonnam National University, Gwangju, Korea). For preparation of the virus inoculum, the virus stock was propagated in confluent monolayers of MARC-145 cells, as described previously [21], and the virus titer was measured using quantitative real-time PCR.

4.3. Animals

Conventional 4-week-old pigs, cross-breeds between Landrace, Yorkshire and Duroc, were obtained from a single healthy herd without any history of PRRS. All pigs were housed in separate air-conditioned rooms and were allowed free access to nutritionally complete antibiotic-free pig feed (Daehan Livestock & Feed, Naju, Jeonnam, Korea) and drinking water. Prior to the experiment, all pigs were tested and confirmed to be seronegative for antibodies to PRRSV by using a commercial enzyme-linked immunosorbent assay kit (IDEXX Laboratories, Westbrook, ME, USA) and tested negative for presence of viremia by real-time PCR [25].

4.4. Experimental Protocol

4.4.1. Experiment 1: Immunostimulatory Effects of HBV on Healthy Pigs

The pigs were randomly allocated to four groups of five pigs each and were injected with HBV subcutaneously via different routes as follows: Group 1, mock control pigs; Group 2, injected with 2.1 mg of HBV into the nasal region (NSI group); Group 3, injected with 2.1 mg of HBV into the neck region (NI group); Group 4, injected with 2.1 mg of HBV into the rectal region (RI group). To measure the immunostimulatory effect of HBV, blood samples were individually collected from the jugular vein 1 day before HBV administration and at 1, 4, 7 and 12 DPA. Body weight was also monitored during the whole study period. All animal procedures were performed in accordance with the guidelines of International Guiding Principles for Biomedical Research Involving Animals by the Council for International Organizations of Medical Sciences. (CIOMS, care of the World Health Organization, Geneva, Switzerland) and approved by the Institutional Animal Care and Use Committee of Chonnam National University (approval number: CNU IACUC-YB-2013-29).

4.4.2. Experiment 2: The Viral Clearance Effect of HBV in the Pigs Experimentally Infected with PRRSV

The pigs were randomly subdivided into groups and injected with HBV, as described in Experiment 1. After 4 days of HBV administration, five milliliters of the viral culture (1×10^6 copies/mL, optimized previously) was inoculated intranasally into both nostril of each pig. After PRRSV infection, blood samples were collected at 0, 4, 7, and 12 DPI for analysis of T lymphocyte $CD4^+/CD8^+$ subsets and cytokine expression profiles. Clinical symptoms and rectal temperature were also monitored every day during the entire study period (12 days), as described previously [23]. At the end of the experiment, all pigs in all groups were euthanized for collection of lung and lymphoid tissues (BLNs and tonsils) for titration of the virus.

4.5. Isolation of PBMCs from Peripheral Blood

PBMCs were isolated as described previously [11]. Briefly, blood samples were collected into lithium-heparin coated tubes and diluted with equal volumes of phosphate buffered saline (PBS). The diluted mixture was layered over a half of the volume of Lymphoprep (Axis-shield, Oslo, Norway) and was separated by gradient centrifugation at $800 \times g$ for 30 min at RT. Contaminating red blood cells were lysed using commercial Red Blood Cell Lysing Buffer Hybri-Max (Sigma-Aldrich, St. Louis, MO, USA). The PBMCs were then washed twice with PBS prior to resuspension in a complete medium consisting of RPMI-1640 medium (Lonza, Basel, Switzerland) containing 10% (v/v) fetal bovine serum (FBS; Gibco, Grand Island, NY, USA) and 2% (v/v) of the antibiotic-antimycotics formulating solution with penicillin, streptomycin, and amphotericin B (Lonza, Basel, Switzerland).

4.6. Analysis of T Lymphocyte $CD4^+/CD8^+$ Subsets by using Flow Cytometry

The PBMCs obtained from the peripheral blood were analyzed to determine the $CD4^+CD8^-$ T lymphocyte and $CD4^-CD8^+$ T lymphocyte subset ratio within the $CD3^+$ subset by means of flow cytometry as described previously [19]. In brief, the cells were stained with a fluorescein isothiocyanate (FITC)-conjugated mouse anti-pig CD3ε antibody (clone BB23-8E6-8C8; BD Biosciences, Franklin Lakes, NJ, USA), a phycoerythrin (PE)/Cy7-conjugated mouse anti-pig CD4a antibody (clone 74-12-4; BD Biosciences, Franklin Lakes, NJ, USA), and a PE-conjugated mouse anti-pig CD8a antibody (clone 76-2-11; BD Biosciences, Franklin Lakes, NJ, USA). After incubation at RT for 30 min in the dark, the cells were washed twice with PBS and the lymphocyte subpopulations were analyzed using a BD Accuri C6 flow cytometer (BD Biosciences, Franklin Lakes, NJ, USA).

4.7. Analysis of Cytokine Expression Levels in PBMCs

We measured the expression levels of IFN-γ, IL-12, TNF-α, and IL-1β in the PBMCs to evaluate the immunostimulatory effect of HBV, especially that related to the pro-inflammatory and Th1 cytokines in the pig immune system. Total RNA was extracted from the PBMCs by using the commercial PureLink RNA Mini Kit (Invitrogen, Grand Island, NY, USA). The RNA concentration was quantified using a NanoDrop ND-1000 instrument (Thermo Fisher Scientific, Waltham, MA, USA), and equal amounts of RNA were reverse-transcribed using the QuantiTect Reverse Transcription Kit (Qiagen, Valencia, CA, USA). To minimize variation in the reverse-transcriptase efficiency, all samples were transcribed simultaneously. Quantitative real-time PCR was performed using a MyiQ2 thermocycler and the SYBR green detection system (Bio-Rad Laboratories, Hercules, CA, USA). The real-time PCR conditions were as follows: 95 °C for 10 min, then 45 cycles at 95 °C, 30 s; 57 °C, 30 s; and 72 °C, 30 s. The oligonucleotide primer pairs that we used for analysis of pig IFN-γ, IL-12, TNF-α, IL-1β, and β-actin expression are shown in Table 1. We confirmed that reaction efficiencies of all primers range between 95% and 105% and R^2 value is over 0.98. The threshold cycle was then determined subsequently. Relative quantitation of pig IFN-γ, IL-12, TNF-α, and IL-1β expression levels were calculated using the comparative Ct method as described previously [26]. The relative quantitation value of the target gene was normalized to an endogenous control β-actin gene, and the differences between control and HBV groups were evaluated as $2^{-\Delta\Delta Ct}$ method (fold change).

Table 1. The real-time PCR primer sequences

Gene		Sequence (5'–3')	Accession number
TNF-α	Forward	CCCCCAGAAGGAAGAGTTTC	JF831365
	Reverse	CGGGCTTATCTGAGGTTTGA	
IL-1β	Forward	GGCCGCCAAGATATAACTGA	NM_214055
	Reverse	GGACCTCTGGGTATGGCTTTC	
IFN-γ	Forward	CAAAGCCATCAGTGAACTCATCA	X53085
	Reverse	TCTCTGGCCTTGGAACATAGTCT	
IL-12	Forward	GGAGTATAAGAAGTACAGAGTGG	U08317
	Reverse	GATGTCCCTGATGAAGAAGC	
β-actin	Forward	CAGGTCATCACCATCGGCAACG	U07786
	Reverse	GACAGCACCGTGTTGGCGTAGAGGT	

4.8. Quantification of PRRSV Genome in Serum and Tissues using Quantitative Real-Time PCR

The serum samples were prepared from the blood samples at 0, 4, 7, and 12 DPI and were collected in EDTA-free tubes after the removal of clots and centrifugation. Lung and lymphoid tissues were homogenized (10%, *w/v*) in minimum essential medium alpha modification (alpha-MEM; Lonza, Basel, Switzerland) containing 2% (*v/v*) of an antibiotic-antimycotics solution consisting of penicillin, streptomycin, and amphotericin B (Lonza, Basel, Switzerland), and were then centrifuged at 4000 × *g* for 30 min at 4 °C. The supernatant was passed through a 0.20-μm non-pyrogenic syringe filter (Pall Corporation, Ann Arbor, MI, USA). RNA extraction from the collected supernatants was performed using the PureLink Viral RNA/DNA Mini Kit (Invitrogen, Grand Island, NY, USA) according to the manufacturer's instructions, and equal amounts of targeted RNA was reverse transcribed into complementary DNA (cDNA) using the QuantiTect reverse transcription kit (Qiagen, Valencia, CA, USA). The cDNA was used for quantification of PRRSV genome by optimized real-time PCR reaction using a standard curve as previously described [27]. Briefly, for the construction of PRRSV genome plasmid, the ORF7 gene of PRRSV was amplified with the following primers: PRRSV forward, 5'-ATAACAACGGCAAGCAG-3'; PRRSV reverse, 5'-CAGTGTAACTTATCCTCCCA-3' (Genebank accession number: AF121131). The resulting PCR product was cloned into the pGEM-T Easy vector (Promega, Madison, WI, USA), which was then transformed into Escherichia coli JM109 competent cells. The plasmid DNA was purified using a Qiaprep Spin Miniprep Kit (Qiagen, Valencia, CA, USA) and quantified using a NanoDrop ND-1000 (Thermo Fisher Scientific, Waltham, MA, USA). According to the above-mentioned quantitation standard curve, real-time PCR for quantitation of the PRRSV genome was performed on a MyiQ2 thermocycler by using the SYBR green detection system (Bio-Rad Laboratories, Hercules, CA, USA). For each assay, a standard curve was constructed using serially diluted plasmid standards with PRRSV at 10^2–10^9 copy numbers/mL. The amplification condition were as follows: 10 min at 95 °C, followed by 50 cycles at 95 °C for 5 s, 57 °C for 15 s and 72 °C for 10 s.

4.9. Histopathological Analysis in Lung Tissue of the Pigs Experimentally Infected with PRRSV

At post-mortem examination, the lung tissue was fixed in 10% neutral-buffered formalin for histopathological examination. The fixed samples were routinely processed, embedded in paraffin, sectioned (at 5 μm thickness), and stained with H&E. Microscopic lesions in the lung tissue were examined by an unbiased certified veterinary pathologist, using previously described scoring systems [28,29] in a blinded fashion. The lung sections were scored for the severity of interstitial pneumonia, ranging from 0–4: 0 = no microscopic lesions; 1 = mild interstitial pneumonia; 2 = moderate multifocal interstitial pneumonia; 3 = moderate diffuse interstitial pneumonia; and 4 = severe interstitial pneumonia.

4.10. Statistical Analysis

The data were expressed as mean ± SD and the means of different parameters were compared among the groups by using one-way analysis of variance (ANOVA) followed by the Tukey-Kramer multiple comparison. All calculations were performed in the GraphPad InStat software, version 3.0 (GraphPad Software, La Jolla, CA, USA). Particularly, the Mann-Whitney U test (nonparametric test; SPSS software, version 17.0; SPSS, Chicago, IL, USA) was used to analyze differences in the microscopic-lesion score (ordinal data) in tissues of the pigs experimentally infected with PRRSV. p value <0.05 was assumed to indicate statistical significance.

Acknowledgments: This study was supported by the Ministry of Agriculture, Food and Rural Affairs, Republic of Korea (Grant No. 311018-02-2-CG000). We would like to thank Kyoung-Oh Cho for providing the LMY strain of PRRSV and Gibeom Kwon, Somin Um, Wooram Bae, Jieun Lee and Soyoung Jeung for their help with the animal experiments. We are also grateful to Gyountae Jo for the assistance with immunological and histological assay.

Author Contributions: Jin-A Lee participated in all experiments and wrote the manuscript. Yun-Mi Kim participated in experiments with RNA isolation and cDNA preparation for real-time PCR and in the histopathological examination. Pung-Mi Hyun, Jong-Woon Jeon and Jin-Kyu Park participated in preparation and analysis of HBV. Bock-Gie Jung performed the animal experiments. Guk-Hyun Suh and Bong-Joo Lee designed the study and participated in data interpretation and manuscript preparation.

Conflicts of Interest: The authors declare no conflict of interest.

References

1. Oršolić, N. Bee venom in cancer therapy. *Cancer Metastasis Rev.* **2012**, *31*, 173–194. [CrossRef] [PubMed]
2. Nam, S.; Ko, E.; Park, S.K.; Ko, S.; Jun, C.Y.; Shin, M.K.; Hong, M.C.; Bae, H. Bee venom modulates murine Th1/Th2 lineage development. *Int. Immunopharmacol.* **2005**, *5*, 1406–1414. [CrossRef] [PubMed]
3. Perrin-Cocon, L.; Agaugué, S.; Coutant, F.; Masurel, A.; Bezzine, S.; Lambeau, G.; André, P.; Lotteau, V. Secretory phospholipase A2 induces dendritic cell maturation. *Eur. J. Immunol.* **2004**, *34*, 2293–2302. [CrossRef] [PubMed]
4. Ramoner, R.; Putz, T.; Gander, H.; Rahm, A.; Bartsch, G.; Schaber, C.; Thurnher, M. Dendritic-cell activation by secretory phospholipase A2. *Blood* **2005**, *105*, 3583–3587. [CrossRef]
5. Mataraci, E.; Dosler, S. *In vitro* activities of antibiotics and antimicrobial cationic peptides alone and in combination against methicillin-resistant *Staphylococcus aureus* biofilms. *Antimicrob. Agents Chemother.* **2012**, *56*, 6366–6371. [CrossRef] [PubMed]
6. Liu, H.; Han, Y.; Fu, H.; Liu, M.; Wu, J.; Chen, X.; Zhang, S.; Chen, Y. Construction and expression of sTRAIL-melittin combining enhanced anticancer activity with antibacterial activity in *Escherichia coli*. *Appl. Microbiol. Biotechnol.* **2013**, *97*, 2877–2884. [CrossRef] [PubMed]
7. Falco, A.; Barrajón-Catalán, E.; Menéndez-Gutiérrez, M.P.; Coll, J.; Micol, V.; Estepa, A. Melittin-loaded immunoliposomes against viral surface proteins, a new approach to antiviral therapy. *Antivir. Res.* **2013**, *97*, 218–221. [CrossRef] [PubMed]
8. Son, D.J.; Lee, J.W.; Lee, Y.H.; Song, H.S.; Lee, C.K.; Hong, J.T. Therapeutic application of anti-arthritis, pain-releasing, and anti-cancer effects of bee venom and its constituent compounds. *Pharmacol. Ther.* **2007**, *115*, 246–270. [CrossRef] [PubMed]
9. Schnitzler, P.; Neuner, A.; Nolkemper, S.; Zundel, C.; Nowack, H.; Sensch, K.H.; Reichling, J. Antiviral activity and mode of action of propolis extracts and selected compounds. *Phytother. Res.* **2010**, *24* (Suppl. 1), 20–28. [CrossRef] [PubMed]
10. Rasul, A.; Millimouno, F.M.; Ali Eltayb, W.; Ali, M.; Li, J.; Li, X. Pinocembrin: A novel natural compound with versatile pharmacological and biological activities. *Biomed. Res. Int.* **2013**. [CrossRef]
11. Jung, B.G.; Lee, J.A.; Park, S.B.; Hyun, P.M.; Park, J.K.; Suh, G.H.; Lee, B.J. Immunoprophylactic effects of administering honeybee (*Apis melifera*) venom spray against *Salmonella* Gallinarum in broiler chicks. *J. Vet. Med. Sci.* **2013**, *75*, 1287–1295. [CrossRef] [PubMed]
12. Nuemann, E.J.; Kliebenstein, J.B.; Johnson, C.D.; Mabry, J.W.; Bush, E.J.; Seitzinger, A.H.; Green, A.L.; Zimmerman, J.J. Assessment of the economic impact of porcine reproductive and respiratory syndrome on swine production in the United States. *JAVMA* **2005**, *227*, 385–392. [CrossRef] [PubMed]

13. Opriessnig, T.; Giménez-Lirola, L.G.; Halbur, P.G. Polymicrobial respiratory disease in pigs. *Anim. Health. Res. Rev.* **2011**, *12*, 133–148. [CrossRef] [PubMed]

14. Gómez-Laguna, J.; Salguero, F.J.; Pallarés, F.J.; Carrasco, L. Immunopathogenesis of porcine reproductive and respiratory syndrome in the respiratory tract of pigs. *Vet. J.* **2013**, *195*, 148–155. [CrossRef] [PubMed]

15. Holmgren, J.; Czerkinsky, C. Mucosal immunity and vaccines. *Nat. Med.* **2005**, *11* (Suppl. 4), S45–S53. [CrossRef] [PubMed]

16. Renukaradhya, G.J.; Dwivedi, V.; Manickam, C.; Binjawadagi, B.; Benfield, D. Mucosal vaccines to prevent porcine reproductive and respiratory syndrome: a new perspective. *Anim. Health. Res. Rev.* **2012**, *13*, 21–37. [CrossRef] [PubMed]

17. Xu, M.; Zhao, M.; Yang, R.; Zhang, Z.; Li, Y.; Wang, J. Effect of dietary nucleotides on immune function in Balb/C mice. *Int. Immunopharmacol.* **2013**, *17*, 50–56. [CrossRef] [PubMed]

18. Dwivedi, V.; Manickam, C.; Binjawadagi, B.; Linhares, D.; Murtaugh, M.P.; Renukaradhya, G.J. Evaluation of immune responses to porcine reproductive and respiratory syndrome virus in pigs during early stage of infection under farm conditions. *Virol. J.* **2012**, *9*, 45. [CrossRef] [PubMed]

19. Dwivedi, V.; Manickam, C.; Binjawadagi, B.; Renukaradhya, G.J. PLGA nanoparticle entrapped killed porcine reproductive and respiratory syndrome virus vaccine helps in viral clearance in pigs. *Vet. Microbiol.* **2013**, *166*, 47–58. [CrossRef] [PubMed]

20. Barranco, I.; Gómez-Laguna, J.; Rodríguez-Gómez, I.M.; Quereda, J.J.; Salguero, F.J.; Pallarés, F.J.; Carrasco, L. Immunohistochemical expression of IL-12, IL-10, IFN-α and IFN-γ in lymphoid organs of porcine reproductive and respiratory syndrome virus-infected pigs. *Vet. Immunol. Immunopathol.* **2012**, *149*, 262–271. [CrossRef] [PubMed]

21. Charerntantanakul, W.; Yamkanchoo, S.; Kasinrerk, W. Plasmids expressing porcine interferon gamma up-regulate pro-inflammatory cytokine and co-stimulatory molecule expression which are suppressed by porcine reproductive and respiratory syndrome virus. *Vet. Immunol. Immunopathol.* **2013**, *153*, 107–117. [CrossRef] [PubMed]

22. Shi, K.C.; Guo, X.; Ge, X.N.; Liu, Q.; Yang, H.C. Cytokine mRNA expression profiles in peripheral blood mononuclear cells from piglets experimentally co-infected with porcine reproductive and respiratory syndrome virus and porcine circovirus type 2. *Vet. Microbiol.* **2010**, *140*, 155–160. [CrossRef] [PubMed]

23. Gómez-Laguna, J.; Salguero, F.J.; Barranco, I.; Pallarés, F.J.; Rodríguez-Gómez, I.M.; Bernabé, A.; Carrasco, L. Cytokine expression by macrophages in the lung of pigs infected with the porcine reproductive and respiratory syndrome virus. *J. Comp. Pathol.* **2010**, *142*, 51–60. [CrossRef] [PubMed]

24. Gimeno, M.; Darwich, L.; Díaz, I.; de la Torre, E.; Pujols, J.; Martín, M.; Inumaru, S.; Cano, E.; Domingo, M.; Montoya, M.; *et al.* Cytokine profiles and phenotype regulation of antigen presenting cells by genotype-I porcine reproductive and respiratory syndrome virus isolates. *Vet. Res.* **2011**, *42*, 9. [CrossRef] [PubMed]

25. Olvera, A.; Sibila, M.; Calsamiglia, M.; Segalés, J.; Domingo, M. Comparison of porcine circovirus type 2 load in serum quantified by a real time PCR in postweaning multisystemic wasting syndrome and porcine dermatitis and nephropathy syndrome naturally affected pigs. *J. Virol. Method.* **2004**, *117*, 75–80. [CrossRef]

26. Livak, K.J.; Schmittgen, T.D. Analysis of relative gene expression data using real-time quantitative PCR and the $2^{-\Delta\Delta Ct}$ Method. *Methods* **2001**, *25*, 402–408. [CrossRef] [PubMed]

27. Chung, W.B.; Chan, W.H.; Chaung, H.C.; Lien, Y.; Wu, C.C.; Huang, Y.L. Real-time PCR for quantitation of porcine reproductive and respiratory syndrome virus and porcine circovirus type 2 in naturally-infected and challenged pigs. *J. Virol. Method.* **2005**, *124*, 11–19. [CrossRef]

28. Opriessnig, T.; Thacker, E.L.; Yu, S.; Fenaux, M.; Meng, X.J.; Halbur, P.G. Experimental reproduction of postweaning multisystemic wasting syndrome in pigs by dual infection with *Mycoplasma hyopneumoniae* and porcine circovirus type 2. *Vet. Pathol.* **2004**, *41*, 624–640. [CrossRef] [PubMed]

29. Manickam, C.; Dwivedi, V.; Patterson, R.; Papenfuss, T.; Renukaradhya, G.J. Porcine reproductive and respiratory syndrome virus induces pronounced immune modulatory responses at mucosal tissues in the parental vaccine strain VR2332 infected pigs. *Vet. Microbiol.* **2013**, *162*, 68–77. [CrossRef] [PubMed]

© 2015 by the authors; licensee MDPI, Basel, Switzerland. This article is an open access article distributed under the terms and conditions of the Creative Commons Attribution (CC-BY) license (http://creativecommons.org/licenses/by/4.0/).

toxins

MDPI

Article

The Effects of Bee Venom Acupuncture on the Central Nervous System and Muscle in an Animal hSOD1^{G93A} Mutant

MuDan Cai [1], Sun-Mi Choi [2] and Eun Jin Yang [1],*

[1] Department of KM Fundamental Research, Korea Institute of Oriental Medicine, 483 Expo-ro, Daejeon, Yuseong-gu 305-811, Korea; mudan126@kiom.re.kr
[2] Executive Director of R&D, Korea Institute of Oriental Medicine, 483 Expo-ro, Daejeon, Yuseong-gu 305-811, Korea; smchoi@kiom.re.kr
* Author to whom correspondence should be addressed; yej4823@hanmail.net; Tel.: +82-42-863-9497; Fax: +82-42-868-9339.

Academic Editor: Sokcheon Pak
Received: 26 November 2014; Accepted: 3 March 2015; Published: 13 March 2015

Abstract: Amyotrophic lateral sclerosis (ALS) is caused by the degeneration of lower and upper motor neurons, leading to muscle paralysis and respiratory failure. However, there is no effective drug or therapy to treat ALS. Complementary and alternative medicine (CAM), including acupuncture, pharmacopuncture, herbal medicine, and massage is popular due to the significant limitations of conventional therapy. Bee venom acupuncture (BVA), also known as one of pharmacopunctures, has been used in Oriental medicine to treat inflammatory diseases. The purpose of this study is to investigate the effect of BVA on the central nervous system (CNS) and muscle in symptomatic hSOD1^{G93A} transgenic mice, an animal model of ALS. Our findings show that BVA at ST36 enhanced motor function and decreased motor neuron death in the spinal cord compared to that observed in hSOD1^{G93A} transgenic mice injected intraperitoneally (i.p.) with BV. Furthermore, BV treatment at ST36 eliminated signaling downstream of inflammatory proteins such as TLR4 in the spinal cords of symptomatic hSOD1^{G93A} transgenic mice. However, i.p. treatment with BV reduced the levels of TNF-α and Bcl-2 expression in the muscle hSOD1^{G93A} transgenic mice. Taken together, our findings suggest that BV pharmacopuncture into certain acupoints may act as a chemical stimulant to activate those acupoints and subsequently engage the endogenous immune modulatory system in the CNS in an animal model of ALS.

Keywords: amyotrophic lateral sclerosis; bee venom acupuncture; acupoint; central nervous system

1. Introduction

Amyotrophic lateral sclerosis (ALS) is caused by the selective and progressive loss of motor neurons, leading to irreversible paralysis; speech, swallowing and respiratory malfunction; and eventually the death of the affected individual after a rapid disease course. ALS is mostly sporadic; 90% of ALS cases occur without a family history of the disease. Mutations in the gene encoding superoxide dismutase-1 (SOD1) cause 15%–20% of familial ALS (fALS) cases, corresponding to 1%–2% of all ALS cases [1]. Recent clinical and electrophysiological data show that the human SOD1-G93A phenotype closely resembles sporadic ALS, indicating comparable disease pathology [2].

Neuroinflammation by glial cells is a normal and necessary process to support neuron cell survival. However, excess neuroinflammation is detrimental, especially in multiple sclerosis (MS), ALS, various types of dementia, Huntington's disease, and other diseases [3,4]. In ALS, non-neuronal cells, such as astrocytes and microglia, release neurotoxic factors and induce neuroinflammatory events causing motor neuron death [5,6]. Several studies related to neuroinflammation by microglia and astrocytes have reported that inflammation is fundamental to the pathogenesis of ALS and have suggested that anti-inflammatory drugs may play an important role in treating ALS patients [7–9]. Anti-inflammatory agents including celastrol, thalidomide, lenalidomide, NDGA, and pioglitazone have delayed the progress of disease in ALS animal models but require further evaluation in clinical studies [10].

Bee venom (BV) or apitoxin is extracted from honeybees and is commonly used in Oriental medicine. BV therapy is used for anticoagulant and anti-inflammation treatments in rheumatism and joint diseases [11]. BV treatment has been shown to reduce pain in patients with chronic rheumatoid arthritis or osteoarthritis [12]. In addition, recent papers have reported that BV treatment has a neuroprotective role against neurodegenerative diseases such as Parkinson's disease (PD) and ALS [13,14].

Pharmacopuncture, injection to acupoints with pharmacological medication or herbal medicine, is a new acupuncture therapy widely available in Korea and China. However, its effectiveness remains unclear. Pharmacopuncture has been used for cancer-related symptoms [15] and an Australian study reported that it induces higher de-qi sensation compared to traditional acupuncture, which may indicate that pharmacopuncture could provide stronger clinical responses than traditional acupuncture [16].

Pharmacopuncture with BV, referred to as bee venom acupuncture (BVA) is applied at specific acupoints to enhance the effect of BV and acupuncture in Korea. Joksamni (ST36) acupoint is known to mediate anti-inflammatory effects [17] and we have demonstrated that Joksamni (ST36) acupoint stimulation enhanced anti-neuroinflammation in symptomatic hSOD1^{G93A} transgenic mice [18]. The purpose of this study is to investigate BVA at ST36 to identify the effects of acupuncture and BV in the CNS and muscle in symptomatic hSOD1^{G93A} transgenic mice. For this study, we established four groups of mice: Non-Tg mice treated with saline acupuncture at ST36 (Non-Tg), hSOD1^{G93A} mice treated with saline (CON) or BVA at ST36 (ST36), and hSOD1^{G93A} mice injected intraperitoneally (i.p.) with BV (IP).

We found that BV treatment significantly improved walking function compared to BV-i.p.-injected hSOD1^{G93A} mice and age-matched hSOD1^{G93A} mice treated with saline acupuncture at ST36. In addition, BVA at ST36 reduced the levels of neuroinflammatory proteins such as Toll-like receptor 4 (TLR4), CD14, and Tumor Necrosis Factor-alpha (TNF-α) in the spinal cord compared with saline acupuncture at ST36-treated hSOD1^{G93A} mice, but BV-i.p. injection in symptomatic hSOD1^{G93A} mice did not. Furthermore, we detected that the nuclear abnormality in the quadriceps femoris muscle was significantly reduced by BVA compared with saline acupuncture at ST36 in ALS mice but was not affected by i.p. injection of BV in hSOD1^{G93A} mice. These findings suggest that BVA at ST36 can be more effective than either ST36 stimulation or BV injection alone in reducing neuroinflammation in the spinal cord of hSOD1^{G93A} transgenic mice.

2. Results

2.1. BV Treatment Improves Motor Functions in Symptomatic hSOD1^{G93A} Mice

To investigate the effect of BV treatment at ST36 or i.p. injection of BV on motor function of symptomatic hSOD1^{G93A} transgenic mice, we conducted the footprint test as a behavioral test. As shown in Figure 1, symptomatic hSOD1^{G93A} transgenic mice (CON group) showed stride lengths of 2.64 ± 0.19 cm and reduced by 2.4-fold compared to that of Non-Tg. However, both BVA at ST36 (ST36 group) and IP group increased stride length by 1.6-fold, and 1.5-fold, respectively, compared to that of saline acupuncture at ST36 (Figure 1B, *, # $p < 0.05$). This result suggests that BV treatment, regardless of the method of administration, could effectively improve motor function in hSOD1^{G93A} transgenic mice.

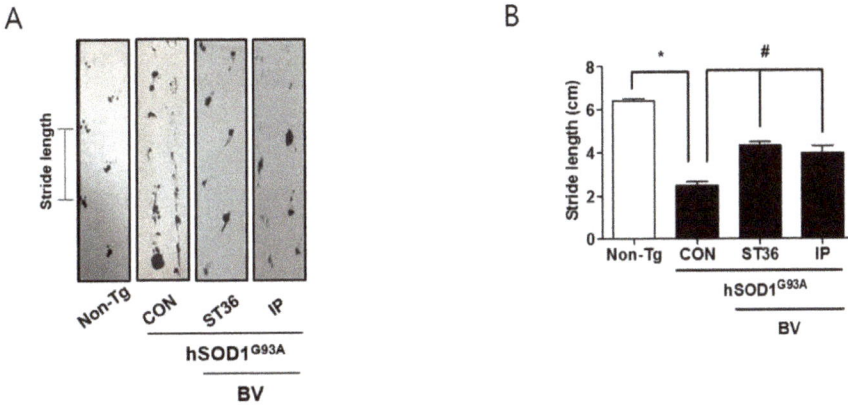

Figure 1. Bee venom (BV) treatment improves stride length in footprints test of hSOD1^{G93A} transgenic mice. Comparison of footprints between groups (**A**). Non-transgenic (Non-Tg) mice are B6SJL mice, and hSOD1^{G93A} transgenic mice were divided into three groups: saline acupuncture at ST36 (CON); BVA at ST36 (ST36); i.p. administration of BV (IP). Quantification of stride length from each group (**B**). (*n* = 9–10/group) Each bar represents the group mean ± SEM (*, # *p* < 0.05).

2.2. BV Treatment Reduces Motor Neuron Cell Death in Symptomatic hSOD1^{G93A} Mice

Based on the result of Figure 1, we determined whether different BV treatment affect motor neuron death in the spinal cord of symptomatic hSOD1^{G93A} mice. Based on Nissl staining, the numbers of motor neurons in the spinal cord of symptomatic hSOD1^{G93A} transgenic mice (CON group) were decreased by 3.5-fold compared to that of Non-Tg (Figure 2A,B, *, # *p* < 0.05). Both BVA at ST36 and i.p. administration of BV were associated with more than twice as many neurons as saline acupuncture at ST36 (Figure 2B). No statistical significance was found between BVA at ST36 and i.p. administration of BV, demonstrating that BV has a neuroprotective role in an animal model of ALS, regardless of the treatment method.

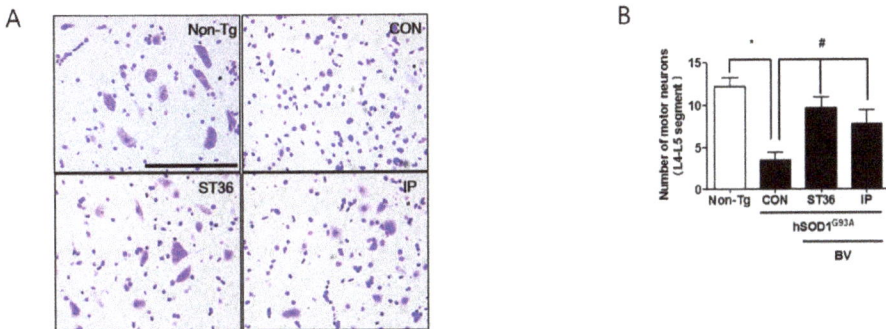

Figure 2. BV treatment alleviates the decrease of motor neurons in hSOD1^{G93A} transgenic mice. Representative photographs of Nissl staining in the ventral horn of the spinal cord of hSOD1^{G93A} transgenic mice (**A**). Quantification of the motor neurons in the L4-L5 segment of the spinal cord in hSOD1^{G93A} transgenic mice (*n* = 4/group) (**B**). Each bar represents the group mean ± SEM (*, # *p* < 0.05). Scale bars = 200 μm.

2.3. BVA at ST36 Augments Anti-Neuroinflammation

To determine the mechanism underlying the effects of BVA at ST36, we studied the effects of BV treatment on neuroinflammation in the spinal cord of symptomatic hSOD1^{G93A} transgenic mice. As expected, BVA at ST36 significantly reduced the levels of the inflammatory proteins TLR4, CD14, and TNF-α, by 1.8-fold, 1.6-fold and 1.7-fold, respectively, in the spinal cords of symptomatic hSOD1^{G93A} transgenic mice compared to those of Non-Tg (Figure 3A–D). However, i.p. administration of BV did not have a similar effect. This suggests that BVA at ST36 prevents motor neuron death by inhibiting neuroinflammation in the spinal cord of symptomatic ALS mice.

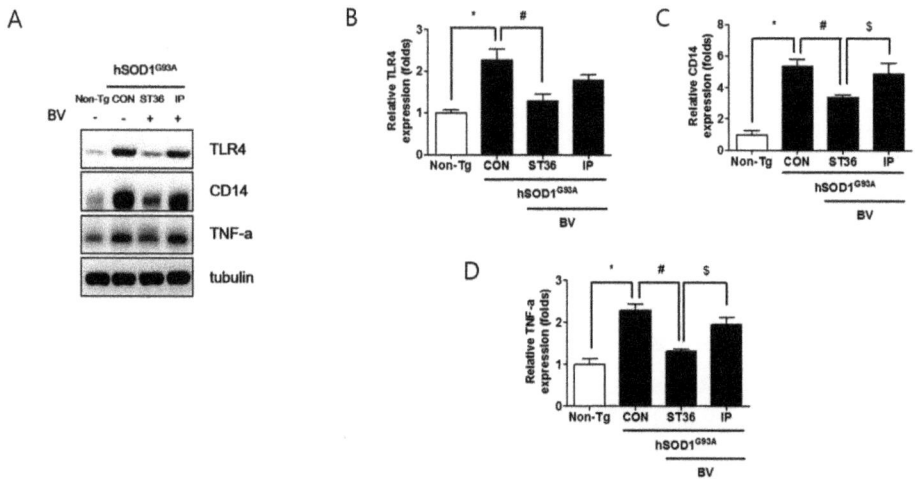

Figure 3. BVA at ST36 eliminates Toll-like receptor 4 (TLR4) signaling-related inflammatory proteins in the spinal cord of ALS mice. Representative images of Western blots for TLR4, CD14, and TNF-α (**A**) (*n* = 5–6/group). Tubulin was used as a loading control. Quantification of immunoblots (**B–D**). Each bar represents the group mean ± SEM (*, #, $ *p* < 0.05).

2.4. BVA at ST36 Prevents Muscle Atrophy in Symptomatic hSOD1^{G93A} Transgenic Mice

To evaluate the effect of BV treatment and the method of administration on the muscle pathology of hSOD1^{G93A} transgenic mice, we examined the effect of BV treatment on fibers of the quadriceps femoris muscle in the hind leg, a target of innervation by the spinal motor neurons counted. As shown in Figure 4A, the fiber diameter was smaller in hSOD1^{G93A} transgenic mice than in Non-Tg. In addition, myofibers containing abnormal nuclei were increased by 10-fold higher than compared to Non-Tg mice. Interestingly, BV treatment at ST36 reduced the number of abnormal nuclei to 3-fold than that of the control (Figure 4B). Intraperitoneal injection of BV reduced the fibers with abnormal nuclei compared to that of saline-treated hSOD1^{G93A} transgenic mice, but the decrease was not significant. To investigate whether BV treatment affected the mitochondrial function of myofibers, we examined the expression level of Bcl-2 protein in the quadriceps femoris muscle. As shown in Figure 4C, the expression level of Bcl-2 was increased by 3.4-fold in CON compared to Non-Tg. However, after BV treatment at ST36 and via i.p. injection, the level of Bcl-2 protein in the quadriceps femoris muscle was 1.8-fold, and 1.5 fold lower, respectively, compared to CON. These findings suggest that BV treatment could ameliorate the mitochondrial dysfunction and muscle structure defects observed in an animal model of ALS. To establish the factor for improvement in walking ability achieved by BV-injected i.p., we examined the expression level of a pro-inflammatory protein, TNF-α, in the quadriceps femoris muscle of hSOD1^{G93A} transgenic mice. As shown in Figure 5, the expression of TNF-α was increased by 4.6-fold in the quadriceps femoris muscle of symptomatic hSOD1^{G93A} mice. However, i.p. injection of BV reduced the level of TNF-α to only 3.1-fold higher than in CON. BVA at ST36 treatment did not reduce the expression level of TNF-α as much as i.p. injected BV in symptomatic hSOD1^{G93A} transgenic mice. Since BV causes inflammation in the skin [19], inflammatory protein such as TNF-α did not reduce in the muscle of BVA at ST36 even though BVA at ST36 induced anti-neuroinflammation in the spinal cord of hSOD1^{G93A} transgenic mice. These data suggest that BV injection i.p. can reduce inflammation in muscle even though it does not reduce neuroinflammation in the spinal cord of hSOD1^{G93A} mice.

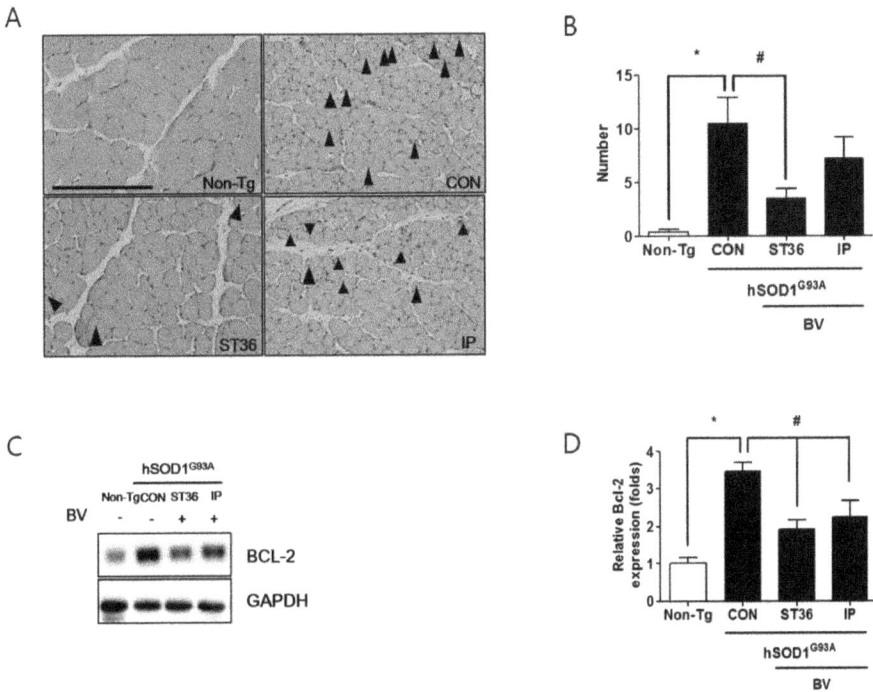

Figure 4. BVA at ST36 reduces nuclear abnormalities in fibers of the quadriceps femoris muscle from hSOD1^{G93A} transgenic mice, viewed in transverse sections. Hematoxylin and eosin staining of quadriceps of Non-Tg, hSOD1^{G93A} transgenic mice (CON), transgenic mice treated with BV at ST36 (ST36), transgenic mice treated with i.p. BV (IP). Arrowhead indicates fiber with abnormal nuclei (**A**); Quantification of mean numbers of abnormal nuclei in fibers of quadriceps femoris muscle of hSOD1^{G93A} mice (**B**); Graph, mean $\pm$ SEM, n = 5–6/group (*, # $p < 0.05$). Representative images of the expression of Bcl-2 in quadriceps femoris muscle from the Non-Tg, CON, ST36, and IP groups (**C**); GAPDH was used as a loading control. One representative experiment of three replicates is shown. Quantification of immunoblots (**D**); Each bar represents the group mean $\pm$ SEM, n = 3/group (* $p < 0.05$, # $p < 0.05$). Scale bars = 200 μm.

A

hSOD1^{G93A}

Non-TgCON ST36 IP

BV   -   -   +   +

TNF-α

GAPDH

B

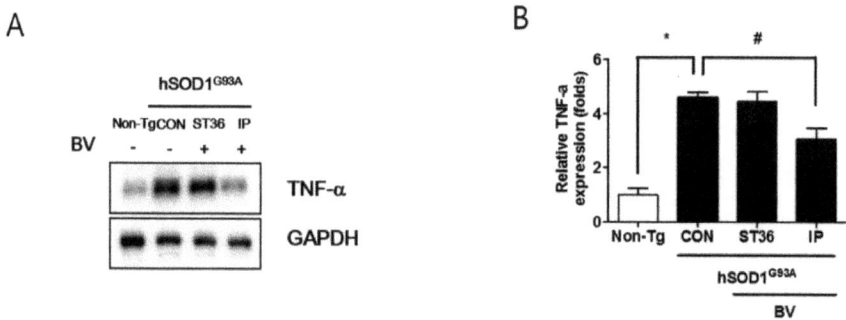

Figure 5. Intraperitoneal injection of BV reduced the levels of inflammatory proteins in quadriceps femoris muscle. Tissue lysates from the quadriceps femoris muscle from the Non-Tg, CON, ST36, and IP groups were immunoblotted with anti-TNF-α (**A**). GAPDH was used as a loading control. Quantification of immunoblots (**B**). Each bar represents the group mean ± SEM ($n = 3$/group) (*, # $p < 0.05$).

3. Discussion

Amyotrophic lateral sclerosis (ALS) is a neurodegenerative disease that causes progressive degeneration of motor neurons in the motor cortex, brainstem, and spinal cord. Familial ALS (fALS), caused by genetic mutations in genes such as Cu^{2+}/Zn^{2+} SOD-1, TAR DNA binding protein-43 (TDP-43), and GGGGCC repeat expansions in the C9orf72 locus, represents 5%–10% of total ALS cases, whereas sporadic ALS (sALS), which is unassociated with any known genetic mutations, represents approximately 90% of ALS cases. To date, the pathological mechanism of ALS remains unclear because ALS is a complex syndrome involving not only motor neurons but also astrocytes, microglia, oligodendrocytes, and muscle cells in both fALS and sALS cases. Because there is no effective treatment for ALS patients, research toward ALS therapy remains ongoing.

Complementary and alternative medicine (CAM), including acupuncture, pharmacoacupuncture, herbal medicine, and massage, is popular due to the significant limitations of conventional therapy. It has previously been reported that 40% of patients with ALS in Germany used CAM therapies, along with 54% of patients in the United Kingdom and an even higher percentage in Asia. Acupuncture is one of the most popular alternative therapies used by patients with ALS [20,21]. Pan *et al.* have reported that ALS patients in Shanghai used CAM therapies including acupuncture, vitamin E, and herbal medicine to reduce the side effects of Riluzole [22]. Wasner *et al.* surveyed the use of CAM by ALS patients and reported that acupuncture, homeopathy, naturopathy, and esoteric treatments were widely used [23].

BV contains several bioactive compounds including melittin, phospholipase A_2, apamin and peptides [24]. BV therapy has been used in Oriental medicine, and BVA is considered to have distinct effects via acupuncture and herbal medicine. However, the mechanism underlying the synergy of BV with acupuncture remains to be identified, even though several papers have reported anti-nociceptive or anti-inflammatory effects of BV in pain, arthritis, rheumatoid diseases, cancer, and skin diseases [11,25]. In addition, recent reports have shown that BV treatment has anti-neuroinflammatory effects in neurodegenerative diseases such as PD and ALS [13,14]. Kim *et al.* showed that BV injection into ST36 increased the modulation of METH-induced hyperactivity and hyperthermia via the central α_2-adrenergic activation compared to other acupoints such as SP9 and GB39 [26]. In an animal model of ALS, BVA protected against motor neuron death through an increase of anti-neuroinflammation in hSOD1^{G93A} transgenic mice [14]. However, it remains to be determined whether the effect of BVA at ST36 is a synergistic effect dependent on acupoint stimulation in combination with BV or is an effect of BV alone. This study tested hSOD1^{G93A} transgenic mice with saline treatment at ST36 for acupoint stimulation, BV treatment at ST36 for acupoint stimulation plus BV effect, and i.p. injection of BV for the effect of BV alone. Interestingly, we found that both BVA at ST36 and i.p. injection of BV enhanced motor function and decreased motor neuron death in the spinal cord compared to saline-treated at ST36 in hSOD1^{G93A} transgenic mice. Furthermore, BVA at ST36 eliminates signaling by inflammatory proteins such as TLR4 in the spinal cord of symptomatic hSOD1^{G93A} transgenic mice. These findings suggest that BV treatment could improve muscle function and reduce motor neuron death, delaying the progress of disease onset in hSOD1^{G93A} transgenic mice.

Studies of ALS have demonstrated functional abnormality and muscle pathology in both fALS and sALS cases [27,28]. In an animal model of ALS, the alteration of myogenesis by the aberrant expression of Pax7 and myogenic regulatory factors induces skeletal muscle defects [29]. Kaspar *et al.* have reported that IGF-1 overexpression in skeletal muscle increased survival and prevented motor neuron death in an animal model of ALS, although a clinical study using IGF-1 therapy did not demonstrate any improvement [30]. To investigate the effect of BV treatment on the quadriceps femoris muscle in the hind limb, we examined myoblast pathology using histochemical and biochemical approaches. BVA at ST36 significantly reduced the number of fibers with abnormal nuclei compared to symptomatic hSOD1^{G93A} mice treated with saline at ST36. In addition, i.p. injection of BV significantly reduced the expression of TNF-α, a pro-inflammatory factor, in the quadriceps femoris muscle of symptomatic hSOD1^{G93A} mice. Taken together, our findings suggest that BV pharmacopuncture into certain acupoints may act as a chemical stimulant that activates the acupoint and subsequently engages the endogenous immune modulatory system in the CNS of an ALS animal model. In addition, the selection of a specific acupoint may be a key factor in producing pharmacological effects of BV acupuncture in an ALS animal model.

Demonstrating the effect of BV on the muscle function requires the study of mitochondria in the muscle of hSOD1^{G93A} transgenic mice. In addition, the effects of BVA at other acupoints should be studied to demonstrate the mechanism underlying the combined effects of BV and acupuncture. Bee venom therapy should consider on sensitization because it has on the risk of anaphylaxis.

4. Materials and Methods

4.1. Animals

Human-SOD1 (hSOD1) G93A transgenic (Tg) mice were purchased from the Jackson Laboratory (Bar Harbor, ME, USA) and maintained as described previously [14]. All mice were allowed access to water and food *ad libitum* and were maintained under a constant temperature (21 ± 2 °C) and humidity (50% ± 10%) under a 12-h light/dark cycle (light on 07:00–19:00). All mice were handled in accordance with the animal care guidelines of the Korea Institute of Oriental Medicine. The 14-week-old transgenic mice were considered symptomatic; mice of this age were used for the study.

4.2. Bee Venom Treatment

Bee venom was purchased from Sigma (St. Louis, MO, USA) and diluted with saline. In this experimental paradigm, the mice were divided into four groups: non-transgenic mice treated with saline at ST36, Non-Tg, *n* = 9; hSOD1^{G93A} mice treated with saline at ST36, CON, *n* = 10; hSOD1^{G93A} mice treated with BVA at ST36, ST36, *n* = 10; and hSOD1^{G93A} mice injected i.p. with BV, IP, *n* = 10. The ST36 acupoint is based on the human acupoint landmark and a mouse anatomical reference [31]; it is anatomically located at 5 mm below and lateral to the anterior tubercle of the tibia. Bee venom was injected into 14-week-old (98-day-old) male hSOD1^{G93A} transgenic mice. The Non-Tg and CON groups were bilaterally injected (subcutaneously) with an equal volume of saline at the ST36 acupoint. BV treatment at 0.1 µg/g was performed three times per week for two weeks. The total amount of BV treatment at ST36-mice or IP-mice was 0.6 µg/g for two weeks.

4.3. Footprint Test

On the second day after the final treatment, we performed a footprint test. Footprints were used to estimate stride length and hind-base width, which reflects the extent of muscle loosening [32,33]. The mice crossed an illuminated alley, 70 cm in length, 12 cm in width, and 16 cm in height, before entering a dark box at the end. Their hindpaws were coated with nontoxic water-soluble ink, and the alley floor was covered with white paper. To obtain clearly visible footprints, at least three trials were conducted. The footprints were then scanned, and stride length was measured using Image J software (NIH, Bethesda, MD, USA).

4.4. Western Blot Analysis

Western blotting was conducted as previously described [34]. Mice were sacrificed immediately after the behavioral testing, and the spinal cords were isolated for Western blotting. The spinal cords and quadriceps femoris muscle were dissected and homogenized in RIPA buffer (50 mM·Tris-HCl, pH 7.4, 1% NP-40, 0.1% SDS, and 150 mM·NaCl) containing a protease inhibitor cocktail. Following homogenization, protein was quantified using the BCA assay. Samples of 20 µg protein were denatured with sodium dodecyl sulfate sampling buffer and then separated using SDS-PAGE, followed by transfer to a PVDF membrane. For the detection of target proteins, the membranes were blocked with 5% non-fat milk in TBS (50 mM Tris-HCl, pH 7.6, 150 mM·NaCl) and subsequently incubated overnight with various primary antibodies: anti-tubulin, anti-TLR4, anti-CD14, anti-TNF-α, anti-GAPDH or anti-Bcl-2. The blots were then probed with peroxidase-conjugated secondary antibodies (Santa Cruz Biotechnology, Santa Cruz, CA, USA) and visualized using enhanced chemiluminescence reagents (Amersham Pharmacia, Piscataway, NJ, USA). Protein bands were detected with the Fusion SL4-imaging system (Fusion, Eberhardzell, Germany). Quantification of the immunoblotting bands was conducted with Image J.

4.5. Tissue Preparation

After the behavioral testing, all mice were anesthetized with an intraperitoneal injection of pentobarbital and perfused with phosphate buffered saline (PBS). The quadriceps femoris muscle and spinal cord tissues were removed and fixed in 4% paraformaldehyde for 3 days at 4 °C until embedding. Briefly, the quadriceps muscle and lumbar 4–5 spinal cord were embedded in paraffin, and the prepared tissues were cut into transverse sections (5 µm thick) and mounted on glass slides. Before staining, sections were deparaffinized in xylene and rehydrated in a graded series of ethanol followed by dH$_2$O.

4.6. Nissl Staining

The staining was used to evaluate the general neuronal morphology and demonstrate the loss of Nissl substances [35] and was performed as previously described [36]. Briefly, after deparaffinizing, the sections of spinal cord were oven-dried, stained with 0.1% cresyl violet, dehydrated through a graded ethanol series (70%, 80%, 90%, and 100% × 2), placed in xylene, and covered with a coverslip after the addition of Histomount media. To quantify the Nissl staining, the cells in the ventral horn of the spinal cord were counted using Image J (version 1.46j) by one researcher blind to the treatment groups. Cells meeting the following criteria were counted: (1) neurons located in the anterior horn ventral to the line tangential to the ventral tip of the central canal; and (2) neurons with a maximum diameter of 20 µm or more.

4.7. H&E Staining

After being deparaffinized, slices of quadriceps muscle were incubated in hematoxylin (Sigma, St. Louis, MO, USA) followed by incubation in eosin. Slices were mounted with Histomount medium (Sigma, St. Louis, MO, USA). Analysis of H&E staining was performed in a blinded test, and fibers with central nuclei were counted in randomly selected areas from each mouse. Three stained sections were counted per hSOD1^{G93A} mouse; tissue samples from 5 to 6 mices per group were stained.

4.8. Statistical Analysis

All data were analyzed using GraphPad Prism 5.0 (GraphPad Software, San Diego, CA, USA) and are presented as the mean ± SEM where indicated. The results of the behavioral test and the histological and Western blot analyses were analyzed using one-way analysis of variance (ANOVA) followed by the Bonferroni's post-hoc tests for multiple comparisons. Statistical significance was set at $p < 0.05$.

5. Conclusions

In summary, our data show that BVA at ST36 and i.p. injection with BV enhanced motor function and decreased motor neuron death in the spinal cord compared to that observed in hSOD1^{G93A} transgenic mice treated saline. Furthermore, BV treatment at ST36 eliminated signaling downstream of inflammatory proteins such as TLR4 in the spinal cords of symptomatic hSOD1^{G93A} transgenic mice. However, i.p. treatment with BV reduced the levels of TNF-α and Bcl-2 expression in the muscle of hSOD1^{G93A} transgenic mice. Taken together, our findings suggest that BV pharmacopuncture into certain acupoints may act as a chemical stimulant to activate acupoints and subsequently engage the endogenous immune modulatory system in the CNS in an animal model of ALS.

Acknowledgments: We thank Sun-Hwa Lee for animal maintenance. This research was supported by grants (K13010, K14012) from the Korea Institute of Oriental Medicine (KIOM).

Author Contributions: E.J.Y. designed the experiments and analyzed the data as well as wrote the manuscript; M.D.C. executed all the experiments and S.M.C. discussed with the manuscript. All authors have read and approved the final manuscript.

Conflicts of Interest: The authors declare no conflict of interest.

References

1. Rosen, D.R.; Siddique, T.; Patterson, D.; Figlewicz, D.A.; Sapp, P.; Hentati, A.; Donaldson, D.; Goto, J.; O'Regan, J.P.; Deng, H.X.; *et al.* Mutations in Cu/Zn superoxide dismutase gene are associated with familial amyotrophic lateral sclerosis. *Nature* **1993**, *362*, 59–62. [CrossRef] [PubMed]

2. Synofzik, M.; Fernandez-Santiago, R.; Maetzler, W.; Schols, L.; Andersen, P.M. The human G93A SOD1 phenotype closely resembles sporadic amyotrophic lateral sclerosis. *J. Neurol. Neurosurg. Psychiatr.* **2010**, *81*, 764–767. [CrossRef] [PubMed]

3. Moller, T. Neuroinflammation in Huntington's disease. *J. Neural Transm.* **2010**, *117*, 1001–1008. [CrossRef] [PubMed]

4. Weydt, P.; Moller, T. Neuroinflammation in the pathogenesis of amyotrophic lateral sclerosis. *Neuroreport* **2005**, *16*, 527–531. [CrossRef] [PubMed]

5. Rojas, F.; Cortes, N.; Abarzua, S.; Dyrda, A.; van Zundert, B. Astrocytes expressing mutant SOD1 and TDP43 trigger motoneuron death that is mediated via sodium channels and nitroxidative stress. *Front. Cell. Neurosci.* **2014**, *8*. [CrossRef] [PubMed]

6. Bowerman, M.; Vincent, T.; Scamps, F.; Perrin, F.E.; Camu, W.; Raoul, C. Neuroimmunity dynamics and the development of therapeutic strategies for amyotrophic lateral sclerosis. *Front. Cell Neurosci.* **2013**, *7*, 214. [CrossRef] [PubMed]

7. McGeer, P.L.; McGeer, E.G. Inflammatory processes in amyotrophic lateral sclerosis. *Muscle Nerve* **2002**, *26*, 459–470. [CrossRef] [PubMed]

8. Meissner, F.; Molawi, K.; Zychlinsky, A. Mutant superoxide dismutase 1-induced IL-1beta accelerates ALS pathogenesis. *Proc. Natl. Acad. Sci. USA* **2010**, *107*, 13046–13050. [CrossRef] [PubMed]

9. Sekizawa, T.; Openshaw, H.; Ohbo, K.; Sugamura, K.; Itoyama, Y.; Niland, J.C. Cerebrospinal fluid interleukin 6 in amyotrophic lateral sclerosis: Immunological parameter and comparison with inflammatory and non-inflammatory central nervous system diseases. *J. Neurol. Sci.* **1998**, *154*, 194–199. [CrossRef] [PubMed]

10. Zhu, Y.; Fotinos, A.; Mao, L.L.; Atassi, N.; Zhou, E.W.; Ahmad, S.; Guan, Y.; Berry, J.D.; Cudkowicz, M.E.; Wang, X. Neuroprotective agents target molecular mechanisms of disease in ALS. *Drug Discov. Today* **2015**, *20*, 65–75. [CrossRef] [PubMed]

11. Kwon, Y.B.; Lee, H.J.; Han, H.J.; Mar, W.C.; Kang, S.K.; Yoon, O.B.; Beitz, A.J.; Lee, J.H. The water-soluble fraction of bee venom produces antinociceptive and anti-inflammatory effects on rheumatoid arthritis in rats. *Life Sci.* **2002**, *71*, 191–204. [CrossRef] [PubMed]

12. Somerfield, S.D.; Brandwein, S. Bee venom and adjuvant arthritis. *J. Rheumatol.* **1988**, *15*, 1878. [PubMed]

13. Kim, J.I.; Yang, E.J.; Lee, M.S.; Kim, Y.S.; Huh, Y.; Cho, I.H.; Kang, S.; Koh, H.K. Bee venom reduces neuroinflammation in the MPTP-induced model of Parkinson's disease. *Int. J. Neurosci.* **2011**, *121*, 209–217. [CrossRef] [PubMed]

14. Yang, E.J.; Jiang, J.H.; Lee, S.M.; Yang, S.C.; Hwang, H.S.; Lee, M.S.; Choi, S.M. Bee venom attenuates neuroinflammatory events and extends survival in amyotrophic lateral sclerosis models. *J. Neuroinflamm.* **2010**, *7*. [CrossRef]

15. Korean Pharmacopuncture Institute. *Pharmacopuncturology*, 2nd ed.; Elsevier Korea: Seoul, Korea, 2011.

16. Strudwick, M.W.; Hinks, R.C.; Choy, S.T. Point injection as an alternative acupuncture technique—An exploratory study of responses in healthy subjects. *Acupunct. Med.* **2007**, *25*, 166–174. [CrossRef] [PubMed]

17. Wilms, H.; Zecca, L.; Rosenstiel, P.; Sievers, J.; Deuschl, G.; Lucius, R. Inflammation in Parkinson's diseases and other neurodegenerative diseases: Cause and therapeutic implications. *Curr. Pharm Des.* **2007**, *13*, 1925–1928. [CrossRef] [PubMed]

18. Yang, E.J.; Jiang, J.H.; Lee, S.M.; Hwang, H.S.; Lee, M.S.; Choi, S.M. Electroacupuncture reduces neuroinflammatory responses in symptomatic amyotrophic lateral sclerosis model. *J. Neuroimmunol.* **2010**, *223*, 84–91. [CrossRef] [PubMed]

19. Cui, X.Y.; Dai, Y.; Wang, S.L.; Yamanaka, H.; Kobayashi, K.; Obata, K.; Chen, J.; Noguchi, K. Differential activation of p38 and extracellular signal-regulated kinase in spinal cord in a model of bee venom-induced inflammation and hyperalgesia. *Mol. Pain.* **2008**, *4*, 17. [CrossRef] [PubMed]

20. Lee, S.; Kim, S. The effects of sa-am acupuncture treatment on respiratory physiology parameters in amyotrophic lateral sclerosis patients: A pilot study. *Evid. Based Complement. Alternat. Med.* **2013**, *2013*. [CrossRef]

21. Liang, S.; Christner, D.; Du Laux, S.; Laurent, D. Significant neurological improvement in two patients with amyotrophic lateral sclerosis after 4 weeks of treatment with acupuncture injection point therapy using enercel. *J. Acupunct. Meridian Stud.* **2011**, *4*, 257–261. [CrossRef] [PubMed]

22. Pan, W.; Chen, X.; Bao, J.; Bai, Y.; Lu, H.; Wang, Q.; Liu, Y.; Yuan, C.; Li, W.; Liu, Z.; *et al.* The use of integrative therapies in patients with amyotrophic lateral sclerosis in Shanghai, China. *Evid. Based Complement. Alternat. Med.* **2013**, *2013*, 613596. [PubMed]

23. Wasner, M.; Klier, H.; Borasio, G.D. The use of alternative medicine by patients with amyotrophic lateral sclerosis. *J. Neurol. Sci.* **2001**, *191*, 151–154. [CrossRef] [PubMed]

24. Lariviere, W.R.; Melzack, R. The bee venom test: A new tonic-pain test. *Pain* **1996**, *66*, 271–277. [CrossRef] [PubMed]

25. Kwon, Y.B.; Lee, J.D.; Lee, H.J.; Han, H.J.; Mar, W.C.; Kang, S.K.; Beitz, A.J.; Lee, J.H. Bee venom injection into an acupuncture point reduces arthritis associated edema and nociceptive responses. *Pain* **2001**, *90*, 271–280. [CrossRef] [PubMed]

26. Kim, K.W.; Kim, H.W.; Li, J.; Kwon, Y.B. Effect of bee venom acupuncture on methamphetamine-induced hyperactivity, hyperthermia and Fos expression in mice. *Brain Res. Bull.* **2011**, *84*, 61–68. [CrossRef] [PubMed]

27. Krasnianski, A.; Deschauer, M.; Neudecker, S.; Gellerich, F.N.; Muller, T.; Schoser, B.G.; Krasnianski, M.; Zierz, S. Mitochondrial changes in skeletal muscle in amyotrophic lateral sclerosis and other neurogenic atrophies. *Brain* **2005**, *128*, 1870–1876. [CrossRef] [PubMed]

28. Echaniz-Laguna, A.; Zoll, J.; Ponsot, E.; N'Guessan, B.; Tranchant, C.; Loeffler, J.P.; Lampert, E. Muscular mitochondrial function in amyotrophic lateral sclerosis is progressively altered as the disease develops: A temporal study in man. *Exp. Neurol.* **2006**, *198*, 25–30. [CrossRef] [PubMed]

29. Manzano, R.; Toivonen, J.M.; Olivan, S.; Calvo, A.C.; Moreno-Igoa, M.; Munoz, M.J.; Zaragoza, P.; Garcia-Redondo, A.; Osta, R. Altered expression of myogenic regulatory factors in the mouse model of amyotrophic lateral sclerosis. *Neurodegener. Dis.* **2011**, *8*, 386–396. [CrossRef] [PubMed]

30. Kaspar, B.K.; Llado, J.; Sherkat, N.; Rothstein, J.D.; Gage, F.H. Retrograde viral delivery of IGF-1 prolongs survival in a mouse ALS model. *Science* **2003**, *301*, 839–842. [CrossRef] [PubMed]

31. Yin, C.S.; Jeong, H.S.; Park, H.J.; Baik, Y.; Yoon, M.H.; Choi, C.B.; Koh, H.G. A proposed transpositional acupoint system in a mouse and rat model. *Res. Vet. Sci.* **2008**, *84*, 159–165. [CrossRef] [PubMed]

32. Filali, M.; Lalonde, R.; Rivest, S. Sensorimotor and cognitive functions in a SOD1(G37R) transgenic mouse model of amyotrophic lateral sclerosis. *Behav. Brain Res.* **2011**, *225*, 215–221. [CrossRef] [PubMed]

33. Mancuso, R.; Olivan, S.; Osta, R.; Navarro, X. Evolution of gait abnormalities in SOD1(G93A) transgenic mice. *Brain Res.* **2011**, *1406*, 65–73. [CrossRef] [PubMed]

34. Cai, M.; Choi, S.M.; Song, B.K.; Son, I.; Kim, S.; Yang, E.J. Scolopendra subspinipes mutilans attenuates neuroinflammation in symptomatic hSOD1(G93A) mice. *J. Neuroinflamm.* **2013**, *10*. [CrossRef]

35. Irugalbandara, Z.E. Simplified differentiation of Nissl granules stained by toluidine blue in paraffin sections. *Stain Technol.* **1960**, *35*, 47–48. [PubMed]

36. Cai, M.; Shin, B.Y.; Kim, D.H.; Kim, J.M.; Park, S.J.; Park, C.S.; Won, H.; Hong, N.D.; Kang, D.H.; Yutaka, Y.; *et al.* Neuroprotective effects of a traditional herbal prescription on transient cerebral global ischemia in gerbils. *J. Ethnopharmacol.* **2011**, *138*, 723–730. [CrossRef] [PubMed]

© 2015 by the authors; licensee MDPI, Basel, Switzerland. This article is an open access article distributed under the terms and conditions of the Creative Commons Attribution (CC-BY) license (http://creativecommons.org/licenses/by/4.0/).

toxins

MDPI

Review

The Protective Effect of Bee Venom on Fibrosis Causing Inflammatory Diseases

Woo-Ram Lee [1], Sok Cheon Pak [2] and Kwan-Kyu Park [1],*

[1] Department of Pathology, College of Medicine, Catholic University of Daegu, 3056-6, Daemyung-4-Dong, Nam-gu, Daegu 705-718, Korea; woolamee@cu.ac.kr

[2] School of Biomedical Sciences, Charles Sturt University, Panorama Avenue, Bathurst, NSW 2795, Australia; spak@csu.edu.au

* Correspondence: kkpark@cu.ac.kr; Tel.: +82-53-650-4149; Fax: +82-53-650-4843

Academic Editor: Ren Lai

Received: 14 August 2015; Accepted: 5 November 2015; Published: 16 November 2015

Abstract: Bee venom therapy is a treatment modality that may be thousands of years old and involves the application of live bee stings to the patient's skin or, in more recent years, the injection of bee venom into the skin with a hypodermic needle. Studies have proven the effectiveness of bee venom in treating pathological conditions such as arthritis, pain and cancerous tumors. However, there has not been sufficient review to fully elucidate the cellular mechanisms of the anti-inflammatory effects of bee venom and its components. In this respect, the present study reviews current understanding of the mechanisms of the anti-inflammatory properties of bee venom and its components in the treatment of liver fibrosis, atherosclerosis and skin disease.

Keywords: bee venom; inflammation; liver fibrosis; atherosclerosis; skin disease

1. Introduction

Bee venom is a natural toxin produced by the honey bee and it has a prime role of defense for the bee colony [1–3]. It has an efficient and complex mixture of substances designed to protect bees against a broad diversity of predators [2]. Bee venom possesses various peptides including melittin, apamin, adolapamin and mast cell degranulating peptide [4,5]. It also contains enzymes, biologically activity amines and non-peptide components. Enzymes are composed of phospholipase A2 (PLA2), hyaluronidase, acid phosphomonesterase, α-D-glucosidase and lysophospholipase, as well as non-peptides such as histamine, dopamine and norepinephrine [6]. Bee venom therapy is a treatment modality that may be thousands of years old and involves the application of live bee stings to the patient's skin or, in more recent years, the injection of bee venom into the skin with a hypodermic needle [7]. Many experiments on the biological and pharmacological activities of bee venom have been carried out [2–8]. Majority of these studies have proven the effectiveness of bee venom in treating pathological conditions such as arthritis [9], pain [10,11] and cancerous tumors [12,13] among others.

The major component of bee venom is melittin, which comprises approximately 50% of the dry weight of bee venom. It is a small linear peptide composed of 26 amino acid residues [14]. Melittin has multiple effects, including anti-bacterial, anti-viral and anti-inflammatory, in various cell types [15,16]. Recent studies have shown that melittin can induce cell cycle arrest, cell growth inhibition and apoptosis in various tumor cells [17–19]. When several melittin peptides accumulate in the cell membrane, phospholipid packing is severely disrupted, thus leading to cell lysis [16]. Melittin triggers not only the lysis of a wide range of plasmatic membranes but also of intracellular ones such as those found in mitochondria. PLA2 and melittin act synergistically, breaking up membranes of susceptible cells and enhancing their cytotoxic effect [20]. However, other paper reported that melittin at concentrations below 2 μM does not disrupt cell membranes of leukocytes [21]. In addition, another paper reported that an optimal dose of melittin protects TGF-β1-induced apoptotic activation of hepatocytes by inhibiting the activation of the Bcl-2 family of proteins, caspases and poly (adenosine diphosphate-ribose) polymerase (PARP) cleavage [22].

Apamin is an integral part of bee venom, accounting for about 2%–3% of its dry weight [3]. It is a peptide neurotoxin comprising 18 amino acid residues that is tightly cross-linked by the presence of two disulphide bonds [5]. Apamin is well known for its pharmacological property of irreversibly blocking Ca^{2+}-activated K^+ (SK) channels [23]. These channels link intracellular calcium transients to changes of the membrane potential by promoting K^+ efflux following increases of intracellular calcium during an action potential [24]. In a previous paper, it was reported that apamin inhibited pro-inflammatory cytokines in lipopolysaccharide (LPS) with fat diet-induced atherosclerotic animal model [25]. Furthermore, a recent study has examined its biological and pharmacological activities [26]. However, little is known about the molecular mechanisms and the levels of gene regulation involved in the anti-inflammatory process. As such, this review focuses on overview of recent research on anti-inflammatory properties of bee venom and its components in liver fibrosis, atherosclerosis and inflammatory skin disease. Moreover, we review possible mechanisms of bee venom for alleviating or preventing the inflammatory diseases.

2. Anti-Inflammatory Effect of Bee Venom on Liver Fibrosis

Liver fibrosis occurs with chronic hepatic damage in a variety of liver diseases including viral hepatitis, alcoholic hepatitis and primary sclerosing cholangitis [27]. In these conditions, fibrotic liver shows changes in tissue architecture and extracellular matrix composition that ultimately compromise organ function [28–31]. The processes of liver repair and of fibrogenesis resemble that of a wound-healing process. Viral infection, alcoholic or drug toxicity, or any other factors that cause damage to hepatocytes, elicit an inflammatory reaction in the liver [14]. Following injury, an acute inflammation response takes place resulting in moderate cell necrosis and extracellular matrix damage [32]. Chronic ethanol consumption is associated with serious and potentially fatal alcohol-related liver diseases such as fatty liver, alcoholic hepatitis and cirrhosis [33]. It is currently understood that the pathogenesis of these diseases is related to apoptosis [34]. Pro-inflammatory cytokine, TNF-α can induce multiple mechanisms that initiate apoptosis in hepatocytes, which leads to liver injury [35]. A paper reported that an optimal dose of bee venom exerts anti-apoptotic effects against ethanol-induced injury to hepatocytes via the mitochondrial pathway [34]. Thus, bee venom protects hepatocyte against TNF-α with actinomycin D induced apoptosis. Low concentrations of bee venom resulted in anti-apoptotic effects that were associated with a decrease in the level of proteolytic fragments of caspases and PARP [36]. Furthermore, a recent study indicates that bee venom inhibits CCL_4-induced hepatic fibrosis through suppression of fibrogenic cytokines in liver fibrosis animal model. This study shows that bee venom down-regulated pro-inflammatory cytokines such as TNF-α and IL-1β. It has been demonstrated that collagen gene expression is regulated by TNF-α at a transcription level and IL-1β exerts a stimulatory effect on the synthesis of extracellular matrix components [37]. Transforming growth factor (TGF)-β is a multifunctional cytokine that mediates cellular differentiation, growth and apoptosis [38]. Park *et al.* reported that TGF-β1 decreased cell viabilities and induced hepatocyte apoptosis. However, adding the 10 ng/mL of bee venom significantly increased the viability of TGF-β1-treated hepatocyte [39]. In addition, Lee and colleagues demonstrated that an optimal dose of melittin exerts anti-apoptotic effects against TGF-β1-induced injury to hepatocytes via the mitochondrial pathway [22]. As such, these papers found that an optimal dose of bee venom and melittin can serve to protect cells against TGF-β1-mediated injury.

The nuclear transcription factor NF-κB is the key player in the development of chronic inflammatory diseases [40]. This transcription factor-involved-pathway is one of the main signaling pathways activated in response to pro-inflammatory cytokines. In addition, activation of this pathway plays a central role in inflammation through the regulation of genes encoding various growth factors [41]. Park *et al.* suggested that melittin attenuates liver injury in thioacetamide-treated mice through modulating inflammation and fibrogenesis [14]. These authors investigated the mechanism for suppression of NF-κB transcription by melittin in TNF-α-treated hepatocytes, examining the effect of melittin on NF-κB promoter activity by transiently transfected luciferase reporter plasmid containing the NF-κB promoter sequence. Melittin significantly inhibited NF-κB promoter activity and NF-κB DNA binding activity in TNF-α-treated hepatocytes. These results suggest that melittin suppresses NF-κB activation, leading to an inhibition of hepatocyte apoptosis [42].

Hepatic stellate cells (HSCs) are perisinusoidal cells residing in the space of Disse. During injury, in response to inflammatory and other stimuli, these cells adopt a myofibroblast-like phenotype and represent the cornerstone of the fibrotic response in the liver [42,43]. Once activated, HSCs up-regulate gene expression of extracellular matrix (ECM) components, matrix-degrading enzymes and their respective inhibitors, resulting in matrix remodeling and accumulation at sites with abundant activated HSCs [31,44]. Park *et al.* reported that melittin inhibited TNF-α secretion in the TNF-α-treated HSCs. Furthermore, melittin inhibited the TNF-α-induced expression of IL-1β and IL-6, especially with 0.5 mg/mL of melittin. This article also showed that melittin protected against thioacetamide-induced liver fibrosis by suppressing liver inflammation and fibrogenesis through the NF-κB signaling pathway. In addition, its anti-fibrotic effect may be attributed to modulation of the inflammatory effect in the activated HSC [14].

Acute hepatic failure is characterized by hepatic encephalopathy, severe coagulopathy, jaundice and hydroperitoneum [45,46]. Administration of a subtoxic dose of D-galactosamine together with LPS has often been used for preparing an animal model with endotoxemic shock and acute liver failure [47]. Upon stimulation with D-galactosamine and LPS, secretion of various pro-inflammatory cytokines and hepatic necrosis occur, which leads to the decreased levels of antioxidant enzymes [48,49]. This liver injury has been associated with significant increases in alanine aminotransferase (ALT) activity and TNF-α level in serum, ultimately leading to extremely high lethality [50]. Park and co-investigators found that melittin prevents D-galactosamine/LPS-induced liver failure by suppressing apoptosis and the inflammatory response in the mouse liver [51]. Melittin decreased the high rate of lethality, alleviated hepatic pathological injury, attenuated hepatic inflammatory responses and inhibited hepatocyte apoptosis. This study provides evidence that melittin may offer an alternative for the prevention of acute hepatic failure.

Some evidence suggests that adult hepatocytes play a role by way of epithelial mesenchymal transition (EMT) in the accumulation of activated fibroblasts [52,53]. EMT is a dynamic cellular program in which polarized epithelial cells lose epithelial properties, undergo morphological changes and acquire mesenchymal characteristics [54]. Hepatocytes can transdifferentiate into mesenchymal cells by EMT and deposit collagen in the liver during chronic injury [55]. A recent study has investigated the anti-fibrosis or anti-EMT mechanism by examining the effect of apamin on TGF-β1-treated hepatocytes or CCl4-injected animal model. This article demonstrated that administration of apamin significantly increased the expression of epithelial marker E-cadherin and decreased mesenchymal marker vimentin in the TGF-β1-treated hepatocytes. In particular, apamin suppressed the expression of Smad-independent and Smad-dependent signaling pathways in hepatocytes. These results demonstrate the potential of apamin for the prevention of EMT progression induced by TGF-β1 *in vitro* [26].

PLA2 from bee venom is a prototypic group III enzyme that hydrolyzes fatty acids and it has been reported that melittin in bee venom enhances the activity of PLA2 [56,57]. In addition, it has been shown that PLA2 prevents neuronal cell death and spinal cord injury [58,59]. Kim *et al.* demonstrated that PLA2 protects against hepatic dysfunction and induces anti-inflammatory cytokine production in acetaminophen-injected mice. This study suggests that PLA2 may have therapeutic potential in preventing acetaminophen-induced hepatotoxicity [60].

3. Anti-Inflammatory Effect of Bee Venom on Atherosclerosis

Atherosclerosis is a chronic inflammatory disease of the arteries resulting from interactions among lipids, monocytes and arterial wall cells [61]. The early stage of atherosclerosis involves the activation of the vascular endothelium in response to many stimuli, such as low-density lipoproteins, free radicals, infectious microorganisms, shear stress, hypertension and toxins from smoking [62]. In the progression of atherosclerosis, the proliferation and migration of vascular smooth muscle cell (VSMC) play an important role in causing stenosis or intimal thickening [63]. The migration and proliferation of VSMC is caused by pathological phenomena such as the accumulation of inflammatory cells and the release of pro-inflammatory cytokines [64,65]. In addition, abundance of macrophages is observed in atherosclerotic lesions, and early lesions of atherosclerosis are characterized by the infiltration of monocyte/macrophages and the presence of macrophage foam cells [63]. Macrophages are multi-potent inflammatory cells with the capacity for synthesis and secretion of pro-inflammatory cytokines such as TNF-α, IL-1β, IL-8 and IL-6 [61,66]. Particularly, TNF-α is reportedly involved in the development of early atherosclerosis by up-regulating vessel wall chemokine and expression of adhesion molecules such as intercellular adhesion molecule (ICAM)-1 and vascular cell adhesion molecule (VCAM)-1 in the aorta [67,68]. The up-regulation of the endothelial adhesion molecules promotes the development of atherosclerotic lesions in rabbits [69], subhuman primates [70] and humans [71,72]. Therefore, the suppression of cell adhesion molecule expression and macrophage accumulation at the level of the endothelium is of particular significance with respect to the management of the vascular inflammatory process. Some of study demonstrated that bee venom inhibits the development of atherosclerosis in C57BL/6 mice induced by injected LPS with feeding of an atherogenic diet. This is likely due to mechanisms involving anti-hypertriglyceridemic and anti-inflammatory effects of bee venom [73]. This study suggested that reduction of adhesion molecules and inflammatory factors by bee venom may be a protection against the atherosclerotic lesion formation.

The increased potential for growth of VSMC is a key abnormality in the development of atherosclerotic lesions [63]. It is well known that, in response to a platelet-derived growth factor (PDGF), VSMC can initiate highly conserved signaling events, which lead to either cell migration or proliferation [74]. Given the nature of VSMC in atherosclerosis, its apoptosis is beneficial in that it offers protection to the walls of arteries against proliferative restenosis induced by arterial injury, including arterial balloon angioplasty or stent implantation [75–80]. Son *et al.* reported that the anti-atherosclerotic effects of melittin were identified by interfering with the induction of apoptosis, inhibiting the proliferation of aortic VSMC and inhibiting downstream molecules of the PDGF receptor [8]. In addition, several studies have investigated the role of type IV collagaenase or gelatinase (MMP-2 and 9) in the regulation of VSMC behavior both *in vitro* and *in vivo* [81,82]. MMP-9 is expressed in the initial stage of the smooth muscle cell (SMC) migration, whereas the MMP-2 activity is observed at a later stage after arterial injury [83]. The synthesis and secretion of MMP-9 can be stimulated by various stimuli, including TNF-α and PDGF, during pathological processes such as atherosclerosis and inflammation [82,84,85]. Jeong *et al.* investigated the effects of melittin on TNF-α-induced migration of human aortic SMCs. The study found that melittin suppresses TNF-α-induced MMP-9 expression by inhibiting its gene transcription, but not by regulating the tissue inhibitor or metalloproteinases. Additionally, suppression of the human aortic SMC migration by melittin appeared to block the MMP-9 expression by inhibiting the NF-κB signal pathway. This study suggested that melittin is a potential agent for the prevention of vascular disorders related to the VSMC migration [65]. Recently, numerous basic research studies have indicated that TNF-α accelerates atherosclerosis in mice. Moreover, IL-1β, which plays an important role in the mediation of inflammatory responses and in the pathogenesis of atherosclerosis, is secreted by macrophages in atherosclerotic lesions [86,87]. Kim *et al.* investigated the protective effects of melittin on serum lipid profiles, pro-inflammatory cytokines, pro-atherosclerotic proteins and adhesion molecule levels in an LPS/high fat-induced mouse model of atherosclerosis and monocyte-derived macrophages. The major finding is that melittin inhibits LPS/high fat-induced expression levels of inflammatory cytokines and adhesion molecules such as TNF-α, IL-1β, ICAM and VCAM. Furthermore, the mechanisms are partly attributable to the inhibition of the NF-κB signaling pathway in LPS-treated monocyte-derived macrophages [88].

Several studies have confirmed that some calcium channel blockers can decrease the area of atherosclerotic lesions, production of oxidative stress and expression of inflammatory cytokines without conspicuously affecting blood lipid levels [89]. Kim *et al.* valuated the anti-atherosclerotic or anti-apoptotic mechanisms of apamin in THP-1-derived macrophages. Treatment of cells with oxLDL significantly promoted the accumulation of lipids and expression of apoptotic proteins. However, treatment of macrophages with apamin inhibited apoptosis through the regulation of Bcl-2 family, caspase-3 and PARP apoptotic pathway. *In vivo*, apamin attenuated apoptotic cell death in atherosclerotic mice [90]. These authors also investigated the protective effect of apamin on LPS/fat-induced atherosclerotic mice. The treated mice showed a large number of atherosclerotic lesions in the aorta. However, treatment with apamin predominantly attenuated atherosclerotic lesions, lipid, Ca^{2+} levels, pro-inflammatory cytokines, adhesion molecules, fibrotic factors and macrophage infiltrations. In regard to mechanism, it was found that treatment with apamin in THP-1-derived macrophages suppresses inflammatory responses by a decrease of the NF-κB signal pathway. Therefore, this study suggests that apamin plays an important role in monocyte/macrophage inflammatory processing and may be of potential value for preventing atherosclerosis [25].

The proliferation of VSMC is governed by the cell cycle, a common convergent point for proliferative signaling cascades [91]. The cell cycle, which consists of three distinct sequential phages (G0/G1, S and G2/M), regulates cellular proliferation [92]. Generally, the cell cycle is tightly regulated by the activity of cycle-dependent kinase (CDK) and the specific regulatory cyclin complex. Specific CDKs are sequentially activated during different phases of the cell cycle [93]. A recent study examined the cellular mechanisms by which apamin inhibits cell cycle progression of the cells exposed to PDGF. This study also investigated the inhibitory effect of apamin on PDGF-induced VSMC proliferation and migration. The results showed that PDGF-treated-VSMC was decreased in cell proliferation and migration through the regulation of cyclin D1, CDK 4, cyclin E and CDK 2. Notably, 2 µg/mL of apamin inhibited the PDGF stimulated proliferation of VSMC through blocking PDGF signaling pathway [94].

4. Anti-Inflammatory Effect of Bee Venom on Skin Disease

Acne vulgaris is the most common skin disease of the pilosebaceous follicle and results in non-inflammatory and inflammatory lesions [95]. *Propionibacterium acnes* (*P. acnes*) is a major contributing factor to the inflammatory component of acne [96]. *P. acnes* contributes to the inflammatory reaction of acne by inducing monocytes and keratinocytes to produce pro-inflammatory cytokines, including IL-1β, IL-8 and TNF-α [97]. The induction of these cytokines by *P. acnes* is mediated by Toll-like receptor (TLR) 2 [98]. Various therapeutic agents, including antibiotics for acne, have been used to inhibit inflammation or bacteria growth. However, antibiotics may lead to the emergence of resistant pathogens and side effects [99]. Thus, research recently focused on the anti-inflammatory property of bee venom. This included the effect of heat-killed *P. acnes* on human keratinocyte and monocyte cell lines. Kim *et al.* investigated the anti-inflammatory effects of bee venom in heat-killed *P. acnes*-treated HaCaT and THP-1 cells, as revealed by ELISA analysis and Western blotting by measuring the pro-inflammatory cytokines and chemokines. Heat-killed *P. acnes* markedly increased the secretion of TNF-α, IL-8 and IFN-γ in HaCaT and THP-1 cells. However, bee venom treatment decreased the secretion of those cytokines. In addition, bee venom inhibited heat-killed *P. acnes*-induced TLR2 expression in HaCaT cells. These results suggest that bee venom blocked TLR2 expression and suppressed the production of pro-inflammatory cytokines induced by *P. acnes* in HaCaT and THP-1 cells [100]. Another recent study conducted by An *et al.* reported that bee venom has a potential anti-bacterial effect against inflammatory skin disease. In this context, *P. acnes* was intradermally injected into ears of ICR mice. Following the injection, bee venom mixed with vaseline was applied to the skin surface of the ear. Histological observation revealed that the *P. acnes* injection induced a considerable increase in the number of infiltrated inflammatory cells and inflammatory cytokines. By contrast, the bee venom treated ears showed noticeably reduced ear thickness. Additionally, bee venom significantly inhibited the number of TNF-α and IL-1β positive cells [101]. Han *et al.* investigated the biological effect of bee venom treatment on keratinocyte migration *in vitro*. Migration assays showed that the distance of cell migration was dramatically increased in the experimental cells exposed to bee venom. This finding suggests that human epidermal keratinocyte migration occurred more rapidly in the bee venom treated cell, indicating that bee venom stimulates keratinocyte migration. Therefore, bee venom could be applied topically to accelerate wound healing by cell regeneration process [6].

During an inflammatory response, TLR activation results in the activation of the MAPK and the transcription factor NF-κB signaling pathways. These pathways then modulate inflammatory gene expression, which is crucial in shaping the innate immune response within the inflammatory skin disease [102]. Lee *et al.* investigated the effects of melittin in the production of inflammatory cytokines in heat-killed *P. acnes*-treated HaCaT cells. Furthermore, the molecular pathogenesis of anti-inflammatory effects of melittin was investigated in living *P. acnes*-induced inflammatory skin disease animal model. Administration of heat-killed *P. acnes* increased expression of IKK, IκB and NF-κB in HaCaT cells. However, the addition of melittin reduced IKK, IκB and NF-κB phosphorylation. These results indicate that treatment with melittin abrogated the effect of *P. acnes* in altering the expression through NF-κB signaling. The same study investigated whether melittin modulates MAPK signaling in heat-killed *P. acnes*-treated HaCaT cells. Findings showed that phosphorylated p38 was markedly increased after treatment with heat-killed *P. acnes*; however, phosphorylated p38 was decreased after treatment with melittin. These results underscore the theory that melittin inhibits pro-inflammatory cytokine expression by suppression of p38 MAPK phosphorylation in heat-killed *P. acnes*-treated HaCaT cells [103].

5. Conclusions

Due to the rising prevalence of side effects from pharmacological approach to inflammatory disease, there is a pressing need for better treatment to alleviate the symptoms of these disorders. The present review is the first to focus on how bee venom and its major components may be incorporated into therapy for inflammatory diseases. We propose that bee venom may serve as an inflammation modulator that subsequently affects the liver fibrosis, atherosclerosis and skin disease. Bee venom and its components regulate pro-inflammatory cytokines in hepatocyte and liver fibrosis animal model. In the atherosclerosis animal model, bee venom appears to inhibit the inflammatory reactions and VSMC proliferation. Furthermore, bee venom seems to accelerate wound healing and antibacterial therapy for the treatment of inflammatory skin disease through the regulation of inflammatory signaling pathway. Collectively, therapy using bee venom and its major components is considered a useful clinical approach for the treatment of inflammatory diseases. In addition, further studies including experimental elucidation of optimal dose, allergic reaction and side effects will lead to a potential therapeutic alternative for inflammatory disease. Since bee venom contains a number of other components, advances in modern sequencing techniques will provide an arsenal of new possibilities to combat other inflammation related diseases.

Acknowledgments: This work was carried out with the support of "Cooperative Research Program for Agriculture Science and Technology Development (Project No. PJ01132501)" Rural Development Administration, Korea.

Author Contributions: All three authors wrote the manuscript and approved the final manuscript.

Conflicts of Interest: The authors declare no conflict of interest.

References

1. An, H.J.; Kim, K.H.; Lee, W.R.; Kim, J.Y.; Lee, S.J.; Pak, S.C.; Han, S.M.; Park, K.K. Anti-fibrotic effect of natural toxin bee venom on animal model of unilateral ureteral obstruction. *Toxins* **2015**, *7*, 1917–1928. [CrossRef] [PubMed]

2. Orsolic, N. Bee venom in cancer therapy. *Cancer Metastasis Rev.* **2012**, *31*, 173–194. [CrossRef] [PubMed]

3. Son, D.J.; Lee, J.W.; Lee, Y.H.; Song, H.S.; Lee, C.K.; Hong, J.T. Therapeutic application of anti-arthritis, pain-releasing, and anti-cancer effects of bee venom and its constituent compounds. *Pharmacol. Ther.* **2007**, *115*, 246–270. [CrossRef] [PubMed]

4. Karimi, A.; Ahmadi, F.; Parivar, K.; Nabiuni, M.; Haghighi, S.; Imani, S.; Afrouzi, H. Effect of honey bee venom on lewis rats with experimental allergic encephalomyelitis, a model for multiple sclerosis. *Iran. J. Pharm. Res.: IJPR* **2012**, *11*, 671–678. [PubMed]

5. Moreno, M.; Giralt, E. Three valuable peptides from bee and wasp venoms for therapeutic and biotechnological use: Melittin, apamin and mastoparan. *Toxins* **2015**, *7*, 1126–1150. [CrossRef] [PubMed]

6. Han, S.M.; Park, K.K.; Nicholls, Y.M.; Macfarlane, N.; Duncan, G. Effects of honeybee (apis mellifera) venom on keratinocyte migration *in vitro*. *Pharm. Mag.* **2013**, *9*, 220–226. [CrossRef] [PubMed]

7. Castro, H.J.; Mendez-Lnocencio, J.I.; Omidvar, B.; Omidvar, J.; Santilli, J.; Nielsen, H.S., Jr.; Pavot, A.P.; Richert, J.R.; Bellanti, J.A. A phase i study of the safety of honeybee venom extract as a possible treatment for patients with progressive forms of multiple sclerosis. *Allergy Asthma Proc.: Off. J. Reg. State Allergy Soc.* **2005**, *26*, 470–476.

8. Son, D.J.; Ha, S.J.; Song, H.S.; Lim, Y.; Yun, Y.P.; Lee, J.W.; Moon, D.C.; Park, Y.H.; Park, B.S.; Song, M.J.; *et al.* Melittin inhibits vascular smooth muscle cell proliferation through induction of apoptosis via suppression of nuclear factor-kappab and akt activation and enhancement of apoptotic protein expression. *J. Pharmacol. Exp. Ther.* **2006**, *317*, 627–634. [CrossRef] [PubMed]

9. Park, H.J.; Lee, S.H.; Son, D.J.; Oh, K.W.; Kim, K.H.; Song, H.S.; Kim, G.J.; Oh, G.T.; Yoon, D.Y.; Hong, J.T. Antiarthritic effect of bee venom: Inhibition of inflammation mediator generation by suppression of nf-kappab through interaction with the p50 subunit. *Arthritis Rheum.* **2004**, *50*, 3504–3515. [CrossRef] [PubMed]

10. Kwon, Y.B.; Ham, T.W.; Kim, H.W.; Roh, D.H.; Yoon, S.Y.; Han, H.J.; Yang, I.S.; Kim, K.W.; Beitz, A.J.; Lee, J.H. Water soluble fraction (<10 kDa) from bee venom reduces visceral pain behavior through spinal alpha 2-adrenergic activity in mice. *Pharmacol. Biochem. Behav.* **2005**, *80*, 181–187. [PubMed]

11. Kim, H.W.; Kwon, Y.B.; Ham, T.W.; Roh, D.H.; Yoon, S.Y.; Lee, H.J.; Han, H.J.; Yang, I.S.; Beitz, A.J.; Lee, J.H. Acupoint stimulation using bee venom attenuates formalin-induced pain behavior and spinal cord fos expression in rats. *J. Vet. Med. Sci./Jpn. Soc. Vete. Sci.* **2003**, *65*, 349–355. [CrossRef]

12. Putz, T.; Ramoner, R.; Gander, H.; Rahm, A.; Bartsch, G.; Thurnher, M. Antitumor action and immune activation through cooperation of bee venom secretory phospholipase a2 and phosphatidylinositol-(3,4)-bisphosphate. *Cancer Immunol. Immunother.: CII* **2006**, *55*, 1374–1383. [CrossRef] [PubMed]

13. Russell, P.J.; Hewish, D.; Carter, T.; Sterling-Levis, K.; Ow, K.; Hattarki, M.; Doughty, L.; Guthrie, R.; Shapira, D.; Molloy, P.L.; *et al.* Cytotoxic properties of immunoconjugates containing melittin-like peptide 101 against prostate cancer: In vitro and *in vivo* studies. *Cancer Immunol. Immunother.: CII* **2004**, *53*, 411–421. [CrossRef] [PubMed]

14. Park, J.H.; Kum, Y.S.; Lee, T.I.; Kim, S.J.; Lee, W.R.; Kim, B.I.; Kim, H.S.; Kim, K.H.; Park, K.K. Melittin attenuates liver injury in thioacetamide-treated mice through modulating inflammation and fibrogenesis. *Exp. Biol. Med. (Maywood)* **2011**, *236*, 1306–1313. [CrossRef] [PubMed]

15. Terra, R.M.; Guimaraes, J.A.; Verli, H. Structural and functional behavior of biologically active monomeric melittin. *J. Mol. Gr. Model.* **2007**, *25*, 767–772. [CrossRef] [PubMed]

16. Raghuraman, H.; Chattopadhyay, A. Melittin: A membrane-active peptide with diverse functions. *Biosci. Rep.* **2007**, *27*, 189–223. [CrossRef] [PubMed]

17. Jeong, Y.J.; Choi, Y.; Shin, J.M.; Cho, H.J.; Kang, J.H.; Park, K.K.; Choe, J.Y.; Bae, Y.S.; Han, S.M.; Kim, C.H.; *et al.* Melittin suppresses egf-induced cell motility and invasion by inhibiting pi3k/akt/mtor signaling pathway in breast cancer cells. *Food Chem. Toxicol.: Int. J. Publ. Br. Ind. Biol. Res. Assoc.* **2014**, *68*, 218–225. [CrossRef] [PubMed]

18. Park, J.H.; Jeong, Y.J.; Park, K.K.; Cho, H.J.; Chung, I.K.; Min, K.S.; Kim, M.; Lee, K.G.; Yeo, J.H.; Chang, Y.C. Melittin suppresses pma-induced tumor cell invasion by inhibiting nf-kappab and ap-1-dependent mmp-9 expression. *Mol. Cells* **2010**, *29*, 209–215. [CrossRef] [PubMed]

19. Shin, J.M.; Jeong, Y.J.; Cho, H.J.; Park, K.K.; Chung, I.K.; Lee, I.K.; Kwak, J.Y.; Chang, H.W.; Kim, C.H.; Moon, S.K.; *et al.* Melittin suppresses hif-1alpha/vegf expression through inhibition of erk and mtor/p70s6k pathway in human cervical carcinoma cells. *PLoS ONE* **2013**, *8*, e69380. [CrossRef] [PubMed]

20. Damianoglou, A.; Rodger, A.; Pridmore, C.; Dafforn, T.R.; Mosely, J.A.; Sanderson, J.M.; Hicks, M.R. The synergistic action of melittin and phospholipase a2 with lipid membranes: Development of linear dichroism for membrane-insertion kinetics. *Protein Pept. Lett.* **2010**, *17*, 1351–1362. [CrossRef] [PubMed]

21. Pratt, J.P.; Ravnic, D.J.; Huss, H.T.; Jiang, X.; Orozco, B.S.; Mentzer, S.J. Melittin-induced membrane permeability: A nonosmotic mechanism of cell death. *In Vitro Cell. Dev. Biol. Anim.* **2005**, *41*, 349–355. [CrossRef] [PubMed]

22. Lee, W.R.; Park, J.H.; Kim, K.H.; Park, Y.Y.; Han, S.M.; Park, K.K. Protective effects of melittin on transforming growth factor-beta1 injury to hepatocytes via anti-apoptotic mechanism. *Toxicol. Appl. Pharmacol.* **2011**, *256*, 209–215. [CrossRef] [PubMed]

23. Thompson, J.M.; Ji, G.; Neugebauer, V. Small-conductance calcium-activated potassium (sk) channels in the amygdala mediate pain-inhibiting effects of clinically available riluzole in a rat model of arthritis pain. *Mol. Pain* **2015**, *11*, 51. [CrossRef] [PubMed]

24. Bond, C.T.; Herson, P.S.; Strassmaier, T.; Hammond, R.; Stackman, R.; Maylie, J.; Adelman, J.P. Small conductance Ca^{2+}-activated k^{+} channel knock-out mice reveal the identity of calcium-dependent afterhyperpolarization currents. *J. Neurosci.: Off. J. Soc. Neurosci.* **2004**, *24*, 5301–5306. [CrossRef] [PubMed]

25. Kim, S.J.; Park, J.H.; Kim, K.H.; Lee, W.R.; Pak, S.C.; Han, S.M.; Park, K.K. The protective effect of apamin on lps/fat-induced atherosclerotic mice. *Evid.-Based Complement. Altern. Med.: eCAM* **2012**, *2012*, 305454. [CrossRef] [PubMed]

26. Lee, W.R.; Kim, K.H.; An, H.J.; Kim, J.Y.; Lee, S.J.; Han, S.M.; Pak, S.C.; Park, K.K. Apamin inhibits hepatic fibrosis through suppression of transforming growth factor beta1-induced hepatocyte epithelial-mesenchymal transition. *Biochem. Biophys. Res. Commun.* **2014**, *450*, 195–201. [CrossRef] [PubMed]

27. Chen, M.H.; Chen, J.C.; Tsai, C.C.; Wang, W.C.; Chang, D.C.; Tu, D.G.; Hsieh, H.Y. The role of tgf-beta 1 and cytokines in the modulation of liver fibrosis by sho-saiko-to in rat's bile duct ligated model. *J. Ethnopharmacol.* **2005**, *97*, 7–13. [CrossRef] [PubMed]

28. Wallace, K.; Burt, A.D.; Wright, M.C. Liver fibrosis. *Biochem. J.* **2008**, *411*, 1–18. [CrossRef] [PubMed]

29. Kim, K.K.; Wei, Y.; Szekeres, C.; Kugler, M.C.; Wolters, P.J.; Hill, M.L.; Frank, J.A.; Brumwell, A.N.; Wheeler, S.E.; Kreidberg, J.A.; *et al.* Epithelial cell alpha3beta1 integrin links beta-catenin and smad signaling to promote myofibroblast formation and pulmonary fibrosis. *J. Clin. Investig.* **2009**, *119*, 213–224. [PubMed]

30. Choi, S.S.; Diehl, A.M. Epithelial-to-mesenchymal transitions in the liver. *Hepatology* **2009**, *50*, 2007–2013. [CrossRef] [PubMed]

31. Bataller, R.; Brenner, D.A. Liver fibrosis. *J. Clin. Investig.* **2005**, *115*, 209–218. [CrossRef] [PubMed]

32. Friedman, S.L. Mechanisms of disease: Mechanisms of hepatic fibrosis and therapeutic implications. *Nat. Clin. Pract. Gastroenterol. Hepatol.* **2004**, *1*, 98–105. [CrossRef] [PubMed]

33. Gao, B.; Bataller, R. Alcoholic liver disease: Pathogenesis and new therapeutic targets. *Gastroenterology* **2011**, *141*, 1572–1585. [CrossRef] [PubMed]

34. Kim, K.H.; Kum, Y.S.; Park, Y.Y.; Park, J.H.; Kim, S.J.; Lee, W.R.; Lee, K.G.; Han, S.M.; Park, K.K. The protective effect of bee venom against ethanol-induced hepatic injury via regulation of the mitochondria-related apoptotic pathway. *Basic Clin. Pharmacol. Toxicol.* **2010**, *107*, 619–624. [CrossRef] [PubMed]

35. Li, J.; Yang, S.; Billiar, T.R. Cyclic nucleotides suppress tumor necrosis factor alpha-mediated apoptosis by inhibiting caspase activation and cytochrome c release in primary hepatocytes via a mechanism independent of akt activation. *J. Biol. Chem.* **2000**, *275*, 13026–13034. [CrossRef] [PubMed]

36. Park, J.H.; Kim, K.H.; Kim, S.J.; Lee, W.R.; Lee, K.G.; Park, K.K. Bee venom protects hepatocytes from tumor necrosis factor-alpha and actinomycin d. *Arch. Pharm. Res.* **2010**, *33*, 215–223. [CrossRef] [PubMed]

37. Kim, S.J.; Park, J.H.; Kim, K.H.; Lee, W.R.; Chang, Y.C.; Park, K.K.; Lee, K.G.; Han, S.M.; Yeo, J.H.; Pak, S.C. Bee venom inhibits hepatic fibrosis through suppression of pro-fibrogenic cytokine expression. *Am. J. Chin. Med.* **2010**, *38*, 921–935. [CrossRef] [PubMed]

38. Imamura, T.; Oshima, Y.; Hikita, A. Regulation of tgf-beta family signalling by ubiquitination and deubiquitination. *J. Biochem.* **2013**, *154*, 481–489. [CrossRef] [PubMed]

39. Park, J.H.; Kim, K.H.; Kim, S.J.; Lee, W.R.; Lee, K.G.; Park, K.K. Effect of bee venom on transforming growth factor-beta1-treated hepatocytes. *Int. J. Toxicol.* **2010**, *29*, 49–56. [CrossRef] [PubMed]

40. Tak, P.P.; Firestein, G.S. Nf-kappab: A key role in inflammatory diseases. *J. Clin. Investig.* **2001**, *107*, 7–11. [CrossRef] [PubMed]

41. De Martin, R.; Hoeth, M.; Hofer-Warbinek, R.; Schmid, J.A. The transcription factor nf-kappa b and the regulation of vascular cell function. *Arterioscler. Thromb. Vasc. Biol.* **2000**, *20*, E83–E88. [CrossRef] [PubMed]

42. Park, J.H.; Lee, W.R.; Kim, H.S.; Han, S.M.; Chang, Y.C.; Park, K.K. Protective effects of melittin on tumor necrosis factor-alpha induced hepatic damage through suppression of apoptotic pathway and nuclear factor-kappa b activation. *Exp. Biol. Med. (Maywood)* **2014**, *239*, 1705–1714. [CrossRef] [PubMed]

43. Sarem, M.; Znaidak, R.; Macias, M.; Rey, R. Hepatic stellate cells: It's role in normal and pathological conditions. *Gastroenterol. Hepatol.* **2006**, *29*, 93–101. [CrossRef] [PubMed]

44. Zhang, Y.; Wang, Y.; Di, L.; Tang, N.; Ai, X.; Yao, X. Mechanism of interleukin-1beta-induced proliferation in rat hepatic stellate cells from different levels of signal transduction. *APMIS: Acta Pathol. Microbiol. Immunol. Scand.* **2014**, *122*, 392–398.

45. Sun, H.; Chen, L.; Zhou, W.; Hu, L.; Li, L.; Tu, Q.; Chang, Y.; Liu, Q.; Sun, X.; Wu, M.; *et al.* The protective role of hydrogen-rich saline in experimental liver injury in mice. *J. Hepatol.* **2011**, *54*, 471–480. [CrossRef] [PubMed]

46. Choi, E.Y.; Hwang, H.J.; Kim, I.H.; Nam, T.J. Protective effects of a polysaccharide from hizikia fusiformis against ethanol toxicity in rats. *Food Chem. Toxicol.: Int. J. Publ. Br. Ind. Biol. Res. Assoc.* **2009**, *47*, 134–139. [CrossRef] [PubMed]

47. Silverstein, R. D-galactosamine lethality model: Scope and limitations. *J. Endotoxin Res.* **2004**, *10*, 147–162. [CrossRef] [PubMed]

48. Gong, X.; Luo, F.L.; Zhang, L.; Li, H.Z.; Wu, M.J.; Li, X.H.; Wang, B.; Hu, N.; Wang, C.D.; Yang, J.Q.; *et al.* Tetrandrine attenuates lipopolysaccharide-induced fulminant hepatic failure in d-galactosamine-sensitized mice. *Int. Immunopharmacol.* **2010**, *10*, 357–363. [CrossRef] [PubMed]

49. Sass, G.; Heinlein, S.; Agli, A.; Bang, R.; Schumann, J.; Tiegs, G. Cytokine expression in three mouse models of experimental hepatitis. *Cytokine* **2002**, *19*, 115–120. [CrossRef] [PubMed]

50. Yamada, I.; Goto, T.; Takeuchi, S.; Ohshima, S.; Yoneyama, K.; Shibuya, T.; Kataoka, E.; Segawa, D.; Sato, W.; Dohmen, T.; *et al.* Mao (ephedra sinica stapf) protects against d-galactosamine and lipopolysaccharide-induced hepatic failure. *Cytokine* **2008**, *41*, 293–301. [CrossRef] [PubMed]

51. Park, J.H.; Kim, K.H.; Lee, W.R.; Han, S.M.; Park, K.K. Protective effect of melittin on inflammation and apoptosis in acute liver failure. *Apop.: Int. J. Program. Cell Death* **2012**, *17*, 61–69. [CrossRef] [PubMed]

52. Zeisberg, M.; Yang, C.; Martino, M.; Duncan, M.B.; Rieder, F.; Tanjore, H.; Kalluri, R. Fibroblasts derive from hepatocytes in liver fibrosis via epithelial to mesenchymal transition. *J. Biol. Chem.* **2007**, *282*, 23337–23347. [CrossRef] [PubMed]

53. Dooley, S.; Hamzavi, J.; Ciuclan, L.; Godoy, P.; Ilkavets, I.; Ehnert, S.; Ueberham, E.; Gebhardt, R.; Kanzler, S.; Geier, A.; *et al.* Hepatocyte-specific smad7 expression attenuates tgf-beta-mediated fibrogenesis and protects against liver damage. *Gastroenterology* **2008**, *135*, 642–659. [CrossRef] [PubMed]

54. Thiery, J.P.; Sleeman, J.P. Complex networks orchestrate epithelial-mesenchymal transitions. *Nat. Rev. Mol. Cell Biol.* **2006**, *7*, 131–142. [CrossRef] [PubMed]

55. Copple, B.L. Hypoxia stimulates hepatocyte epithelial to mesenchymal transition by hypoxia-inducible factor and transforming growth factor-beta-dependent mechanisms. *Liver Int.: Off. J. Int. Assoc. Study Liv.* **2010**, *30*, 669–682. [CrossRef] [PubMed]

56. Zhao, H.; Kinnunen, P.K. Modulation of the activity of secretory phospholipase a2 by antimicrobial peptides. *Antimicrob. Agents Chemother.* **2003**, *47*, 965–971. [CrossRef] [PubMed]

57. Monti, M.C.; Casapullo, A.; Santomauro, C.; D'Auria, M.V.; Riccio, R.; Gomez-Paloma, L. The molecular mechanism of bee venom phospholipase a2 inactivation by bolinaquinone. *Chembiochem* **2006**, *7*, 971–980. [CrossRef] [PubMed]

58. Lopez, F.O.; Lopez O'Rourke, V.J.; Fernandez Mariscal, E.; Vilarrasa Sauquet, R.; Sanudo Martin, I. [c3 spinal cord ependymoma c03]. *Med. Clin.* **2011**, *136*, 605.

59. Jeong, J.K.; Moon, M.H.; Bae, B.C.; Lee, Y.J.; Seol, J.W.; Park, S.Y. Bee venom phospholipase a2 prevents prion peptide induced-cell death in neuronal cells. *Int. J. Mol. Med.* **2011**, *28*, 867–873. [PubMed]

60. Kim, H.; Keum, D.J.; Kwak, J.; Chung, H.S.; Bae, H. Bee venom phospholipase a2 protects against acetaminophen-induced acute liver injury by modulating regulatory t cells and il-10 in mice. *PLoS ONE* **2014**, *9*, e114726. [CrossRef] [PubMed]

61. Ross, R. Atherosclerosis–an inflammatory disease. *N. Engl. J. Med.* **1999**, *340*, 115–126. [CrossRef]

62. Stoll, G.; Bendszus, M. Inflammation and atherosclerosis: Novel insights into plaque formation and destabilization. *Stroke A J. Cereb. Circ.* **2006**, *37*, 1923–1932. [CrossRef] [PubMed]

63. Ross, R. The pathogenesis of atherosclerosis: A perspective for the 1990s. *Nature* **1993**, *362*, 801–809. [CrossRef] [PubMed]

64. Gerthoffer, W.T. Mechanisms of vascular smooth muscle cell migration. *Circ. Res.* **2007**, *100*, 607–621. [CrossRef] [PubMed]

65. Jeong, Y.J.; Cho, H.J.; Whang, K.; Lee, I.S.; Park, K.K.; Choe, J.Y.; Han, S.M.; Kim, C.H.; Chang, H.W.; Moon, S.K.; *et al.* Melittin has an inhibitory effect on tnf-alpha-induced migration of human aortic smooth muscle cells by blocking the mmp-9 expression. *Food Chem. Toxicol.: An Int. J. Publ. Br. Ind. Biol. Res. Assoc.* **2012**, *50*, 3996–4002. [CrossRef] [PubMed]

66. Cipollone, F.; Iezzi, A.; Fazia, M.; Zucchelli, M.; Pini, B.; Cuccurullo, C.; de Cesare, D.; de Blasis, G.; Muraro, R.; Bei, R.; *et al.* The receptor rage as a progression factor amplifying arachidonate-dependent inflammatory and proteolytic response in human atherosclerotic plaques: Role of glycemic control. *Circulation* **2003**, *108*, 1070–1077. [CrossRef] [PubMed]

67. Zhang, L.; Peppel, K.; Sivashanmugam, P.; Orman, E.S.; Brian, L.; Exum, S.T.; Freedman, N.J. Expression of tumor necrosis factor receptor-1 in arterial wall cells promotes atherosclerosis. *Arterioscler. Thromb. Vasc. Biol.* **2007**, *27*, 1087–1094. [CrossRef] [PubMed]

68. Ohta, H.; Wada, H.; Niwa, T.; Kirii, H.; Iwamoto, N.; Fujii, H.; Saito, K.; Sekikawa, K.; Seishima, M. Disruption of tumor necrosis factor-alpha gene diminishes the development of atherosclerosis in apoe-deficient mice. *Atherosclerosis* **2005**, *180*, 11–17. [CrossRef] [PubMed]

69. Li, H.; Cybulsky, M.I.; Gimbrone, M.A., Jr.; Libby, P. An atherogenic diet rapidly induces vcam-1, a cytokine-regulatable mononuclear leukocyte adhesion molecule, in rabbit aortic endothelium. *Arterioscler. Thromb.: A J. Vasc. Biol./Am. Heart Assoc.* **1993**, *13*, 197–204. [CrossRef]

70. Shi, Q.; Vandeberg, J.F.; Jett, C.; Rice, K.; Leland, M.M.; Talley, L.; Kushwaha, R.S.; Rainwater, D.L.; Vandeberg, J.L.; Wang, X.L. Arterial endothelial dysfunction in baboons fed a high-cholesterol, high-fat diet. *Am. J. Clin. Nutr.* **2005**, *82*, 751–759. [PubMed]

71. Poston, R.N.; Haskard, D.O.; Coucher, J.R.; Gall, N.P.; Johnson-Tidey, R.R. Expression of intercellular adhesion molecule-1 in atherosclerotic plaques. *Am. J. Pathol.* **1992**, *140*, 665–673. [PubMed]

72. Hwang, S.J.; Ballantyne, C.M.; Sharrett, A.R.; Smith, L.C.; Davis, C.E.; Gotto, A.M., Jr.; Boerwinkle, E. Circulating adhesion molecules vcam-1, icam-1, and e-selectin in carotid atherosclerosis and incident coronary heart disease cases: The atherosclerosis risk in communities (aric) study. *Circulation* **1997**, *96*, 4219–4225. [CrossRef] [PubMed]

73. Lee, W.R.; Kim, S.J.; Park, J.H.; Kim, K.H.; Chang, Y.C.; Park, Y.Y.; Lee, K.G.; Han, S.M.; Yeo, J.H.; Pak, S.C.; *et al.* Bee venom reduces atherosclerotic lesion formation via anti-inflammatory mechanism. *Am. J. Chin. Med.* **2010**, *38*, 1077–1092. [CrossRef] [PubMed]

74. Jung, F.; Haendeler, J.; Goebel, C.; Zeiher, A.M.; Dimmeler, S. Growth factor-induced phosphoinositide 3-oh kinase/akt phosphorylation in smooth muscle cells: Induction of cell proliferation and inhibition of cell death. *Cardiovasc. Res.* **2000**, *48*, 148–157. [CrossRef]

75. Yang, H.M.; Kim, H.S.; Park, K.W.; You, H.J.; Jeon, S.I.; Youn, S.W.; Kim, S.H.; Oh, B.H.; Lee, M.M.; Park, Y.B.; *et al.* Celecoxib, a cyclooxygenase-2 inhibitor, reduces neointimal hyperplasia through inhibition of akt signaling. *Circulation* **2004**, *110*, 301–308. [CrossRef] [PubMed]

76. Lesauskaite, V.; Ivanoviene, L.; Valanciute, A. Programmed cellular death and atherogenesis: From molecular mechanisms to clinical aspects. *Medicina (Kaunas)* **2003**, *39*, 529–534. [PubMed]

77. Hofmann, C.S.; Sonenshein, G.E. Green tea polyphenol epigallocatechin-3 gallate induces apoptosis of proliferating vascular smooth muscle cells via activation of p53. *FASEB J.: Off. Publ. Fed. Am. Soc. Exp. Biol.* **2003**, *17*, 702–704. [CrossRef] [PubMed]

78. Curcio, A.; Torella, D.; Cuda, G.; Coppola, C.; Faniello, M.C.; Achille, F.; Russo, V.G.; Chiariello, M.; Indolfi, C. Effect of stent coating alone on *in vitro* vascular smooth muscle cell proliferation and apoptosis. *Am. J. Physiol. Heart Circ. Physiol.* **2004**, *286*, H902–H908. [CrossRef] [PubMed]

79. Chen, J.H.; Wu, C.C.; Hsiao, G.; Yen, M.H. Magnolol induces apoptosis in vascular smooth muscle. *Naunyn-Schmied. Arch. Pharmacol.* **2003**, *368*, 127–133. [CrossRef] [PubMed]

80. Perlman, H.; Sata, M.; Krasinski, K.; Dorai, T.; Buttyan, R.; Walsh, K. Adenovirus-encoded hammerhead ribozyme to bcl-2 inhibits neointimal hyperplasia and induces vascular smooth muscle cell apoptosis. *Cardiovasc. Res.* **2000**, *45*, 570–578. [CrossRef]

81. Newby, A.C.; Zaltsman, A.B. Molecular mechanisms in intimal hyperplasia. *J. Pathol.* **2000**, *190*, 300–309. [CrossRef]

82. Cho, A.; Reidy, M.A. Matrix metalloproteinase-9 is necessary for the regulation of smooth muscle cell replication and migration after arterial injury. *Circ. Res.* **2002**, *91*, 845–851. [CrossRef] [PubMed]

83. Bendeck, M.P.; Zempo, N.; Clowes, A.W.; Galardy, R.E.; Reidy, M.A. Smooth muscle cell migration and matrix metalloproteinase expression after arterial injury in the rat. *Circ. Res.* **1994**, *75*, 539–545. [CrossRef] [PubMed]

84. Chen, Q.; Jin, M.; Yang, F.; Zhu, J.; Xiao, Q.; Zhang, L. Matrix metalloproteinases: Inflammatory regulators of cell behaviors in vascular formation and remodeling. *Med. Inflamm.* **2013**, *2013*, 928315. [CrossRef] [PubMed]

85. Moon, S.K.; Cha, B.Y.; Kim, C.H. Erk1/2 mediates tnf-alpha-induced matrix metalloproteinase-9 expression in human vascular smooth muscle cells via the regulation of nf-kappab and ap-1: Involvement of the ras dependent pathway. *J. Cell. Physiol.* **2004**, *198*, 417–427. [CrossRef] [PubMed]

86. Branen, Ł.; Hovgaard, L.; Nitulescu, M.; Bengtsson, E.; Nilsson, J.; Jovinge, S. Inhibition of tumor necrosis factor-alpha reduces atherosclerosis in apolipoprotein e knockout mice. *Arterioscler. Thromb. Vasc. Biol.* **2004**, *24*, 2137–2142. [CrossRef] [PubMed]

87. Canault, M.; Peiretti, F.; Mueller, C.; Kopp, F.; Morange, P.; Rihs, S.; Portugal, H.; Juhan-Vague, I.; Nalbone, G. Exclusive expression of transmembrane tnf-alpha in mice reduces the inflammatory response in early lipid lesions of aortic sinus. *Atherosclerosis* **2004**, *172*, 211–218. [CrossRef] [PubMed]

88. Kim, S.J.; Park, J.H.; Kim, K.H.; Lee, W.R.; Kim, K.S.; Park, K.K. Melittin inhibits atherosclerosis in lps/high-fat treated mice through atheroprotective actions. *J. Atheroscler. Thromb.* **2011**, *18*, 1117–1126. [CrossRef] [PubMed]

89. Mancini, G.B. Antiatherosclerotic effects of calcium channel blockers. *Prog. Cardiovasc. Dis.* **2002**, *45*, 1–20. [CrossRef] [PubMed]

90. Kim, S.J.; Park, J.H.; Kim, K.H.; Lee, W.R.; An, H.J.; Min, B.K.; Han, S.M.; Kim, K.S.; Park, K.K. Apamin inhibits thp-1-derived macrophage apoptosis via mitochondria-related apoptotic pathway. *Exp. Mol. Pathol.* **2012**, *93*, 129–134. [CrossRef] [PubMed]

91. Braun, K.; Ehemann, V.; Waldeck, W.; Pipkorn, R.; Corban-Wilhelm, H.; Jenne, J.; Gissmann, L.; Debus, J. Hpv18 e6 and e7 genes affect cell cycle, prb and p53 of cervical tumor cells and represent prominent candidates for intervention by use peptide nucleic acids (pnas). *Cancer Lett.* **2004**, *209*, 37–49. [CrossRef] [PubMed]

92. Elledge, S.J. Cell cycle checkpoints: Preventing an identity crisis. *Science* **1996**, *274*, 1664–1672. [CrossRef] [PubMed]

93. Fuster, J.J.; Fernandez, P.; Gonzalez-Navarro, H.; Silvestre, C.; Nabah, Y.N.; Andres, V. Control of cell proliferation in atherosclerosis: Insights from animal models and human studies. *Cardiovasc. Res.* **2010**, *86*, 254–264. [CrossRef] [PubMed]

94. Kim, J.Y.; Kim, K.H.; Lee, W.R.; An, H.J.; Lee, S.J.; Han, S.M.; Lee, K.G.; Park, Y.Y.; Kim, K.S.; Lee, Y.S.; *et al.* Apamin inhibits pdgf-bb-induced vascular smooth muscle cell proliferation and migration through suppressions of activated akt and erk signaling pathway. *Vasc. Pharmacol.* **2015**, *70*, 8–14. [CrossRef] [PubMed]

95. Leyden, J.J. The evolving role of propionibacterium acnes in acne. *Semin. Cutan. Med. Surg.* **2001**, *20*, 139–143. [CrossRef] [PubMed]

96. Jung, M.K.; Ha, S.; Son, J.A.; Song, J.H.; Houh, Y.; Cho, E.; Chun, J.H.; Yoon, S.R.; Yang, Y.; Bang, S.I.; *et al.* Polyphenon-60 displays a therapeutic effect on acne by suppression of tlr2 and il-8 expression via down-regulating the erk1/2 pathway. *Arch. Dermatol. Res.* **2012**, *304*, 655–663. [CrossRef] [PubMed]

97. Vowels, B.R.; Yang, S.; Leyden, J.J. Induction of proinflammatory cytokines by a soluble factor of propionibacterium acnes: Implications for chronic inflammatory acne. *Infect. Immun.* **1995**, *63*, 3158–3165. [PubMed]

98. Kim, J. Review of the innate immune response in acne vulgaris: Activation of toll-like receptor 2 in acne triggers inflammatory cytokine responses. *Dermatology* **2005**, *211*, 193–198. [CrossRef] [PubMed]

99. Aslam, I.; Fleischer, A.; Feldman, S. Emerging drugs for the treatment of acne. *Expert Opin. Emerg. Drugs* **2015**, *20*, 91–101. [CrossRef] [PubMed]

100. Kim, J.Y.; Lee, W.R.; Kim, K.H.; An, H.J.; Chang, Y.C.; Han, S.M.; Park, Y.Y.; Pak, S.C.; Park, K.K. Effects of bee venom against propionibacterium acnes-induced inflammation in human keratinocytes and monocytes. *Int. J. Mol. Med.* **2015**, *35*, 1651–1656. [CrossRef] [PubMed]

101. An, H.J.; Lee, W.R.; Kim, K.H.; Kim, J.Y.; Lee, S.J.; Han, S.M.; Lee, K.G.; Lee, C.K.; Park, K.K. Inhibitory effects of bee venom on propionibacterium acnes-induced inflammatory skin disease in an animal model. *Int. J. Mol. Med.* **2014**, *34*, 1341–1348. [CrossRef] [PubMed]

102. Grange, P.A.; Raingeaud, J.; Calvez, V.; Dupin, N. Nicotinamide inhibits propionibacterium acnes-induced il-8 production in keratinocytes through the nf-kappab and mapk pathways. *J. Dermatol. Sci.* **2009**, *56*, 106–112. [CrossRef] [PubMed]

103. Lee, W.R.; Kim, K.H.; An, H.J.; Kim, J.Y.; Chang, Y.C.; Chung, H.; Park, Y.Y.; Lee, M.L.; Park, K.K. The protective effects of melittin on propionibacterium acnes-induced inflammatory responses *in vitro* and *in vivo*. *J. Investig. Dermatol.* **2014**, *134*, 1922–1930. [CrossRef] [PubMed]

© 2015 by the authors; licensee MDPI, Basel, Switzerland. This article is an open access article distributed under the terms and conditions of the Creative Commons Attribution (CC-BY) license (http://creativecommons.org/licenses/by/4.0/).

toxins

MDPI

Review

Pharmacological Alternatives for the Treatment of Neurodegenerative Disorders: Wasp and Bee Venoms and Their Components as New Neuroactive Tools

Juliana Silva, Victoria Monge-Fuentes, Flávia Gomes, Kamila Lopes, Lilian dos Anjos, Gabriel Campos, Claudia Arenas, Andréia Biolchi, Jacqueline Gonçalves, Priscilla Galante, Leandro Campos and Márcia Mortari *

Neuropharmacology Laboratory, Department of Physiological Sciences, Institute of Biological Sciences, University of Brasília, Brasília 70910-900, Brazil; ju.castroesilva@gmail.com (J.S.); victorananobio@gmail.com (V.M.-F.); flaviia.medeiros@hotmail.com (F.G.); kamila_farm@yahoo.com.br (K.L.); lilian.dosanjos@gmail.com (L.A.); gabriel_avohay@hotmail.com (G.C.); clauji55@gmail.com (C.A.); andreia.biolchi@gmail.com (A.B.); jacq.coimbra@gmail.com (J.G.); prigalante@yahoo.com.br (P.G.); leandro.ambrosio@gmail.com (L.C.)

* Correspondence: mmortari@unb.br; Tel.: +55-61-3107-3123; Fax: +55-61-3107-2904

Academic Editor: Sokcheon Pak

Received: 15 May 2015; Accepted: 5 August 2015; Published: 18 August 2015

Abstract: Neurodegenerative diseases are relentlessly progressive, severely impacting affected patients, families and society as a whole. Increased life expectancy has made these diseases more common worldwide. Unfortunately, available drugs have insufficient therapeutic effects on many subtypes of these intractable diseases, and adverse effects hamper continued treatment. Wasp and bee venoms and their components are potential means of managing or reducing these effects and provide new alternatives for the control of neurodegenerative diseases. These venoms and their components are well-known and irrefutable sources of neuroprotectors or neuromodulators. In this respect, the present study reviews our current understanding of the mechanisms of action and future prospects regarding the use of new drugs derived from wasp and bee venom in the treatment of major neurodegenerative disorders, including Alzheimer's Disease, Parkinson's Disease, Epilepsy, Multiple Sclerosis and Amyotrophic Lateral Sclerosis.

Keywords: neurological disease; bee venom; wasp venom; polyamine toxins; Melittin; Apamin; AvTx-7; Wasp Kinin; Mastoparan; Pompilidotoxins

1. Introduction

Insect venoms have been used by traditional Chinese and Korean medicine as well as ancient Egyptian and Greek civilizations since 1000–3000 BC to control a number of diseases, including neurological disorders [1–3]. Moreover, religious texts such as the Vedas, the Bible and the Koran report the use of bee products to treat diseases [3,4].

The diversity of biologically active molecules from animal venoms is well-known and has long garnered the interest of toxinologists. However, progress is more evident in recent years due to advances in the fields of proteomics, transcriptomics and genomics [5]. The area of venom-based drugs in particular has benefited from these advances along with high throughput screening techniques, which have accelerated the discovery of useful venom-derived drugs.

Bee and wasp venoms are known to be rich in neuroactive molecules that may be valuable in the development of new drugs or act as pharmacological tools to study the normal and pathological functioning of the nervous system [6,7]. As such, this review focuses on the main results obtained for the use of wasp and bee venoms in the treatment of the most prevalent neurodegenerative disorders.

It is important to note that several of these compounds could become important new sources for the development of more effective medication with fewer adverse effects. The bioprospection of these compounds is vital since the drugs currently used to treat major neurological disorders (*i.e.*, Epilepsy, Parkinson's Disease (PD) and Alzheimer's Disease (AD)) provide only symptomatic relief, and the incidence of serious adverse effects remains high [8–11].

The nervous system is an important target for these toxins, which can modulate synapses as well as generate and propagate action potentials by selectively acting on different ion channels and receptors [12]. Interestingly, evolution has fine-tuned venoms for optimal activity, providing us with a vast array of potential therapeutic drugs, which can be used to design pharmacological agents for the treatment of several diseases, including central nervous system (CNS) disorders [12,13] (Figure 1).

2. General Profile of the Main Neurodegenerative Diseases

According to the World Health Organization (WHO), neurological disorders include Epilepsy, Alzheimer's Disease (AD) and other dementias, Parkinson's Disease (PD), Multiple Sclerosis (MS), Migraine, Cerebrovascular Disease, Poliomyelitis, Tetanus, Meningitis and Japanese Encephalitis, among others. These diseases are major causes of mortality, accounting for 12% of total deaths worldwide [11]. They are frequently stigmatized, since they are socially incapacitating and can cause cognitive impairment, behavioral disorders, depression and suicide [14,15].

The effectiveness of wasp and bee venom against neurodegenerative diseases has only been investigated for a select group of disorders. Thus, we have performed a brief epidemiological, symptomatic and histopathological summary of the following target diseases: Alzheimer's Disease, Parkinson's Disease, Epilepsy, Multiple Sclerosis and Amyotrophic Lateral Sclerosis (ALS).

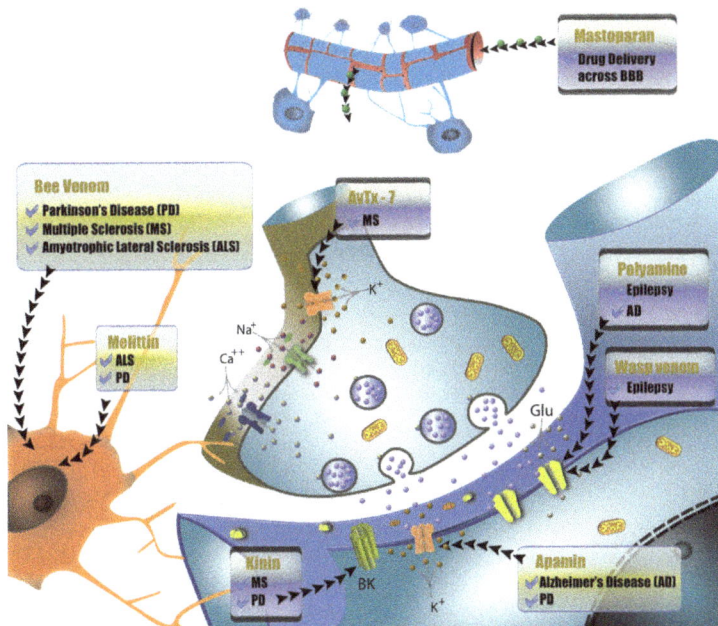

Figure 1. Main targets for wasp and bee venoms in the nervous system according to the type of neurodegenerative disorder treated.

Among these neurological disorders, neurodegenerative conditions significantly impact not only individuals, but also caregivers and society. The most prevalent neurodegenerative disease is AD, followed by PD and Epilepsy. Neurodegenerative diseases are a heterogeneous group with relentless progression, where aging is a major risk factor in the development [16]. Despite their heterogeneity, all of these diseases are characterized by cognitive impairment, motor alterations and personality changes. Unfortunately, the specific etiology of neuronal death and protein deposition in these diseases remains unknown [16,17].

2.1. Alzheimer's Disease and Other Dementias

Dementia is one of the most frequent causes of cognitive impairment in older adults, with forecasts indicating a worldwide increase from 25 million in 2000 to 115.4 million by 2050. Alzheimer's alone is responsible for over half of these cases [18–20].

Alzheimer's is symptomatically characterized by memory deficits, cognitive impairments and personality changes [17]. In general, the first clinical signs are impaired short-term memory accompanied by attention and verbal fluency difficulties. Other cognitive functions also deteriorate with the evolution of the disease, including the ability to make calculations, visual-spatial skills and the ability to use everyday objects and tools [17,20].

Estimates indicate the disease will affect more than 80 million people by 2040 and increased life expectancy will see the number of people with AD grow by 300% in developing countries. Since the disease is progressive, patients require prolonged special care after diagnosis, with annual costs estimated at nearly EUR 20,000 per person, exceeding that of patients with cancer [17].

Major contributors to neurodegeneration in brains affected by AD are the deposition of senile plaques, composed primarily of Aβ peptide, and neurofibrillary tangles formed largely by tau protein, which accumulate in neuropils from the cerebral cortex and hippocampus. Moreover, mitochondrial alterations such as fission-fusion abnormalities, defects in electron transport chain proteins, cytoskeletal abnormalities, calcium metabolism, intrinsic apoptosis pathways and caspase activation, as well as free radical generation are also involved in AD pathology [21,22].

More than 100 years after identifying the hallmark lesions in AD, there is still no minimally effective disease modifying therapy available [22]. From 2002 to 2012, of 221 agents submitted to trials for disease-modifying potential, none was different from the placebo in terms of positively affecting primary outcomes [23]. Alzheimer's treatment is symptomatic and relies on the administration of cholinesterase inhibitors (AChEI) (only tacrine, donepezil, rivastigmine and galantamine are currently approved for AD treatment) and NMDA receptor antagonists (only memantine is approved) [17]. Intervention with AChEI decreases acetylcholine metabolism and enhances neurotransmission, which is associated with memory and cognition reduction in AD [24]. NMDA antagonists act by compensating abnormal tonic activation by glutamate and are more efficient in moderate to severe stages of the disease [25]. Given that these drugs merely provide symptomatic relief, there is an urgent need to develop neuroprotective treatments for AD.

2.2. Parkinson's Disease

Parkinson's Disease is a universal, incurable, multifactorial and neurodegenerative disorder characterized by gradual degeneration and loss of dopaminergic neurons in the *substantia nigra* (SN). This leads to nigrostriatal pathway denervation, with the presence of Lewy body cytoplasmatic inclusions, predominantly resulting in motor symptomatology. In addition, non-motor symptoms are often identified in PD patients and may precede motor signs [26]. The disorder affects 1% of the population during the fifth or sixth decade of life and is primarily related to aging, with no definitive biomarker available for PD diagnosis [27,28].

Although PD etiology is not yet fully understood, it is possible that a large set of environmental and genetic factors in association with intrinsic neuronal vulnerability in the SN could be involved in the neuronal death typically observed in PD, primarily by inducing oxidative stress and mitochondrial dysfunction [29]. These factors include pesticide exposure, glutamate excitotoxicity, protein misfolding and aggregation, an imbalance in calcium homeostasis and neuroinflammation by microglial activation [30,31]. However, no drug has been clinically proven to modify disease progression, either by protecting surviving dopaminergic cells from degeneration or by restoring lost cells.

In this context, pharmacological treatment for PD remains focused on motor symptoms, mostly by restoring striatal dopamine levels through the administration of dopamine agonists. L-DOPA, a dopamine precursor, is the gold standard for this approach and is often associated with an inhibitor of peripheral degradation (carbidopa and benserazide). Despite its efficiency, long-term L-DOPA treatment is linked to side effects such as motor fluctuations (shorter duration of action) and dyskinesias (abnormal involuntary movements), both of which can significantly reduce quality of life in patients [32,33].

2.3. Epilepsy

Epilepsy is an enduring predisposition of the brain to generate epileptic seizures along with the neurobiological, cognitive, psychological and social consequences that the condition causes [34]. More recently it has been defined according to events such as the occurrence of at least two unprovoked (or reflex) seizures in a 24 h period, one unprovoked (or reflex) seizure with the likelihood of further similar seizures, or diagnosis of an epileptic syndrome [35].

Estimates suggest that approximately 65 million people of all ages may be affected by epilepsy [36] and that the majority face treatment problems due to pharmacoresistance to antiepileptic drug (AED) therapy [37,38]. AEDs are classified into three generations, according to their introduction into the market. The first generation of these drugs was sold in the USA and Europe from 1857 to 1958, followed by the second generation between 1960 and 1975. Drugs introduced in the 1960s are potent enzymatic inducers of cytochrome P450 that lead to clinically significant adverse drug interactions and hypersensitive reactions [39]. The 1980s saw the introduction of 15 additional AEDs (third generation), providing more appropriate drug alternatives for patients. However, it is important to underscore that each drug has its advantages and limitations, making treatment a difficult process [40]. Furthermore, these drugs are still inefficient in drug resistant epilepsy, challenging our understanding of the underlying mechanisms of this phenomenon and how to overcome or prevent them. Recent progress in understanding the molecular and cellular events that cause this disease have allowed better management of strategies for the discovery and development of more effective AEDs [41].

2.4. Multiple Sclerosis

Multiple sclerosis (MS) is a chronic inflammatory, demyelinating and neurodegenerative disorder of the CNS that begins in young adulthood and may be the result of the interaction between genetic and environmental factors, together with certain pathological hallmarks of an autoimmune disease [42–44]. According to the National Multiple Sclerosis Society, the disease affects around 2.1 million people worldwide [45]. MS has a significant socioeconomic impact that is comparable to other neurological conditions. This is because mean disease duration is approximately 38 years, thus affecting individuals at a time when they are entering, developing, or consolidating their professional careers [42].

The pathogenesis of MS is complex and only partially understood, hampering diagnosis and thus the choice of appropriate treatment. Nevertheless, a group of experts recently revised the MS phenotypic classification that includes the five MS subtypes: Relapsing-remitting MS (RRMS), clinically isolated syndrome (CIS), radiologically isolated syndrome (RIS), primary-progressive MS (PPMS) and secondary-progressive MS (SPMS) [46]. Considering the complexity of MS pathophysiology and diagnosis, only a brief description will be given of the main phenotypes included since MS classification began (RRMS, PPMS, and SPMS).

Relapsing-remitting multiple sclerosis (RRMS) represents about 80% of all cases, lasts for about 15 years and is characterized by acute exacerbations from which patients completely or partially recover, with periods of relative clinical stability in between [43,44]. When neurological function declines, the disease progresses to the following stage and is known as primary-progressive multiple sclerosis (PPMS). This type affects 10% of patients, who often present with progressive cerebellar syndrome and myelopathy, or other progressive symptoms [44,47]. Secondary-progressive multiple sclerosis (SPMS) is characterized by a progressive loss of motor function after an initial relapse, occurring about 20 years after the initial event [48]. Furthermore, RRMS is best characterized by an intense focal inflammatory component, whereas PPMS and SPMS exhibit more neurodegenerative features with concomitant chronic inflammation and axon loss [49].

Similar to other neurodegenerative disorders, the limitations of current therapies for MS include lack of superior treatment efficacy, serious adverse effects and long-term safety [43]. Significant advances in the treatment of RRMS are observed when the main goal is to target inflammation and modify the course of the disease; however, the same cannot be said about progressive forms of MS [47,50]. In addition, halting or reversing disease progression is only possible by using remyelinating and neuroprotecting agents, which does not occur in current treatments.

2.5. Amyotrophic Lateral Sclerosis

Amyotrophic lateral sclerosis (ALS) is a devastating, progressive and incurable adult-onset neurodegenerative disease characterized by the loss of upper and lower motor neurons in the primary motor cortex, brainstem and spinal cord. The disease affects motor functioning, resulting in paralysis and eventual death, typically from respiratory failure [51–56]. Average survival is 3 years after the first symptoms emerge and 5%–10% of patients survive beyond 10 years [57].

The worldwide average incidence rate for ASL is 2.1/100,000 person-years and a point prevalence of 5.4/100,000 persons, strongly linked to increased age [57]. Although little is known about the etiology of ALS, some studies indicate that 10% of cases are familial ALS and 85%–90% are classified as sporadic [53,58,59].

There is increasing evidence that patients with familial and sporadic forms of ALS exhibit signs of multi-modal dysfunction, even in early stages. Previous population-based studies estimated that around 35% of patients exhibit these impairments, including behavioral changes and executive and cognitive function deficits. Furthermore, about 15% of those affected with ALS may also suffer from frontotemporal dementia (ALS-FTD). This leads to reduced quality of life, caregiver stress, clinical effects from ventilator use and gastrostomies, negatively influencing survival time [51,55,60–62].

The mechanisms responsible for disease onset and progression remain unknown, hindering the development of targeted therapies for ALS [59]. Given the multifaceted nature of the disease, most of the current approaches employed in clinical trials focus on the emerging concept of stem cell-based therapeutics [59,63]. Riluzole is the only Food and Drug Administration (FDA) approved treatment for ALS and prolongs survival by only a few months [59,64].

3. Venoms and Toxins from Wasps and Bees to Combat Neurodegenerative Disorders

The biological capacity to develop a secretion with highly specialized functions and a venomous apparatus is limited to certain groups, including cnidarians, some mollusk families, arthropods, certain reptiles and fish [12]. All insects that can sting are members of the order Hymenoptera, which includes ants, wasps and bees. The most extensively characterized venoms are bee venoms, mainly from the *Apis* genus, as well as some social and solitary wasp genera [4,65].

3.1. Bee Venom

Apitherapy is the medicinal therapeutic use of honeybee products, consisting of honey, propolis, royal jelly, pollen, beeswax and, in particular, bee venom (BV). Depending on the disease being treated, BV therapy can be used by applying a cream, liniment, or ointment, via injection, acupuncture or even directly through a live bee sting [4]. However, the most commonly used method is bee venom acupuncture (BVA), which involves the injection of diluted bee venom into acupuncture points. It can be employed as an alternative medicine in patients with PD, pain and other inflammatory diseases, such as rheumatoid arthritis and osteoarthritis [66–68].

Bee venom therapy is based on the fact that these crude extracts exhibit a wide variety of pharmacologically active molecules. This pool of chemical compounds is formed by biogenic amine, enzymes (phospholipase A2), basic peptides and proteins (melittin and apamin) and a mixture of water-soluble and nitrogen-containing substances [5].

One of the main biological activities identified in the venom of *Apis mellifera*, the most widely studied honeybee, is the inhibition of inflammatory and nociceptive responses [68]. Studies have shown that inhibition can occur in multiple aspects, making apitherapy the most common application for the treatment of inflammatory diseases such as arthritis, bursitis, tendinitis, rheumatoid arthritis and Lyme Disease [68,69].

Interestingly, BV has also been used in humans to treat neurological diseases with neuroinflammatory aspects, such as multiple sclerosis and Parkinson's Disease [66,67,70] (Figure 1). Furthermore, several studies on neuroinflammatory diseases in animal models have increasingly supported the effectiveness of this treatment [71–74] (Table 1).

In regard to anti-neuroinflammatory activity, crude honeybee venom and its components are important tools for the treatment of diseases accompanied by microglial activation [75,76]. Microglia are a population of macrophage cells in the brain that play an important role in immune defense and CNS tissue repair and are vital in controlling normal homeostatic functions in the brain [77].

Under pathogenic conditions, microglia are rapidly overactivated in response to neuronal injury and migrate to the affected sites of the CNS, significantly contributing to neuronal death in specific brain regions [78]. Resting microglia are generally benign to the brain; however, once activated through injury or during removal of unwanted cellular debris, they produce inflammatory cytokines, glutamate, quinolinic acid, superoxide radicals (O_2^-) and nitric oxide (NO), undermining cerebral homeostasis.

In this context, the suppression of microglial activation and the neuroprotective effect of BV were observed in several *in vitro* and *in vivo* studies, as well as in clinical trials. Studies in humans have shown that BV may be beneficial in the treatment of diseases that trigger cell death by microglial activation, particularly PD [79,80]. Parkinson's patients treated with BV acupuncture obtained promising results in idiopathic Parkinson's Disease Rating Scale Tests [79], demonstrating the remarkable ability of BV acupuncture (BVA) to interfere with PD progression.

Table 1. Use of Bee Venom and its components for the treatment of neurodegenerative diseases in *in vivo* models.

Venom or Compound	Neurological Disease	Model Tested	Administration via	Dose	Reference
Bee venom	Parkinson's Disease	1-methyl-4-phenyl-1,2,4,5-tetrahydropyridine (MPTP) in mice	s.c. acupuncture (point GB34)	0.02 mL bee venom (1:2000 w/v) once every 3 days for 2 weeks	[81]
Bee venom	Parkinson's Disease	MPTP in mice	s.c. acupuncture (bilateral point ST36)	A single injection (0.6 mg/kg)	[82]
Bee venom	Parkinson's Disease	MPTP/probenecid in mice	i.p.	Two injections 3.5 days apart for 5 weeks Low—12 µg/kg/BW High—120 µg/kg/BW	[83]
Bee venom	Parkinson's Disease	MPTP in mice	i.p.	one i.p. injection BV (1 mg/kg) every day for 6 days	[84]
Bee venom	Parkinson's Disease	Rotenone-induced oxidative stress and apoptosis	s.c. acupuncture (point GB34)	0.02 mL bee venom (1:2000 w/v) once every 3 days for 2 weeks	[85]
Bee venom	Multiple Sclerosis	Experimental allergic encephalomyelitis model in rats	-	2 mg/kg or 5 mg/kg	[86]
Bee venom	Amyotrophic Lateral Sclerosis	hSOD1^{G93A} transgenic mice	s.c. acupuncture (bilateral point ST36)	0.1 µg/g—3 times/week for 2 weeks	[87]
Bee venom	Amyotrophic Lateral Sclerosis	hSOD1^{G93A} transgenic mice	s.c. acupuncture (bilateral point ST36) i.p.	0.1 µg/g—3 times/week for 2 weeks	[88]
Apamin	Parkinson's Disease	MPTP/probenecid mice	i.p.	Two injections 3.5 days apart for 5 weeks Low—0.5 µg/kg/BW High—1.0 µg/kg/BW	[83]
Melittin	Amyotrophic Lateral Sclerosis	hSOD1^{G93A} transgenic mice	s.c. acupuncture (bilateral point ST36)	0.1 µg/g twice a week	[89]

In vivo models for BVA and PD have also been tested. Bilateral acupoint stimulation of lower hind limbs prevented the loss of dopaminergic (DA) neurons in the striatum and SN for MPTP-induced PD (1-methyl-4-phenyl-1,2,4,5-tetrahydropyridine) and increased striatal dopamine levels [81–83]. MPTP mimics PD in rodents, involving the progressive loss of neurons in SN and causing behavioral alterations typical of PD, making it the most widely used model to study the disease. Chung and colleagues (2012) corroborated the results previously recorded for dopaminergic neuroprotection and observed a reduction in the infiltration of CD4T cells and microglial deactivation in an MPTP-induced PD mouse model [85]. In addition, BVA suppressed neuroinflammatory responses by MAC-1 and iNOS, microglial activation and loss of neurons in SN in the same mouse model [82]. It is important to note that the protective effect of BV on DA neurons of the SN is not restricted to acupoint stimulation, since it is also observed when using intraperitoneal injections [83].

Recently, an extensive and important study indicated that BV was capable of normalizing neuroinflammatory and apoptotic markers and restoring brain neurochemistry after simulated PD injury in mice [85], revealing the significant potential of BV application for PD therapy. Moreover, BV exhibited no signs of toxicity on general physiological functions when administered subcutaneously within a higher therapeutic range (100–200 fold) [90].

In *in vitro* tests, BV reduced the production of NO, COX-2, PGE2 and pro-inflammatory cytokines (IL-1β, IL-6 and TNF-α) in murine microglia cultures stimulated by lipopolysaccharides (BV-2 cell line) [75,91–94]. Additionally, tests using SH-SY5Y human neuroblastoma cells and MPTP demonstrated an increase in cell viability, reduced apoptosis by DNA fragmentation assays, and inhibited cell death cascade activation after pre-treatment with BV [95].

Bee venom has also been investigated in the treatment of MS and ALS (Table 1). In 2007 and 2008, two reviews summarized relevant findings regarding the therapeutic potential of venoms and other non-conventional approaches in MS treatment [80,96]. An interesting cross-sectional study involving 154 patients with MS investigated how often they used complementary and alternative medicine (CAM), including apitherapy [97]. The authors concluded that about 61% used CAM, and more than 90% of these used it as an adjunct to allopathic treatments. Furthermore, 65.8% of the interviewees reported an improvement. Given its importance and the growing interest in BV therapy, the American Apitherapy Society began to track patients who receive this treatment regularly, enrolling over 6000 members who take BV for MS or rheumatoid arthritis [98].

An FDA—approved investigational new drug trial involving nine patients with progressive MS evaluated the safety of BV [70]. Intradermal injections of gradually increasing doses were administered for 17 weeks until treatment reached 2.0 mg/week. A questionnaire, functional neurological tests and changes in measurement of somatosensory-evoked potentials were used to assess responses to therapy. None of the subjects displayed severe allergic reactions, although four reported worsening neurological symptoms and had to discontinue treatment. Two other patients showed objective improvement and three exhibited subjective symptom improvement. This was a preliminary study performed on a small number of patients and, despite the few positive results obtained, it was difficult to establish definitive conclusions regarding the efficacy of apitherapy.

In the same year, a high quality clinical trial for apitherapy in MS [99] evaluated the effectiveness of BV in 26 relapsing-remitting or secondary progressive MS patients [100]. This crossover study tested two groups; one received bee sting therapy for 24 weeks and placebo for another 24 weeks, while the other was given the same treatments in reverse order. Live bees were used to administer BV three times a week, with an increasing number of stings in each session to a maximum of 20 bee stings. Although it was well tolerated with no serious adverse events, the therapy failed to reduce fatigue, disease activity or disability, or improve quality of life. By contrast, phase II of the study assessed the efficacy of BV in patients with either RRMS or chronic progressive MS and found that BV intradermal injections decreased functional debilitation [101]. Treatment was administered until positive clinical effects reached a plateau, with an initial dose of one bee sting. In general, more than 68% of patients experienced some beneficial effects from BV therapy, including better balance, coordination, bladder and bowel control, as well as improved extremity strength, fatigue, endurance, spasticity and numbness, providing important evidence for the use of BV in MS. The authors attributed most of the positive findings to patients suffering from chronic-progressive MS when compared to relapsing-remitting MS, largely due to inherent variability among these MS patients, hindering result assessment.

A more recent study showed significant positive effects attributed to BV treatment in an experimental allergic encephalomyelitis animal model for MS induced by guinea pig spinal cord homogenate [86]. The results indicate that BV significantly decreases clinical symptoms and immunization effects in Lewis rats, as well as penetration of inflammatory cells and serum TNF-α and nitrate levels.

Considering all the findings reported on BV therapy for MS and according to Namaka and collaborators (2008), the different results reported to date may be due to the therapeutic protocols used, type of animal model and/or type of challenged cell line, in addition to potential time and dose-dependent properties [96].

Bee venom has also been studied for the treatment of ALS. A study using a symptomatic animal model for ALS with mutant hSOD1^{G93A} transgenic mice showed an improvement in motor activity in the rotarod test and prolonged life span for mice treated with BV acupoint stimulation [87]. The results obtained were substantiated by reduced levels of cytokines, typically released by activated microglia and astrocytes, leading to the neuroprotective effect observed. Moreover, by contributing to the reduction of motor neuron degeneration, BV prevented mitochondrial disruption and activated cell survival signal transduction pathways.

Research using the same animal model found that transgenic mice that received BV exhibited reduced expression of α-synuclein modifications, ubiquitinated α-synuclein and recovered spinal cord proteasomal activity [102]. It is important to underscore that animals received only two subcutaneous injections of 0.1 μg/g of BV at an acupoint, which was sufficient to induce positive effects.

Interestingly, another recently published study compared the effects of BV treatment using different administration routes for the same symptomatic model of ALS [88]. It was noted that BV treatment through an acupoint was more effective than intraperitoneal (i.p.) BV administration and acupoint stimulation alone. The results demonstrated an improvement in walking function, lower levels of neuroinflammatory proteins (TLR4, CD14 and TNF-α) in the spinal cord and reduced nuclear abnormality in the quadriceps femoris muscle.

In a study evaluating the ability of BV to act on the impaired ubiquitin-proteasome system [103], NSC34 motor neuronal cells expressing the mutant gene hSOD1^{G85R} were used and stimulated with 2.5 μg/mL of BV for 24 h. Once again the results showed restored proteasome activity and a reduction in the amount of misfolded SOD1. However, BV did not activate the autophagic pathway in these cells, a process frequently impaired in ALS that results in the aberrant accumulation of misfolded and/or aggregated proteins within spinal cord cells. This BV effect is remarkable because it reduces protein aggregation by targeting the ubiquitin system as opposed to activating the autophagy pathway.

Thus, when taken together, these findings reinforce the therapeutic potential of BV treatment, demonstrating an antineuroinflammatory effect, reduced neuronal loss caused by misfolded protein aggregates and glutamate neurotoxicity, restoration of the ubiquitin-proteasome system and motor improvement. These results could have important clinical implications for BV use as a coadjuvant treatment in both ALS and other neurodegenerative disorders.

3.2. Wasp Venom

With respect to wasps, important studies reveal the pharmacological potential of these venoms, present primarily in the *Polybia* genus (Table 2). In 2005, Cunha and colleagues described the effects on rats of an intracerebroventricular (i.c.v.) injection of crude and denatured venom of the social wasp *Polybia ignobilis* [104]. Interestingly, crude venom provoked severe generalized tonic-clonic seizures, respiratory depression and death. On the other hand, denatured venom had an antiepileptic effect on acute seizures induced by i.c.v. injection of bicuculline, picrotoxin and kainic acid, but not on pentylenetetrazole(PTZ)-induced seizures. In addition, the denatured venom inhibited [³H]-glutamate binding in membranes from the rat cerebral cortex at lower concentrations than those used for [³H]-GABA binding [105]. These results indicate that specific components in the venom of *P. ignobilis* may interact with GABA and glutamate receptors, representing a significant source of neuroactive molecules (Figure 1).

Table 2. Use of Wasp Venom and its components for the treatment of neurodegenerative diseases in *in vivo* models.

Venom or Compound	Neurological Disease	Model Tested	Route of Administration	Dose	Reference
Denatured venom—*P. ignobilis*	Epilepsy	Acute seizures model induced by chemoconvulsants in rats	i.c.v.	400 μg/animal	[104]
Denatured venom—*P. occidentalis*	Epilepsy	Acute seizures model induced by chemoconvulsants in rats	i.c.v.	120, 240 and 300 μg/animal	[105]
Low molecular weight compounds—*P. occidentalis*	Epilepsy	Acute seizures model induced by PTZ	i.c.v.	70, 210 and 350 μg/animal	[106]
Bradykinin	Stroke	Transient forebrain ischemia in rats	i.p.	150 μg/kg 48 h after ischemia	[107]
Bradykinin	Stroke	Transient forebrain ischemia in rats	i.p.	150 μg/kg 48 h after ischemia	[108]

Similarly, i.c.v. administration of the denatured venom of *P. occidentalis* inhibited epileptic seizures caused by the same chemical convulsants previously described and was ineffective against PTZ-induced seizures [105]. A subsequent study with low molecular weight compounds (LMWC) from *P. paulista* wasps demonstrated their ability to block PTZ-induced seizures [106]. This effect is likely due to the presence of different compounds that act on GABA receptors.

Finally, research on crude venom from the social wasp *Agelaia vicina* revealed its ability to competitively inhibit high- and low-affinity GABA and glutamate uptake [109]. This is an important result since diseases such as Stroke, Epilepsy and PD involve abnormalities in GABA and glutamate uptake systems [110,111].

4. Compounds Isolated from Wasp and Bee Venom for the Treatment of Neurodegenerative Diseases

In addition to crude venom, several venom components have been widely used in Oriental medicine to relieve pain and treat inflammatory diseases such as rheumatoid arthritis and tendinitis [68,69,112]. Other potential venom-related treatments include the inhibition of neuroinflammatory responses, useful in the treatment of PD, AD and MS. This section of the review highlights the most recent and innovative therapeutic and biological applications of bee venom compounds: Melittin and Apamin (Table 1); and wasp venom compounds: Pompilidotoxins, Mastoparans, Kinins and Polyamine toxins (Table 2).

4.1. Peptides from Bee Venom as Therapeutic Sources

4.1.1. Melittin

Melittin is the main component found in BV, accounting for 40% to 60% of dry venom, and is the best characterized peptide in BV. This linear peptide has 26 amino acid residues, alkaline characteristics, a predominantly hydrophobic *N*-terminal region and a hydrophilic *C*-terminal, resulting in amphiphilic properties [113] (Figure 2A). It appears to be primarily responsible for intense local pain, inflammation, itching and irritation in higher doses. On the other hand, in very small doses Melittin can cause a wide range of central and systemic effects, including anti-inflammatory effects, increased capillary permeability and lower blood pressure, among others [114].

The effect of Melittin on the CNS has been documented since 1973, when studies showed its marked effect on inhibiting general behavior, exploratory activity and "emotionality", in addition to disrupting spontaneous and evoked bioelectric activity in the brain. Moreover, high doses of this peptide can induce a depressant effect evaluated by electroencephalography in anesthetized cats. This effect was associated with reduced systemic blood pressure [115,116].

In 2011, Yang and collaborators studied the therapeutic effect of Melittin in a transgenic mouse model for ALS. In this model, Melittin-treated animals exhibited a decline in the number of activated microglia and expression of proinflammatory factor TNF-α, inhibiting the increased neuroinflammation responsible for neuronal death in this disease. Moreover, Melittin regulates the production of misfolded proteins by activating chaperones and alleviating α-synuclein post-translational modification, an important mechanism for PD and ALS pathologies. Melittin also restored proteasome activity in the brainstem and spinal cord. Interestingly, treatment with this alkaline peptide in a symptomatic ALS animal model improved motor function and reduced neuronal death [117].

Additionally, *in vitro* assays revealed the potential in Melittin as an agent for the prevention of neurodegenerative diseases, considering its ability to inhibit the apoptotic factor and cell death in neuroblastoma SH-SY5Y cells [118]. Melittin also demonstrated a potent suppressing effect on proinflammatory responses for BV2 microglia by reducing proinflammatory mediators and production of NO, PGE2 and cytokines [89]. Thus, it is suggested that this compound may have significant therapeutic potential for the treatment of neurodegenerative diseases accompanied by microglial activation, such as PD (Figure 1).

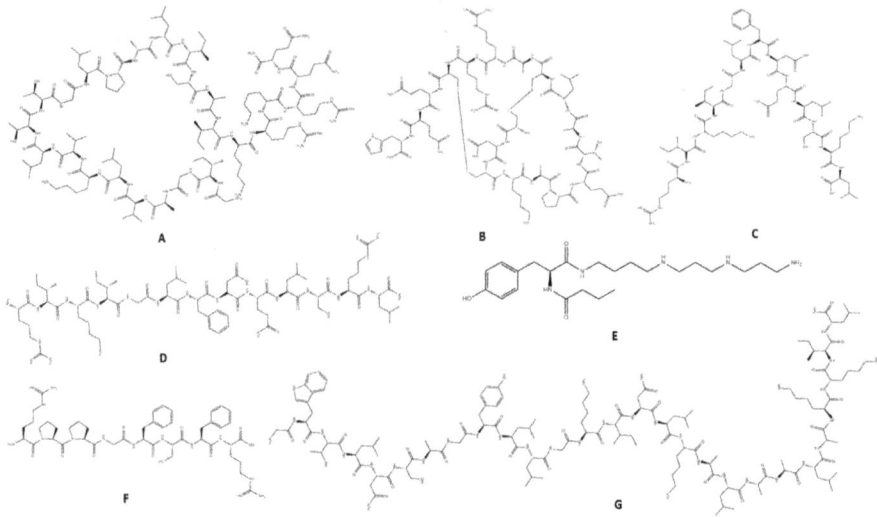

Figure 2. Chemical structures of compounds found in bee and wasp venoms. (**A**) Melittin [119]; (**B**) Apamin [120]; (**C**) Alpha-pompilidotoxin [121]; (**D**) Beta-pompilidotoxin [122]; (**E**) Philanthotoxin [123]; (**F**) Bradykinin [124]; (**G**) Transportan [125].

Recently, Dantas and colleagues (2014) investigated the pharmacological effects of Melittin on the nervous system of mice [126]. The animals were submitted to behavioral tests, including the catalepsy test, open field and apomorphine rotation tests. The results showed that mice treated with Melittin displayed no cataleptic effects or changes in motor activity, although there was a reduction in the effects induced by the apomorphine test. As such, the authors found that Melittin exhibited antipsychotic properties and may be an alternative for the treatment of psychotic diseases, reducing the classic side effects caused by conventional neuroleptic drugs.

4.1.2. Apamin

Neurotoxin Apamin is the smallest peptide, accounting for less than 2% of BV, with 18 amino acids residues, a high cysteine content and alkalinity (Figure 2B). Moreover, it is well known for its pharmacological property of irreversibly blocking Ca^+ activated K^+ channels (SK channels) and is considered the most widely used blocker for this type of channel [113,127].

Small-conductance Ca^{2+}-activated K^+ (SK) channels control the firing frequency of neurons, especially at AMPA and NMDA glutamatergic synapses, and are responsible for hyperpolarization following action potentials [128]. These channels can be positively or negatively modulated. Positive modulation involves binding the compound, which then facilitates channel activity, thus impairing memory and learning. The opposite is true for negative modulation, where memory and learning improve and calcium channel sensibility declines [129]. Apamin acts through the second mechanism described. In neurons, this SK channel blockage decreases hyperpolarizing effects, modulating synaptic plasticity and memory encoding [130,131]. In addition, when compared to other arthropod neurotoxins, Apamin has an unusual ability to cross the blood brain barrier (BBB) and acts mainly in the CNS, where SK channels are extensively expressed [130].

Alvarez-Fischer *et al.* (2013) studied the protective effect of this peptide on dopaminergic neurons in a chronic mouse model of MPTP-induced PD [83]. The animals received i.p. injections in two different dosages of Apamin (low: 0.5 μg/kg; high: 1.0 μg/kg) in order to assess brain lesions and behavioral effects in mice. Results showed that Apamin protected nigral DA neurons and increased striatal DA levels in the nerve terminals. In the behavioral test, data were paradoxical, indicating that mice treated with Apamin spent significantly less time on the spindle in comparison to saline-treated animals with MPTP brain lesions, despite the authors' suggestion that Apamin may improve neuroprotection of dopaminergic neurons [83]. In this context, cell cultures that mimic the selective demise of mesenphalic dopaminergic neurons showed a lower degeneration rate after Apamin treatment [132]. Furthermore, Apamin has also been evaluated for the treatment of PD using the motor score from the Unified Parkinson's Disease Rating Scale. In this study, Apamin exhibited primarily neurorestorative activity in PD, as well as symptomatic and neuroprotective activity [133].

Several behavioral and electrophysiological studies have suggested Apamin in the treatment of AD, indicating that the blockage of SK channels by this compound may enhance neuronal excitability, synaptic plasticity, and long-term potentiation in the CA1 hippocampal region (Figure 1) [134]. Likewise, Apamin is a valuable tool in the investigation of physiological mechanisms involved in higher brain functions, such as cognitive processes or mood control, and there is already a patented method for early diagnosis of AD using Apamin [135–139]. However, it is important to underscore that SK blockage may accelerate neurodegenerative processes, making additional research in this field imperative.

4.2. Wasp Venom Peptides as Therapeutic Sources

4.2.1. Pompilidotoxins

Pompilidotoxins are a group of neuroactive molecules that were first described by Konno *et al.* [140,141]. They consist of two neurotoxins known as α- and β-pompilidotoxin (PMTX), derived from solitary wasps *Anoplius samariensis* and *Batozonelus maculifrons*, respectively. These molecules are peptides composed of 13 amino acid residues, differing solely in the presence of an amino acid at position 12, corresponding to lysine in α-PMTX and Arginine in β-PMTX (Figure 2C,D, respectively). This minimal structural difference appears to be responsible for the significant potency of β-PMTX, approximately five times higher than α-PMTX, when tested in the lobster neuromuscular junction. Moreover, both peptides act on mammalian central neurons, primarily by blocking Na^+ current inactivation [142].

It has been demonstrated that α-PMTX interrupts synchronous firing of rat cortical neurons, facilitates synaptic transmission in hippocampal slices and decelerates the inactivation of tetrodotoxin-sensitive voltage-gated sodium channels (VGSCs) from rat trigeminal neurons [143,144]. In turn, β-PMTX modulated spontaneous rhythmic activity in spinal networks [145] and acted on hippocampal CA1 neurons by interfering with postsynaptic potential, increasing excitatory potential and interrupting rapid inhibitory potential [146]. Given that the main action of Pompilidotoxins is to slow the inactivation of VGSCs, these peptides may provide a better understanding of the molecular determinants associated with alterations in these channels involved in neuropathological conditions. The alteration of sodium channels has been described as a contributor to the events involved in several neurological disorders, especially persistent sodium currents that can participate in the physiopathology of some types of epilepsy and MS [147,148]. It is important to note that finely orchestrated activation and inactivation is essential for the correct maintenance of neuronal excitability and the slightest change in this equilibrium can result in serious consequences for the individual.

4.2.2. AvTx-7

Research by Pizzo *et al.* (2004) showed that neuroactive peptide Avtx7, isolated from the venom of social wasp *Agelaia vicina*, acted on the blockage of tetraethylammonium and 4-aminopyridine (4-AP)-sensitive K^+ channels (Figure 1) [149]. As such, this novel neurotoxin may be a valuable tool in better understanding how K^+ channels work on neurological diseases, such as dementia and MS. These results were obtained using cortical brain synaptosomes and by assessing glutamate release as a response to different potassium blockers. K^+ channels are critically involved in the nervous system, consequently, alterations in their function can lead to important perturbations in membrane excitability and neuronal function. For instance, the dysfunction of a subfamily or subtype of K^+ channels might induce AD or PD [150]. Thus, K^+ channel blockade, for instance by 4-AP, has been linked to an action potential extension with a consequent increase in duration, which is relevant for the treatment of MS. Since 1990, the use of 4-AP in patients with MS has been described to reduce fatigue and improve visual field defects [151]. However, despite its therapeutic effects, drawbacks include low selectivity, causing severe adverse effects and difficulty determining individual therapeutic dose. In this respect, research targets more selective pharmaceuticals to treat MS by using these blockers, though with fewer side effects [152].

In regard to potassium blockers, an important line of research proposes their use as a meaningful non-dopaminergic alternative for the treatment of neurodegenerative diseases, such as advanced-stage PD. The use of these blockers is favorable in three mechanisms: Increased neurotransmitter release (*i.e.*, glutamate), modulation of neuronal network oscillation and greater cortical excitation. In relation to 4-AP, advanced clinical trials have shown satisfactory results, leading to FDA approval in 2013 for the treatment of movement dysfunction in patients with MS [153]. In this field, the discovery and identification of AvTx7 provides new pharmacological options, since its mechanism seems to be related to 4-AP.

4.2.3. Mastoparan

Mastoparan is a class of multifunctional peptides found in solitary and social wasp venom, with its primary activity described in mast cell degranulation, giving the peptide its name [154]. Thus, these peptides exhibit a number of remarkable pharmacological activities, such as antimicrobial, antitumor, insulinotropic and neurological effects [114,155–159].

The first Mastoparan was identified and chemically characterized by Hirai *et al.* in 1979, when this molecule was isolated from the social wasp *Vespula lewisii*. Mastoparans are short cationic peptides with 10 to 14 amino acid residues, two to four lysine residues and C-terminal amidation, characteristics that are essential for proper peptide action [160,161]. These peptides can interact and penetrate biological membranes via the positively charged side-chains of their amphipathic α-helical structures [161]. In light of this property, Mastoparans were recently classified as cell-penetrating peptides (CPP) [162].

Crossing the BBB is a significant challenge in neuropharmacology. The BBB is responsible for regulating brain homeostasis through selective permeability that protects the CNS. However, these characteristics also affect drug delivery and bioavailability to the CNS. Advances in the fields of pharmacokinetics, molecular biology, nanotechnology and toxinology have resulted in strategies to facilitate the crossing of drugs through the BBB, thus, increasing drug concentration in the brain [163]. Cell permeable peptides (CPP), particularly Mastoparans, serve as vehicles for the delivery of different molecules and particles into the brain and neurons and have been studied in combination with compounds that act on the CNS [164].

With the aim of enabling neuroactive compounds to permeate the BBB, researchers have created new chimeric peptides (Transportan), connecting Mastoparans and the neuropeptide Galanin in two different ways. The first compound, named Transportan, is formed by 12 residues of Galanin and a full length Mastoparan connected by a lysine, resulting in a chimera with 27 residues [164] (Figure 2G). The second compound, called Transportan 10, consists of seven terminal residues of Galanin and a full Mastoparan connected by a lysine residue [165].

Galanin, discovered in 1983, is a neuropeptide that in humans contains 30 amino acid residues and 29 in other species, for revision see [166]. Its name originates from the fusion of Glycin and Alanin, the *N*-terminal and *C*-terminal amino acids, respectively. Widely distributed in the peripheral and central nervous systems, Galanin has been associated with the pathophysiology of neurodegenerative diseases such as AD and Epilepsy [166]. Several studies report that the overexpression of Galanin detected in AD can preserve cholinergic striatal neuron function, which in turn may slow AD symptoms [167]. The chimeric construction of Transportan and Transportan 10 has been used as a drug delivery system for Galanin in the CNS and as treatment for neurodegenerative diseases, acting as a neuroprotective agent (Figure 1).

Another important function of Mastoparans is that they act as an antidote to one of the most powerful neurotoxins in the world, Botulinum toxin A (BoTx-A). If inhaled, only one gram of crystallized BoTx-A dispersed in the air can kill a million people [168]. Intoxication is so rapid and severe that some countries developed biological weapons containing BoTx for use in World War II. Intoxicated patients are treated with serum therapy. However, this does not reverse the toxic effects already induced in the organism [169]. As such, in an effort to treat this intoxication, a group of researchers employed Mastoparan 7 as a CPP in a chimeric construction denominated Drug Delivery Vehicle-Mas 7 (DDV-Mas 7). Consisting of a non-toxic heavy chain fragment of BoTx-A and Mastoparan 7, this chimeric peptide induced neurotransmitter release in a culture of mice spinal cord neurons, reversing the effect of the BoTx-A and allowing Ach liberation, followed by muscular contraction [160].

Mastoparans also modulate G-protein activity without receptor interaction, currently considered a preeminent tool for the study and understanding of this complex intracellular signaling system [170–173]. Several neurological disorders, including Mood Disorders, Epilepsy, AD, and PD are related to G protein-coupled receptors [174–176]. Thus, over the last decade, natural, modified or chimeric Mastoparans have been used as a potential treatment for a number of neurological conditions.

4.2.4. Wasp Kinin

Another class of peptide frequently encountered in wasp venom is Kinin, composed of Bradykinin (BK) and its analogues, largely responsible for the pain caused after a wasp sting and the paralyzing action used for prey capture [177–179]. Naturally present in different animals, BK was first described in 1949 by Rocha and Silva as consisting of nine amino acid residues (Figure 2F), with its primary activity described in mammal platelets [180]. This small peptide plays an important role in controlling blood pressure, renal and cardiac function, and inflammation [181]. It is important to note that Kinin was the first neurotoxin component isolated from wasp venom. In addition, Kinin acts on the insect CNS, where it irreversibly blocks the synaptic transmission of nicotinic acetylcholine receptors [179–182]. Furthermore, Kinin components, produced via the kallikrein-kinin system, have been found in abundance throughout both the rat and human CNS attracting interest in neuroprotective research [183] (Figure 1). Two major Kinin receptor families have been identified: B2 and B1 receptors. Their expression is low under normal conditions, but is up-regulated following injury, infection and inflammation [184].

Although several studies report that BK likely triggers a specific cascade of inflammatory events in the CNS, it has also been shown to possess anti-inflammatory (neuroprotective) properties, suppressing the release of inflammatory cytokines (TNF-α and IL-1β) from microglia in *in vitro* assays [183]. According to these authors, BK modulated microglial function by negative feedback for cytokine production, increasing prostaglandin synthesis and causing greater microglial cAMP production [183].

BK can also be beneficial after ischemic stroke, particularly if administered in the latter stages as opposed to the initial phases, where its harmful effects include inflammatory response and neurogenic inflammation [185]. It is noteworthy that molecular and functional evidence has suggested that interaction with B1 receptors may provide a new therapeutic approach in MS, primarily by reducing the infiltration of immune cells (lymphocytes T) into the brain [184]. Additionally, treatment with BK applied two days after transient forebrain ischemia in rats in post-conditioning studies provided 97% neuroprotection for the particularly vulnerable CA1 hippocampal neurons, as well as a decrease in Caspase3 expression and iNOS-positive cells, and also a suppression in the release of cytosolic cytochrome *c* and MnSOD [107,108]. This indicates that the neuroprotective mechanism initiated by BK may also inhibit the mitochondria-mediated apoptotic pathway [108]. The neuroprotective role of BK has also been reinforced by evidence of its action in the retina, protecting against neuronal loss induced by glutamatergic toxicity. This BK-induced protection caused a downstream reaction in NO generation and an upstream reaction in radical oxygen generation [186].

As observed, BK agonists may provide a new platform for drugs designed to treat neurodegenerative disorders that involve microglial activation, such as PD and acute brain damage. In this respect, wasp venom contains a multitude of Kinins with different activity potency profiles. A good example is Thr^6-Bradykinin, a compound isolated from several wasp venom samples. The single substitution of serine for threonine in this compound results in enhanced action when compared to BK. According to Mortari *et al.* (2007), this peptide displays remarkable anti-nociceptive effects when injected directly into the rat CNS; it is approximately three times more potent and remains active longer than BK [187]. These results can be explained by a more stable conformation in its secondary structure and/or the modification may protect against hydrolysis through neuronal kininases, preserving the effect of the peptide on B2 receptors [187,188].

4.3. Polyamine Toxins as Therapeutic Sources

Polyamine toxins are a group of low molecular weight (<1 kDa), non-oligomeric compounds isolated primarily from the venom of wasps, followed by spider venoms [189,190] (Figure 2E). The first polyamine toxin described, Philanthotoxin-433 (PhTX-433), was isolated from the venom of the wasp *Philanthus triangulum* [191]. These small natural molecules exhibit a number of biological activities and have been used as tools in the study of ionotropic glutamate (iGLU; AMPA) and nicotinic acetylcholine (nACh) receptors since the 1980s [190,192,193]. Interest is centered on its action as a non-selective and potent antagonist of glutamate receptors in the invertebrate and vertebrate nervous system (Figure 1) [192–194]. Moreover, it is believed that the abnormal activation of iGLU receptors is involved in neurological and psychiatric diseases such as AD, PD, Stroke, Depression, Epilepsy, Neuropathic Pain and Schizophrenia [195,196].

With respect to iGLU, current polyamine toxins (PhTXs) and their derivatives have the ability to differentiate which AMPA receptors are in fact permeable to Ca^{2+} ion, acting as a non-selective open-channel blocker [190,197]. As a result, PhTXs can control the excessive opening of overactivated ion channels (due to pathological conditions) and block the exaggerated influx of calcium, culminating in neuroprotection [193,198]. Interestingly, this mechanism of action is similar to that of Memantine, a drug used in the symptomatic treatment of moderate to severe AD [199]. Thus, the existence of a drug that has obtained good clinical results and its similarity with polyamine toxins illustrates the potentially promising role of these molecules and highlights the need for further research.

Recently, a computational model approach was devised to better understand how polyamine toxins interact with ion channels coupled with glutamate receptors [200]. This study found that these molecules could bind to the narrowest central region of the ion channel and block local ion flow. Membrane potential is important in toxin-receptor interaction, and as such, polyamine toxins are generally highly voltage-dependent blockers of iGLU [200]. In this regard, Nørager *et al.* recently developed fluorescent templates using polyamine toxin analogues to visualize these ligands in iGLU of living tissue [201].

5. Conclusions

Due to the rising prevalence of neurodegenerative diseases among the elderly, there is a pressing need for better treatment to alleviate the social and financial burden of these disorders. There are multiple targets for treating neurodegenerative diseases, considered complex syndromes that are difficult to control in a stable and lasting manner. Effective treatment of these diseases may require that the different pathogenic events associated with neurodegenerative diseases, such as the clearance of disaggregated proteins targeted in conjunction with neuroprotective and immunomodulatory strategies. In this respect, therapy using bee and wasp venoms is considered a psychoneurological approach for autoimmune and neurodegenerative diseases. Since these venoms contain a number of compounds, mainly peptides, advances in modern identification and sequencing techniques have facilitated and subsidized the elucidation of their full composition, thus providing an arsenal of new possibilities to combat a series of neurodegenerative diseases, using different neuroactive mechanisms of action.

Acknowledgments: Our research group was supported by the National Council for Scientific and Technological Development (CNPq), the Coordination for the Improvement of Higher Education Personnel (CAPES) and the Federal District Research Foundation (FAPDF).

Author Contributions: All authors actively participated in the writing of this review. J.S. contributed with the design and coordination of this work, and along with M.M. wrote the introduction, the sections on Neurodegenerative disease as target for the action of wasp and bee venom, Venoms and toxins from wasp and bee to combat neurodegenerative disorders, and the subsections on Multiple Sclerosis, Bee Venom, Wasp Venom, and Wasp Kinin. V.M.F. reviewed the manuscript and wrote the abstract and the conclusions. F.G. collaborated on diagramming and organizing all references, and wrote the subsection on Epilepsy and Polyamine toxins. K.L. contributed to the writing of the subsections Pompilidotoxins and AvTx-7 and conceived and collaborated with the figures. L.A. participated in the writing of the subsection Amyotrophic Lateral Sclerosis and also contributed to the Polyamines toxins as therapeutic sources. G.C. wrote about Alzheimer's Disease and other dementias and Parkinson's Disease. C.A. wrote the subsection about Melittin and collaborated on creating Table 2. A.M.B contributed to the writing of the subsections Bee Venom and Wasp Venom. J.G. described the subsection on Mastoparan. P.G. wrote the subsection about Apamin and contributed to Table 1. L.C. designed the figures. M.M organized, conceived and helped with the writing in general, and reviewed the article and citations for content. All authors read and approved the final manuscript.

Conflicts of Interest: The authors declare no conflict of interest.

References

1. Bogdanov, S. Bee venom: Composition, health, medicine: A review. *Peptides* **2015**, *1*, 1–20.
2. Pemberton, R.W. Insects and other arthropods used as drugs in Korean traditional medicine. *J. Ethnopharmacol.* **1999**, *65*, 207–216. [CrossRef]
3. Adewole, A.M.; Ileke, K.D.; Oluyede, P.O. Perception and knowledge of bee venom therapy as an alternative treatment for common ailments in southwestern Nigeria. *FUTA J. Res. Sci.* **2013**, *9*, 235–240.
4. Ali, M.A. Studies on bee venom and its medical uses. *Int. J. Adv. Res. Technol.* **2012**, *1*, 69–83.
5. Santos, L.D.; Pieroni, M.; Menegasso, A.R.S.; Pinto, J.R.A.S.; Palma, M.S. A new scenario of bioprospecting of Hymenoptera venoms through proteomic approach. *J. Venom. Anim. Toxins Incl. Trop. Dis.* **2011**, *17*, 364–377.

6. Mortari, M.R.; Cunha, A.O.S.; Ferreira, L.B.; dos Santos, W.F. Neurotoxins from invertebrates as anticonvulsants: From basic research to therapeutic application. *Pharmacol. Ther.* **2007**, *114*, 171–183. [CrossRef] [PubMed]

7. Mortari, M.R.; Cunha, A.O.S. New perspectives in drug discovery using neuroactive molecules from the venom of Arthropods. In *An Integrated View of the Molecular Recognition and Toxinology—From Analytical Procedures to Biomedical Applications*; Radis-Baptista, G., Ed.; InTech: Rijeka, Croatia, 2013.

8. Bialer, M.; White, H.S. Key factors in the discovery and development of new antiepileptic drugs. *Nat. Rev. Drug Discov.* **2010**, *9*, 68–82. [CrossRef] [PubMed]

9. Calabresi, P.; Di Filippo, M.; Ghiglieri, V.; Tambasco, N.; Picconi, B. Levodopa-induced dyskinesias in patients with Parkinson's Disease: Filling the bench-to-bedside gap. *Lancet Neurol.* **2010**, *9*, 1106–1117. [CrossRef]

10. Hung, A.Y.; Schwarzschild, M.A. Treatment of Parkinson's Disease: What's in the non-dopaminergic pipeline? *Neurotherapeutics* **2014**, *11*, 34–46. [CrossRef] [PubMed]

11. World Health Organization. Neurological Disorders: Public Health Challengers. 2006. Available online: http://www.who.int/mental_health/neurology/neurological_disorders_report_web.pdf (accessed on 10 March 2015).

12. Escoubas, P.; Quinton, L.; Nicholson, G.M. Venomics: Unravelling the complexity of animal venoms with mass spectrometry. *J. Mass Spectrom.* **2008**, *43*, 279–295. [CrossRef] [PubMed]

13. Ménez, A.; Stöcklin, R.; Mebs, D. 'Venomics' or: The venomous systems genome project. *Toxicon* **2006**, *47*, 255–259. [CrossRef] [PubMed]

14. De Boer, H.M.; Mula, M.; Sander, J.W. The global burden and stigma of epilepsy. *Epilepsy Behav.* **2008**, *12*, 540–546. [CrossRef] [PubMed]

15. Jacoby, A. Stigma, epilepsy, and quality of life. *Epilepsy Behav.* **2002**, *3*, 10–20. [CrossRef]

16. Beal, M.F. Aging, energy, and oxidative stress in neurodegenerative diseases. *Ann. Neurol.* **1995**, *38*, 357–366. [CrossRef] [PubMed]

17. Hampel, H.; Prvulovic, D.; Teipel, S.; Jessen, F.; Luckhaus, C.; Frölich, L.; Riepe, M.W.; Dodel, R.; Leyhe, T.; Bertram, L.; *et al.* The future of Alzheimer's Disease: The next 10 years. *Prog. Neurobiol.* **2011**, *95*, 718–728. [CrossRef] [PubMed]

18. World Health Organization. Dementia: A Public Health Priority. 2012. Available online: http://whqlibdoc. who.int/publications/2012/9789241564458_eng.pdf (accessed on 10 March 2015).

19. Wimo, A.; Winblad, B.; Aguero-Torres, H.; von Strauss, E. The magnitude of dementia occurrence in the world. *Alzheimer Dis. Assoc. Disord.* **2003**, *17*, 63–67. [CrossRef] [PubMed]

20. Davey, D.A. Alzheimer's Disease and vascular dementia: One potentially preventable and modifiable disease? Part II: Management, prevention and future perspective. *Neurodegener. Dis. Manag.* **2014**, *4*, 261–270. [CrossRef] [PubMed]

21. Kumar, A.; Ekavali, A.S. A review on Alzheimer's Disease pathophysiology and its management: An update. *Pharmacol. Rep.* **2015**, *67*, 195–203. [CrossRef] [PubMed]

22. Castellani, R.J.; Perry, G. Pathogenesis and disease-modifying therapy in Alzheimer's Disease: The flat line of progress. *Arch. Med. Res.* **2012**, *43*, 694–698. [CrossRef] [PubMed]

23. Cummings, J.L.; Morstorf, T.; Zhong, K. Alzheimer's Disease drug-development pipeline: Few candidates, frequent failures. *Alzheimers Res. Ther.* **2014**, *6*, 1–7. [CrossRef] [PubMed]

24. Frölich, L. The cholinergic pathology in Alzheimer's Disease—Discrepancies between clinical and pathophysiological findings. *J. Neural. Transm.* **2002**, *109*, 1003–1014. [PubMed]

25. Schneider, L.S.; Dagerman, K.S.; Higgins, J.P.; McShane, R. lack of evidence for the efficacy of memantine in mild Alzheimer disease. *Arch. Neurol.* **2011**, *68*, 991–998. [CrossRef] [PubMed]

26. Olanow, C.W.; Obeso, J.A. The significance of defining preclinical or prodromal Parkinson's Disease. *Mov. Disord.* **2012**, *27*, 666–669. [CrossRef] [PubMed]

27. Sharma, S.; Moon, C.S.; Khogali, A.; Haidous, A.; Chabenne, A.; Ojo, C.; Jelebinkov, M.; Kurdi, Y.; Ebadi, M. Biomarkers in Parkinson's disease (recent update). *Neurochem. Int.* **2013**, *63*, 201–229. [CrossRef] [PubMed]

28. Rodriguez, M.; Rodriguez-Sabate, C.; Morales, I.; Sanchez, A.; Sabate, M. Parkinson's Disease as a result of aging. *Aging Cell* **2015**, *14*, 293–308. [CrossRef] [PubMed]

29. Savitt, J.M.; Dawson, V.L.; Dawson, T.M. Diagnosis and treatment of Parkinson disease: Molecules to medicine. *J. Clin. Investig.* **2006**, *116*, 1744–1754. [CrossRef] [PubMed]

30. Tansey, M.G.; McCoy, M.K.; Frank-Cannon, T.C. Neuroinflammatory mechanisms in Parkinson's Disease: Potential environmental triggers, pathways, and targets for early therapeutic intervention. *Exp. Neurol.* **2007**, *208*, 1–25. [CrossRef] [PubMed]

31. Dong, X.X.; Wang, Y.; Qin, Z.H. Molecular mechanisms of excitotoxicity and their relevance to pathogenesis of neurodegenerative diseases. *Acta Pharmacol. Sin.* **2009**, *30*, 379–387. [CrossRef] [PubMed]

32. Sharma, S.; Singh, S.; Sharma, V.; Singh, V.P.; Deshmukh, R. Neurobiology of l-DOPA induced dyskinesia and the novel therapeutic strategies. *Biomed. Pharmacother.* **2015**, *70*, 283–293. [CrossRef] [PubMed]

33. Kakkar, A.K.; Dahiya, N. Management of Parkinson's Disease: Current and future pharmacotherapy. *Eur. J. Pharmacol.* **2015**, *750*, 74–81. [CrossRef] [PubMed]

34. Fisher, R.S.; van Emde Boas, W.; Blume, W.; Elger, C.; Genton, P.; Lee, P.; Engel, J.J. Epileptic seizures and epilepsy: Definitions proposed by the International League Against Epilepsy (ILAE) and the International Bureau for Epilepsy (IBE). *Epilepsia* **2005**, *46*, 470–472. [CrossRef] [PubMed]

35. Fisher, R.S.; Acevedo, C.; Arzimanoglou, A.; Bogacz, A.; Cross, J.H.; Elger, C.E.; Engel, J.J.; Forsgren, L.; French, J.A.; Glynn, M.; *et al.* A practical clinical definition of epilepsy. *Epilepsia* **2014**, *55*, 475–482. [CrossRef] [PubMed]

36. Ngugi, A.K.; Bottomley, C.; Kleinschmidt, I.; Sander, J.W.; Newton, C.R. Estimation of the burden of active and life-time epilepsy: A meta-analytic approach. *Epilepsia* **2010**, *51*, 883–890. [CrossRef] [PubMed]

37. Kwan, P.; Arzimanoglou, A.; Berg, A.T.; Brodie, M.J.; Allen Hauser, W.; Mathern, G.; Moshé, S.L.; Perucca, E.; Wiebe, S.; Frech, J. Definition of drug resistant epilepsy: Consensus proposal by the ad hoc Task Force of the ILAE Commission on Therapeutic Strategies. *Epilepsia* **2010**, *51*, 1069–1077. [CrossRef] [PubMed]

38. Kwan, P.; Schachter, S.C.; Brodie, M.J. Drug-resistant epilepsy. *N. Engl. J. Med.* **2011**, *365*, 919–926. [CrossRef] [PubMed]

39. Schmidt, D.; Schachter, S.C. Drug treatment of epilepsy in adults. *BMJ* **2014**, *348*, g254. [CrossRef] [PubMed]

40. Simonato, M.; Löscher, W.; Cole, A.J.; Dudek, F.E.; Engel, J., Jr.; Kaminski, R.M.; Loeb, J.A.; Scharfman, H.; Staley, K.J.; Velisek, L.; *et al.* Finding a better drug for epilepsy: Preclinical screening strategies and experimental trial design. *Epilepsia* **2012**, *53*, 1860–1867. [CrossRef] [PubMed]

41. Löscher, W.; Klitgaard, H.; Twyman, R.E.; Schmidt, D. New avenues for anti-epileptic drug discovery and development. *Nat. Rev. Drug Discov.* **2013**, *12*, 757–776. [CrossRef] [PubMed]

42. Moore, P.; Harding, K.E.; Clarkson, H.; Pickersgill, T.P.; Wardle, M.; Robertson, N.P. Demographic and clinical factors associated with changes in employment in multiple sclerosis. *Mult. Scler.* **2013**, *19*, 1647–1654. [CrossRef] [PubMed]

43. Pawate, S.; Bagnato, F. Newer agents in the treatment of multiple sclerosis. *Neurologist* **2015**, *19*, 104–117. [CrossRef] [PubMed]

44. Sand, I.K. Classification, diagnosis, and differential diagnosis of multiple sclerosis. *Curr. Opin. Neurol.* **2015**, *28*, 1–13. [CrossRef] [PubMed]

45. National Multiple Sclerosis Society. Available online: http://www.nationalmssociety.org/What-is-MS/Who-Gets-MS (accessed on 10 April 2015).

46. Lublin, F.D. New multiple sclerosis phenotypic classification. *Eur. Neurol.* **2014**, *72*, 1–5. [CrossRef] [PubMed]

47. Ontaneda, D.; Fox, R.J. Progressive multiple sclerosis. *Curr. Opin. Neurol.* **2015**, *28*, 1–7. [CrossRef] [PubMed]

48. Rovaris, M.; Confavreux, C.; Furlan, R.; Kappos, L.; Comi, G.; Filippi, M. Secondary progressive multiple sclerosis: Current knowledge and future challenges. *Lancet Neurol.* **2006**, *5*, 343–354. [CrossRef]

49. Lassman, H. Multiple sclerosis: Is there neurodegeneration independent from inflammation? *J. Neurol. Sci.* **2007**, *259*, 3–6. [CrossRef] [PubMed]

50. Ontaneda, D.; Hyland, M.; Cohen, J.A. Multiple sclerosis: New insights in pathogenesis and novel therapeutics. *Annu. Rev. Med.* **2012**, *63*, 389–404. [CrossRef] [PubMed]

51. Giordana, M.T.; Ferrero, P.; Grifoni, S.; Pellerino, A.; Naldi, A.; Montuschi, A. Dementia and cognitive impairment in amyotrophic lateral sclerosis: A review. *Neurol. Sci.* **2011**, *32*, 9–16. [CrossRef] [PubMed]

52. Al-Chalabi, A.; Jones, A.; Troakes, C.; King, C.; Al-Sarraj, S.; van den Berg, L.H. The genetics and neuropathology of amyotrophic lateral sclerosis. *Acta Neuropathol.* **2012**, *124*, 339–352. [CrossRef] [PubMed]

53. Endo, F.; Komine, O.; Fujimori-Tonou, N.; Katsuno, M.; Jin, S.; Watanabe, S.; Sobue, G.; Dezawa, M.; Wyss-Coray, T.; Yamanaka, K. Astrocyte-Derived TGF-β1 Accelerates Disease Progression in ALS Mice by Interfering with the Neuroprotective Functions of Microglia and T Cells. *Cell Rep.* **2015**, *11*, 1–13. [CrossRef] [PubMed]

54. Freer, C.; Hylton, T.; Jordan, H.M.; Kaye, W.E.; Singh, S.; Huang, Y. Results of Florida's Amyotrophic Lateral Sclerosis Surveillance Project, 2009–2011. *BMJ Open* **2015**, *5*, 1–6. [CrossRef] [PubMed]

55. Jelsone-Swain, L.; Persad, C.; Burkard, D.; Welsh, R.C. Action Processing and Mirror Neuron Function in Patients with Amyotrophic Lateral Sclerosis: An fMRI Study. *PLoS ONE* **2015**, *10*, 1–22. [CrossRef] [PubMed]

56. Mazzini, L.; Gelati, M.; Profico, D.C.; Sgaravizzi, G.; Pensi, M.P.; Muzi, G.; Ricciolini, C.; Nodari, L.R.; Carletti, S.; Giorgi, C.; *et al.* Human neural stem cell transplantation in ALS: Initial results from a phase I trial. *J. Transl. Med.* **2015**, *13*, 1–16. [CrossRef] [PubMed]

57. Chiò, A.; Logroscino, G.; Traynor, B.J.; Collins, J.; Simeone, J.C.; Goldstein, L.A.; White, L.A. Global epidemiology of amyotrophic lateral sclerosis: A systematic review of the published literature. *Neuroepidemiology* **2013**, *41*, 118–130. [CrossRef] [PubMed]

58. Byrne, S.; Walsh, C.; Lynch, C.; Bede, P.; Elamin, M.; Kenna, K.; McLaughlin, R.; Hardiman, O. Rate of familial amyotrophic lateral sclerosis: A systematic review and meta-analysis. *J. Neurol. Neurosurg. Psychiatry* **2011**, *82*, 623–627. [CrossRef] [PubMed]

59. Lunn, J.S.; Sakowski, S.A.; Feldman, E.L. Concise review: Stem cell therapies for amyotrophic lateral sclerosis: Recent advances and prospects for the future. *Stem Cells* **2014**, *32*, 1099–1109. [CrossRef] [PubMed]

60. Achi, E.Y.; Rudnicki, S.A. ALS and Frontotemporal Dysfunction: A Review. *Neurol. Res. Int.* **2012**, *2012*, 1–9. [CrossRef] [PubMed]

61. Oh, S.I.; Park, A.; Kim, H.J.; Oh, K.W.; Choi, H.; Kwon, M.J.; Ki, C.S.; Kim, H.T.; Kim, S.H. Spectrum of cognitive impairment in Korean ALS patients without known genetic mutations. *PLoS ONE* **2014**, *9*, 1–9. [CrossRef] [PubMed]

62. Montuschi, A.; Iazzolino, B.; Calvo, A.; Moglia, C.; Lopiano, L.; Restagno, G.; Brunetti, M.; Ossola, I.; lo Presti, A.; Cammarosano, S.; *et al.* Cognitive correlates in amyotrophic lateral sclerosis: A population-based study in Italy. *J. Neurol. Neurosurg. Psychiatry* **2015**, *86*, 168–173. [CrossRef] [PubMed]

63. Nicaise, C.; Mitrecic, D.; Falnikar, A.; Lepore, A.C. Transplantation of stem cell-derived astrocytes for the treatment of amyotrophic lateral sclerosis and spinal cord injury. *World J. Stem Cells* **2015**, *7*, 380–398. [CrossRef] [PubMed]

64. Borasio, G.D.; Miller, R.G. Clinical characteristics and management of ALS. *Semin. Neurol.* **2001**, *21*, 155–166. [CrossRef] [PubMed]

65. De Lima, P.R.; Brochetto-Braga, M.R. Hymenoptera venom review focusing on Apis mellifera. *J. Venom. Anim. Toxins incl. Trop. Dis.* **2003**. [CrossRef]

66. Kim, H.J.; Jeon, B.S. Is acupuncture efficacious therapy in Parkinson's Disease? *J. Neurol. Sci.* **2014**, *341*, 1–7. [CrossRef] [PubMed]

67. Ezzo, J.; Hadhazy, V.; Birch, S.; Lao, L.; Kaplan, G.; Hochberg, M.; Berman, B. Acupuncture for osteoarthritis of the knee: A systematic review. *Arthritis Rheum.* **2001**, *44*, 819–825. [CrossRef]

68. Lee, M.S.; Pittler, M.H.; Shin, B.C.; Kong, J.C.; Ernst, E. Bee venom acupuncture for musculoskeletal pain: A review. *J. Pain* **2008**, *9*, 289–297. [CrossRef] [PubMed]

69. Lee, J.D.; Park, H.J.; Chae, Y.; Lim, S. An overview of bee venom acupuncture in the treatment of arthritis. *Evid. Based Complement. Alternat. Med.* **2005**, *2*, 79–84. [CrossRef] [PubMed]

70. Castro, H.J.; Mendez-Lnocenio, J.I.; Omidvar, B.; Omidvar, J.; Santilli, J.; Nielsen, H.S., Jr.; Pavot, A.P.; Richert, J.R.; Bellanti, J.A. A phase I study of the safety of honeybee venom extract as a possible treatment for patients with progressive forms of multiple sclerosis. *Allergy Asthma Proc.* **2005**, *26*, 470–476. [PubMed]

71. Kwon, Y.B.; Lee, J.H.; Han, H.J.; Mar, W.C.; Beitz, A.J.; Lee, H.J. Bee venom injection into an acupunture point reduces arthritis associated edema and nociceptive responses. *Pain* **2001**, *90*, 271–280. [CrossRef]

72. Kwon, Y.B.; Kim, H.W.; Ham, T.W.; Yoon, S.Y.; Roh, D.H.; Jan, H.J.; Beitz, H.J.; Yang, I.S.; Lee, J.H. The anti-inflammatory effect of bee venom stimulation in a mouse air pouch model is mediated by adrenal medullary activity. *J. Neuroendocrinol.* **2003**, *15*, 93–96. [CrossRef] [PubMed]

73. Kang, S.S.; Pak, S.C.; Choi, S.H. The effect of whole bee venom on arthritis. *Am. J. Chin. Med.* **2002**, *30*, 73–80. [CrossRef] [PubMed]

74. Suh, S.J.; Kim, K.S.; Kim, M.J.; Chang, Y.C.; Lee, S.D.; Kim, M.S.; Kim, C.H. Effects of bee venom on protease activities and free radical damages in synovial fluid from type II collagen-induced rheumatoid arthritis rats. *Toxicol. In Vitro* **2006**, *20*, 1465–1471. [CrossRef] [PubMed]

75. Moon, D.O.; Park, S.Y.; Lee, K.J.; Heo, M.S.; Kim, K.C.; Kim, M.O.; Lee, J.D.; Choi, Y.H.; Kim, G.Y. Bee venom and melittin reduce proinflammatory mediators in lipopolysaccharide-stimulated BV2 microglia. *Int. Immunopharmacol.* **2007**, *7*, 1092–1101. [CrossRef] [PubMed]

76. Mahomoodally, M.F.; Bhugun, V.; Chutterdharry, G. Complementary and alternative medicines use against neurodegenerative diseases. *Adv. Pharmacol. Pharm.* **2013**, *1*, 103–123.

77. Kim, S.U.; Vellis, J. Microglia in health and disease. *J. Neurosci. Res.* **2005**, *81*, 302–313. [CrossRef] [PubMed]

78. Gonzalez-Scarano, F.; Baltuch, G. Microglia as mediators of inflammatory and degenerative diseases. *Annu. Rev. Neurosci.* **1999**, *22*, 219–240. [CrossRef] [PubMed]

79. Cho, S.; Shim, S.; Rhee, H.; Park, H.; Jung, W.; Moon, S.; Park, J.; Ko, C.; Cho, K.; Park, S. Effectiveness of acupuncture and bee venom acupuncture in idiopathic Parkinson's Disease. *Pakinsonism Relat. Disord.* **2012**, *18*, 948–952. [CrossRef] [PubMed]

80. Mirshafiey, A. Venom therapy in multiple sclerosis. *Neuropharmacology* **2007**, *53*, 353–361. [CrossRef] [PubMed]

81. Doo, A.R.; Kim, S.T.; Kim, S.N.; Moon, W.; Yin, C.S.; Chae, Y.; Park, H.J. Neuroprotective effects of bee venom pharmaceutical acupuncture in acute 1-methyl-4-phenyl-1,2,3,6-tetrahydropyridine-induced mouse model of Parkinson's Disease. *Neurol. Res.* **2010**, *32*, 88–91. [CrossRef] [PubMed]

82. Kim, S.; Park, J.; Kim, K.; Lee, W.; Kim, K.; Park, K. Melittin inhibits atherosclerosis in LPS/High-Fat treated mice through atheroprotective actions. *J. Atheroscler. Thromb.* **2011**, *18*, 1117–1126. [CrossRef] [PubMed]

83. Alvarez-Fisher, D.; Noelker, C.; Vulinovic, F.; Grünewald, A.; Chevarin, C.; Klein, C.; Oertel, W.H.; Hirsch, E.C.; Michel, P.P.; Hartmann, A. Bee venom and its component Apamin as neuroprotetive agents in Parkinson disease mouse model. *PLoS ONE* **2013**, *8*, e61700. [CrossRef] [PubMed]

84. Chung, E.S.; Kim, H.; Lee, G.; Park, S.; Kim, H.; Bae, H. Neuro-protective effects of bee venom by suppression of neuroinflammatory responses in a mouse model of Parkinson's Disease: Role of regulatory T cells. *Brain Behav. Immun.* **2012**, *26*, 1322–1330. [CrossRef] [PubMed]

85. Khalil, W.K.; Assaf, N.; ElShebiney, S.A.; Salem, N.A. Neuroprotective effects of bee venom acupuncture therapy against rotenone-induced oxidative stress and apoptosis. *Neurochem. Int.* **2014**, *80*, 79–86. [CrossRef] [PubMed]

86. Karimi, A.; Ahmadi, F.; Parivar, K.; Nabiuni, M.; Haghighi, S.; Imani, S.; Afrouzi, H. Effect of honey bee venom on lewis rats with experimental allergic encephalomyelitis, a model for multiple sclerosis. *Iran. J. Pharm. Res.* **2012**, *11*, 671–678. [PubMed]

87. Yang, E.J.; Jiang, J.H.; Lee, S.M.; Yang, S.C.; Hwang, H.S.; Lee, M.S.; Choi, S.M. Bee venom attenuates neuroinflammatory events and extends survival in amyotrophic lateral sclerosis models. *J. Neuroinflammation* **2010**, *15*, 7–69. [CrossRef] [PubMed]

88. Cai, M.; Choi, S.M.; Yang, E.J. The effects of bee venom acupuncture on the central nervous system and muscle in an animal hSOD1G93A mutant. *Toxins* **2015**, *7*, 846–858. [CrossRef] [PubMed]

89. Han, S.M.; Kim, J.M.; Park, K.K.; Chang, Y.C.; Pak, S.C. Neuroprotective effects of Melittin on hydrogen peroxide-induced apoptotic cell death in neuroblastoma SH-SY5Y cells. *BMC Complement. Altern. Med.* **2014**, *14*, 1–8. [CrossRef] [PubMed]

90. Kim, H.W.; Ham, T.W.; Yoon, S.Y.; Yang, I.S.; Lee, H.J.; Lee, J.H. General pharmacological profiles of bee venom and its water soluble fractions in rodent models. *J. Vet. Sci.* **2004**, *5*, 309–318. [PubMed]

91. Han, S.; Lee, K.; Yeo, J.; Kweoh, H.; Woo, S.; Lee, M.; Baek, H.; Kim, S.; Park, K. Effect of honey bee venom on microglial cells nitric oxide and tumor necrosis factor-alpha production stimulated by LPS. *J. Ethnopharmacol.* **2007**, *111*, 176–181. [CrossRef] [PubMed]

92. Rekka, E.; Kourounakis, L.; Kourounakis, P. Antioxidant activity of and interleukin production affected by honey bee venom. *Arzneimittelforschung* **1990**, *40*, 912–913. [PubMed]

93. Nam, K.W.; Je, K.H.; Lee, J.H.; Han, H.J.; Lee, H.J.; Kang, S.K.; Mar, W. Inhibition of COX-2 activity and proinflammatory cytokines (TNFalpha and IL-1beta) production by water-soluble sub-fractionated parts from bee (Apis mellifera) venom. *Arch. Pharm. Res.* **2003**, *26*, 383–388. [CrossRef] [PubMed]

94. Jang, H.S.; Kim, S.K.; Han, J.B.; Ahn, H.J.; Bae, H.; Min, B.I. Effects of bee venom on the pro-inflammatory responses in RAW264.7 macrophage cell line. *J. Ethnopharmacol.* **2005**, *99*, 157–160. [CrossRef] [PubMed]

95. Doo, A.R.; Kim, S.N.; Kim, S.T.; Park, J.Y.; Chung, S.H.; Choe, B.Y.; Chae, Y.; Lee, H.; Yin, C.S.; Park, H.J. Bee venom protects SH-SY5Y human neuroblastoma cells from 1-methyl-4-phenylpyridinium-induced apoptotic cell death. *Brain Res.* **2012**, *1429*, 106–115. [CrossRef] [PubMed]

96. Namaka, M.; Crook, A.; Doupe, A.; Kler, K.; Vaconcelos, M.; Klowak, M.; Gong, Y.; Wojewnik-Smith, A.; Melanson, M. Examining the evidence: Complementary adjunctive therapies for multiple sclerosis. *Neurol. Res.* **2008**, *30*, 710–719. [CrossRef] [PubMed]

97. Apel, A.; Greim, B.; Zetti, U.K. How frequently do patients with multiple sclerosis use complementary and alternative medicine? *Complement. Ther. Med.* **2005**, *13*, 258–263. [CrossRef] [PubMed]

98. The American Apitherapy Society Inc. Available online: http://www.apitherapy.org (accessed on 13 April 2015).

99. Bowling, A.C. Complementary and alternative medicine and multiple sclerosis. *Neurol. Clin.* **2011**, *29*, 465–480. [CrossRef] [PubMed]

100. Wesselius, T.; Heersema, D.J.; Mostert, J.P.; Heerings, M.; Admiraal-Behloul, F.; Talebian, A.; van Buchem, M.A.; de Keyser, J. A randomized crossover study of bee sting therapy for multiple sclerosis. *Neurology* **2005**, *65*, 1764–1768. [CrossRef] [PubMed]

101. Hauser, R.A.; Daguio, M.; Wester, D.; Hauser, M.; Kirchman, A.; Skinkis, C. Bee-venom therapy for treating multiple sclerosis - a clinical trial. *Altern. Complement. Ther.* **2001**, *7*, 37–45. [CrossRef]

102. Yang, E.J.; Choi, S.M. Synuclein modification in an ALS animal model. *Evid. Based Complement Altern. Med.* **2013**, *2013*, 1–7. [CrossRef] [PubMed]

103. Kim, S.H.; Jung, S.Y.; Lee, K.W.; Lee, S.H.; Cai, M.; Choi, S.M.; Yang, E.J. Bee venom effects on ubiquitin proteasome system in hSOD1(G85R)-expressing NSC34 motor neuron cells. *BMC Complement. Altern. Med.* **2013**, *13*, 1–9.

104. Cunha, A.O.S.; Mortari, M.R.; Oliveira, L.; Carolino, R.O.G.; Coutinho-Netto, J.; Santos, W.F. Anticonvulsant effects of the wasp Polybia ignobilis venom on chemically induced seizures and action on GABA and glutamate receptors. *Comp. Biochem. Physiol. C Toxicol. Pharmacol.* **2005**, *141*, 50–57. [CrossRef] [PubMed]

105. Mortari, M.R.; Cunha, A.O.S.; Oliveira, L.; Vieira, E.B.; Gelfuso, E.A.; Coutinho-Netto, J.; Santos, W.F. Anticonvulsant and behavioural effects of the denatured venom of the social wasp Polybia occidentalis (Polistinae, Vespidae). *Basic Clin. Pharmacol. Toxicol.* **2005**, *97*, 289–295. [CrossRef] [PubMed]

106. Couto, L.L.; dos Anjos, L.C.; Araujo, M.A.F.; Mourão, C.A.; Schwartz, C.A.; Ferreira, L.B.; Mortari, M.R. Anticonvulsant and anxiolytic activity of the peptide fraction isolated from the venom of the social wasp Polybia paulista. *Pharmacogn. Mag.* **2012**, *8*, 292–299. [PubMed]

107. Danielisová, V.; Gottlieb, M.; Némethová, M.; Burda, J. Effects of bradykinin postconditioning on endogenous antioxidant enzyme activity after transient forebrain ischemia in rat. *Neurochem. Res.* **2008**, *33*, 1057–1064. [CrossRef] [PubMed]

108. Danielisová, V.; Gottlieb, M.; Némethová, M.; Kravcuková, P.; Domoráková, I.; Mechírová, E.; Burda, J. Bradykinin postconditioning protects pyramidal CA1 neurons against delayed neuronal death in rat hippocampus. *Cell. Mol. Neurobiol.* **2009**, *29*, 871–878. [CrossRef] [PubMed]

109. Pizzo, A.B.; Fontana, A.C.K.; Coutinho-Netto, J.; Santos, W.F. Effects of the crude venom of social wasp Agelaia vicina on γ-Aminobutyric acid and glutamate uptake in synapsosomes from rat cerebral cortex. *J. Biochem. Mol. Toxicol.* **2000**, *14*, 88–94. [PubMed]

110. Krzyzanowska, W.; Pomierny, B.; Filip, M.; Pera, J. Glutamate transporters in brain ischemia: To modulate or not? *Acta Pharmacol. Sin.* **2014**, *35*, 444–462. [CrossRef] [PubMed]

111. Soni, N.; Reddy, B.V.; Kumar, P. GLT-1 transporter: An effective pharmacological target for various neurological disorders. *Pharmacol. Biochem. Behav.* **2014**, *127*, 70–81. [CrossRef] [PubMed]

112. Lee, J.; Son, M.; Choi, J.; Jun, J.; Kim, J.; Lee, M. Bee venom acupuncture for rheumatoid arthritis: A systematic review of randomised clinical trials. *BMJ* **2014**, *4*, e006140. [CrossRef] [PubMed]

113. Moreno, M.; Giralt, E. Three valuable peptides from bee and wasp venoms for therapeutic and biotechnological use: Mellitin, Apamin and Mastoparan. *Toxins* **2015**, *7*, 1126–1150. [CrossRef] [PubMed]

114. Raghuraman, H.; Chattopadhyay, A. Melittin: A membrane-active peptide with diverse functions. *Biosci. Rep.* **2007**, *27*, 189–223. [CrossRef] [PubMed]

115. Son, D.; Lee, J.; Lee, Y.; Song, H.; Lee, C.; Hong, J. Therapeutic application of anti-arthritis, pain-releasing, anti-cancer effects of bee venom and its constituent compounds. *Pharmacol. Ther.* **2007**, *115*, 246–270. [CrossRef] [PubMed]

116. Vyatchannikov, N.K.; Sinka, A.Y. The effect of mellitin—The major constituent of the bee venom-on the central nervous system. *Farmakol. Toksikol.* **1973**, *36*, 526–530.

117. Ishay, J.; Ben-Shachrar, D.; Elazar, Z.; Kaplinsky, E. Effects of melittin on the central nervous system. *Toxicon* **1975**, *13*, 277–283. [CrossRef]

118. Yang, E.J.; Kim, S.H.; Yang, S.C.; Lee, S.M.; Choi, S.M. Melittin restores proteasome function in an animal model of ALS. *J. Neuroinflammation* **2011**, *8*, 1–9. [CrossRef] [PubMed]

119. ChemSpider, Melittin. Available online: http://www.chemspider.com/Chemical-Structure.26567345.html (accessed on 11 August 2015).

120. ChemSpider, Apamin. Available online: http://www.chemspider.com/Chemical-Structure.21169555.html (accessed on 11 August 2015).

121. Toxin and Toxin Target Database, Alpha-pompilidotoxin. Available online: http://www.t3db.ca/toxins/T3D2490 (accessed on 11 August 2015).

122. Toxin and Toxin Target Database, Beta-pompilidotoxin. Available online: http://www.t3db.ca/toxins/T3D2491 (accessed on 11 August 2015).

123. ChemSpider, Philanthotoxin. Available online: http://www.chemspider.com/Chemical-Structure.103077.html (accessed on 11 August 2015).

124. ChemSpider, Bradykinin. Available online: http://www.chemspider.com/Chemical-Structure.388341.html?rid=15413dd8-4d59-4d52-aac6-45e35a46f78d (accessed on 11 August 2015).

125. ChemSpider, Transportan. Available online: http://www.chemspider.com/Chemical-Structure.17290614.html?rid=5ed04597-4113-4665-98c5-cef78b91243 (accessed on 11 August 2015).

126. Dantas, C.G.; Nunes, T.L.G.M.; Paixão, A.O.; Reis, F.P.; Júnior, W.L.; Cardoso, J.C.; Gramacho, K.P.; Gomes, M.Z. Pharmacological evaluation of bee venom and Melittin. *Rev. Bras. Farmacogn.* **2014**, *24*, 67–72. [CrossRef]

127. Lamy, C.; Goodchild, S.J.; Weatherall, K.L.; Jane, D.E.; Liégeois, J.F.; Seutin, V.; Marrion, N.V. Allosteric block of $K_{Ca}2$ channels by Apamin. *J. Biol. Chem.* **2010**, *285*, 27067–27077. [CrossRef] [PubMed]

128. Lam, J.; Coleman, N.; Garing, A.L.; Wulff, H. The therapeutic potential of small-conductance $K_{Ca}2$ channels in neurodegenerative and psychiatric diseases. *Expert Opin. Ther. Targets* **2013**, *17*, 1203–1220. [CrossRef] [PubMed]

129. Stackman, R.W.; Hammond, R.S.; Linardatos, E.; Gerlach, A.; Maylie, J.; Adelman, J.P.; Tzounopoulos, T. Small conductance Ca^{2+}-activated K^+ channels modulate synaptic plasticity and memory encoding. *J. Neurosci.* **2002**, *22*, 10163–10171. [PubMed]

130. Gati, C.; Mortari, M.; Schwartz, E. Towards Therapeutic Applications of Arthropod Venom K(+)-channel blockers in CNS neurologic diseases involving memory acquisition and storage. *J. Toxicol.* **2012**, *2012*. [CrossRef] [PubMed]

131. Adelman, J.P.; Maylie, J.; Sah, P. Small-conductance Ca^{2+}-activated K^+ channels: Form and function. *Annu. Rev. Physiol.* **2012**, *74*, 245–269. [CrossRef] [PubMed]

132. Salthun-Lassalle, B.; Hirsch, E.C.; Wolfart, J.; Ruberg, M.; Michel, P.P. Rescue of Mesencephalic Dopaminergic Neurons in Culture by Low-Level Stimulation of Voltage-Gated. *J. Neurosci.* **2004**, *24*, 5922–5930. [CrossRef] [PubMed]

133. Hartmann, A.; Bonnet, A.M.; Schüpbach, M. Medicament for Treating Parkinson Disease. U.S. Patent N° US 8357658 B2, 22 January 2013.

134. Tzounopoulos, T.; Stackman, R. Enhancing Synaptic Plasticity and Memory: A role for small-conductance Ca^{2+}-activated K^+ channels. *Neuroscientist* **2003**, *9*, 434–439. [CrossRef] [PubMed]

135. Alkon, D.L.; Etcheberrigaray, R.; Ito, E.; Gibson, G.E. Diagnostic Tests for Alzheimers Disease. U.S. Patent N° 5580748 A, 3 December 1996. Available online: http://www.google.com.ar/patents/US5580748 (accessed on 13 April 2015).

136. Masters, C.L.; Bush, A.I.; Beyreuther, K.T. Method of Assaying for Alzheimer's Disease. U.S. Patent N° 5705401 A, 6 January 1998. Available online: http://www.google.com.ar/patents/US5705401 (accessed on 13 April 2015).

137. Potter, H. Method of Diagnosing and Monitoring a Treatment for Alzheimer's Disease. Patent N° 5778893 A, 14 July 1998. Available online: http://www.google.com/patents/US5778893 (accessed on 13 April 2015).

138. Garcia, M.L.; Galvez, A.; Garcia-Calvo, M.; King, V.F.; Vazquez, J.; Kaczorowski, G.J. Use of toxins to study potassium channels. *J. Bioenerg. Biomembr.* **1991**, *23*, 615–646. [CrossRef] [PubMed]

139. Van der Staay, F.J.; Fanelli, R.J.; Blokland, A.; Schmidt, B.H. Behavioral effects of Apamin, a selective inhibitor of the SK_{Ca}-channel, in mice and rats. *Neurosci. Biobehav. Rev.* **1999**, *23*, 1087–1110. [CrossRef]

140. Konno, K.; Miwa, A.; Takayama, H.; Hisada, M.; Itagaki, Y.; Naoki, H.; Yasuhada, T.; Kawai, N. α-Pompilidotoxin (α-PMTX), a novel neurotoxin from the venom of a solitary wasp, facilitates transmission in the crustacean neuromuscular synapse. *Neurosc. Lett.* **1997**, *238*, 99–102. [CrossRef]

141. Konno, K.; Hisada, M.; Itagaki, Y.; Naoki, H.; Kawai, N.; Miwa, A.; Yasuhara, T.; Takayama, H. Isolation and structure of Pompilidotoxins, novel peptide neurotoxins in solitary wasp venoms. *Biochem. Biophys. Res. Commun.* **1998**, *250*, 612–616. [CrossRef] [PubMed]

142. Schiavon, E.; Stevens, M.; Zaharenko, A.J.; Konno, K.; Tytgat, J.; Wanke, E. Voltage-gated sodium channel isoform-specific effects of Pompilidotoxins. *FEBS J.* **2010**, *277*, 918–930. [CrossRef] [PubMed]

143. Harsch, A.; Konno, K.; Takayama, H.; Kawai, N.; Robinson, H. Effects of α-pompilidotoxin on synchronized firing in networks of rat cortical neurons. *Neurosc. Lett.* **1998**, *252*, 49–52. [CrossRef]

144. Sahara, Y.; Gotoh, M.; Konno, K.; Miwa, A.; Tsubokawa, H.; Robinson, H.P.; Kawai, N. A new class of neurotoxin from wasp venom slows inactivation of sodium current. *Eur. J. Neurosci.* **2000**, *12*, 1961–1970. [CrossRef] [PubMed]

145. Magloire, V.; Czarnecki, A.; Anwander, H.; Streit, J. β-pompilidotoxin modulates spontaneous activity and persistent sodium currents in spinal networks. *Neuroscience* **2011**, *172*, 129–138. [CrossRef] [PubMed]

146. Yokota, H.; Tsubokawa, H.; Miyawaki, T.; Konno, K.; Nakayama, H.; Masuzawa, T.; Kawai, N. Modulation of synaptic transmission in hippocampal CA1 neurons by a novel neurotoxin (β-pompilidotoxin) derived from wasp venom. *Neurosci. Res.* **2001**, *41*, 365–371. [CrossRef]

147. Mantegazza, M.; Curia, G.; Biagini, G.; Ragsdale, D.S.; Avoli, M. Voltage-gated sodium channels as therapeutic targets in epilepsy and other neurological disorders. *Lancet Neurol.* **2010**, *9*, 413–424. [CrossRef]

148. Eijkelkamp, N.; Linley, J.E.; Baker, M.D.; Minett, M.S.; Cregg, R.; Werdehausen, R.; Rugiero, F.; Wood, J.N. Neurological perspectives on voltage-gated sodium channels. *Brain* **2012**, *135*, 2585–2612. [CrossRef] [PubMed]

149. Pizzo, A.B.; Beleboni, R.O.; Fontana, A.C.; Ribeiro, A.M.; Miranda, A.; Coutinho-Netto, J.; dos Santos, W.F. Characterization of the actions of AvTx 7 isolated from Agelaia vicina (Hymenoptera: Vespidae) wasp venom on synaptosomal glutamate uptake and release. *J. Biochem. Mol. Toxicol.* **2004**, *18*, 61–68. [CrossRef] [PubMed]

150. Tian, C.; Zhu, R.; Zhu, L.; Qiu, T.; Cao, Z.; Kang, T. Potassium channels: Structures, diseases and modulators. *Chem. Biol. Drug Des.* **2014**, *83*, 1–26. [CrossRef] [PubMed]

151. Jensen, H.B.; Ravnborg, M.; Dalgas, U.; Stenager, E. 4-Aminopyridine for symptomatic treatment of multiple sclerosis: A systematic review. *Ther. Adv. Neurol. Disord.* **2014**, *7*, 97–113. [CrossRef] [PubMed]

152. Davis, F.A.; Stefoski, D.; Rush, J. Orally administered 4-aminopyridine improves clinical signs in multiple sclerosis. *Ann. Neurol.* **1990**, *27*, 186–192. [CrossRef] [PubMed]

153. Luca, C.C.; Singer, C. Can 4-aminopyridine modulate dysfunctional gait networks in Parkinson's Disease? *Parkinsonism Relat. Disord.* **2013**, *19*, 777–782. [CrossRef] [PubMed]

154. Hirai, Y.; Yasuhara, T.; Yoshida, H.; Nakajima, T.; Fujino, M.; Kitada, C. A new mast cell degranulatin peptide "mastoparano" in the venom of Vespula lewisii. *Chem. Pharm. Bull.* **1979**, *27*, 1942–1944. [CrossRef] [PubMed]

155. Blazquez, P.S.; Garzon, J. Mastoparan reduces the supraspinal analgesia mediated by IX/6-opioid receptors in mice. *Eur. J. Pharm.* **1994**, *258*, 159–162. [CrossRef]

156. Blazquez, P.S.; Garzon, J. αN-Acetyl-β-Endorphin-(1_31) Disrupts the Diminishing Effect of Mastoparan on Opioid- and Clonidine-Evoked Supraspinal Antinociception in Mice. *JPET* **1995**, *273*, 787–792.

157. Yandek, L.E.; Pokorny, A.; Floren, A.; Knoelke, K.; Langel, U.; Almeida, P.F.F. Mechanism of the Cell-Penetrating Peptide Transportan 10 permeation of lipid bilayers. *Biophys. J.* **2007**, *92*, 2434–2444. [CrossRef] [PubMed]

158. Rocha, T.; Souza, B.M.; Palma, M.S.; Cruz-Höfling, M.A.; Harris, J.B. The neurotoxicological effects of mastoparano Polybia-MPII at the murine neuromuscular junction: An ultrastructural and immunocytochemical study. *Histochem. Cell Biol.* **2009**, *132*, 395–404. [CrossRef] [PubMed]

159. Zhang, P.; Ray, R.; Singh, B.R.; Ray, P. Mastoaparan-7 rescues botulinum toxin-A poisoned neurons in a mause spinal cord cell culture model. *Toxicon* **2013**, *76*, 37–43. [CrossRef] [PubMed]

160. Souza, B.M.; Cabrera, M.P.S.; Neto, J.R.; Palma, M.S. Investigating the effect of different positioning of lysine residues along the peptide chain of mastoparans for their secondary structures and biological activities. *Amino Acids* **2011**. [CrossRef]

161. Silva, A.V.R.; Souza, B.M.; Cabrera, M.P.S.; Dias, N.B.; Gomes, P.C.; Neto, J.R.; Stabeli, R.G.; Palma, M.S. The effects of C-terminal amidation of mastoparans on their biological actions and interactions with membrane-mimetic systems. *Biochim. Biophys. Acta* **2014**, *1838*, 2357–2368. [CrossRef] [PubMed]

162. Fanghänel, S.; Wadhwani, P.; Strandberg, E.; Verdurmen, W.P.; Bürck, J.; Ehni, S.; Mykhailiuk, P.K.; Afonin, S.; Gerthsen, D.; Komarov, I.V.; *et al.* Structure analysis and conformational transitions of the cell penetrating peptide transportan 10 in the membrane-bound state. *PLoS ONE* **2014**, *9*, e99653. [CrossRef] [PubMed]

163. Chen, Y.; Liu, L. Modern methods for delivery drugs across the blood-brain barrier. *Adv. Drug Deliv. Rev.* **2012**, *64*, 640–655. [CrossRef] [PubMed]

164. Pooga, M.; Hällbrink, M.; Zorko, M.; Langel, U. Cell penetration by transportan. *FASEB J.* **1998**, *12*, 67–77. [PubMed]

165. Yandek, L.E.; Pokorny, A.; Florén, A.; Knoelke, K.; Langel, U.; Almeida, P.F. Mechanism of the cell-penetrating peptide transportan 10 permeation of lipid bilayers. *Biophys. J.* **2007**, *92*, 2434–2444. [CrossRef] [PubMed]

166. Webling, K.E.; Runesson, J.; Bartfai, T.; Langel, U. Galanin receptors and ligands. *Front. Endocrinol. (Lausanne)* **2012**, *3*, 1–14. [CrossRef] [PubMed]

167. Counts, S.E.; He, B.; Che, S.; Ginsberg, S.D.; Mufson, E.J. Galanin hyperinnervation upregulates choline acetyltransferase expression in cholinergic basal forebrain neurons in Alzheimer's disease. *Neurodegener. Dis.* **2008**, *5*, 228–231. [CrossRef] [PubMed]

168. Montecucco, C.; Schiavo, G. Mechanism of action of tetanus and botulinum neurotoxins. *Mol. Microbiol.* **1995**, *13*, 1–8. [CrossRef]

169. Simpson, L. The life story of a botulinum toxin molecule. *Toxicon* **2013**, *68*, 40–59. [CrossRef] [PubMed]

170. Jones, S.; Howl, J. Biological applications of the receptor mimetic peptide Mastoparan. *Curr. Protein Pept. Sci.* **2006**, *7*, 501–508. [CrossRef] [PubMed]

171. Todokoro, Y.; Yumen, I.; Fukushima, K.; Kamg, S.; Park, J.; Kohno, T.; Wakamatsu, K.; Akutsu, H.; Fujiwara, T. Structure of tightly membrane-bound Mastoparan-X, a G-protein-activating peptide, determined by solid-state NMR. *Biophys. J.* **2006**, *91*, 1368–1379. [CrossRef] [PubMed]

172. Higashijima, T.; Uzu, S.; Nakajima, T.; Ross, E.M. Mastoparan, a peptide toxin from wasp venom, mimics receptors by activating GTP-biding regulatory proteins (G proteins). *JBC* **1988**, *263*, 6491–6494.

173. Lagerström, M.C.; Schiöth, H.B. Structural diversity of G protein-coupled receptors and significance for drug discovery. *Nat. Rev. Drug Discov.* **2008**, *7*, 339–357. [CrossRef] [PubMed]

174. Thathiah, A.; De Strooper, B. The role of G protein-coupled receptors in the pathology of Alzheimer's Disease. *Nat. Rev. Neurosci.* **2011**, *12*, 73–87. [CrossRef] [PubMed]

175. Guixa-Gonzalez, R.; Bruno, A.; Marti-Solano, M.; Selent, J. Crosstalk within GPCR heteromers in schizophrenia and Parkinson's Disease: Physical or just functional? *Curr. Med. Chem.* **2012**, *19*, 1119–1134. [CrossRef] [PubMed]

176. González-Maeso, J.; Rodríguez-Puertas, R.; Meana, J.J.; García-Sevilla, J.A.; Guimón, J. Neurotransmitter receptor-mediated activation of G-proteins in brains of suicide victims with mood disorders: Selective supersensitivity of alpha(2A)-adrenoceptors. *Mol. Psychiatry* **2002**, *7*, 755–767. [CrossRef] [PubMed]

177. Nakajima, T.; Uzu, S.; Wakamatsu, K.; Saito, K.; Miyazawa, T.; Yasuhara, T.; Tsukamoto, Y.; Fujino, M. Amphiphilic peptides in wasp venom. *Biopolymers* **1986**, *25*, S115–S121. [PubMed]

178. Konno, K.; Palma, M.S.; Hitara, I.Y.; Juliano, M.A.; Juliano, L.; Yasuhara, T. Identification of bradykinins in solitary wasp venoms. *Toxicon* **2002**, *40*, 309–312. [CrossRef]

179. Picolo, G.; Hisada, M.; Moura, A.B.; Machado, M.F.; Conceição, I.M.; Melo, R.L.; Oliveira, V.; Lima-Landman, M.T.; Cury, Y.; Konno, K.; *et al.* Bradykinin-related peptides in the venom of the solitary wasp Cyphononyx fulvognathus. *Biochem. Pharmacol.* **2010**, *79*, 478–486. [CrossRef] [PubMed]

180. Rocha e Silva, M.; Beraldo, W.T.; Rosenfield, G. Bradykinin, a hypotensive and smooth muscle stimulating factor released from plasma globulin by snake venoms and by trypsin. *Am. J. Physiol.* **1949**, *156*, 261–273. [PubMed]

181. Moreau, M.E.; Garbacki, N.; Molinaro, G.; Brown, N.J.; Marceau, F.; Adam, A. The kallikrein-kinin system: Current and future pharmacological targets. *J. Pharmacol. Sci.* **2005**, *99*, 6–38. [CrossRef] [PubMed]

182. Piek, T.; Hue, B.; Mony, L.; Nakajima, T.; Pelhate, M.; Yasuhara, T. Block of synaptic transmission in insect CNS by toxins from the venom of the wasp Megascolia flavifrons (Fab.). *Comp. Biochem. Physiol. C* **1987**, *87*, 287–295. [CrossRef]

183. Noda, M.; Kariura, Y.; Pannasch, U.; Nishikawa, K.; Wang, L.; Seike, T.; Ifuku, M.; Kosai, Y.; Wang, B.; Nolte, C.; *et al.* Neuroprotective role of bradykinin because of the attenuation of pro-inflammatory cytokine release from activated microglia. *J. Neurochem*. **2007**, *101*, 397–410. [CrossRef] [PubMed]

184. Golias, Ch.; Charalabopoulos, A.; Stagikas, D.; Charalabopoulos, K.; Batistatou, A. The kinin system—Bradykinin: Biological effects and clinical implications. Multiple role of the kinin system—Bradykinin. *Hippokratia* **2007**, *11*, 124–128. [PubMed]

185. Thornton, E.; Ziebell, J.M.; Leonard, A.V.; Vink, R. Kinin receptor antagonists as potential neuroprotective agents in central nervous system injury. *Molecules* **2010**, *15*, 6598–6618. [CrossRef] [PubMed]

186. Yasuyoshi, H.; Kashii, S.; Zhang, S.; Nishida, A.; Yamauchi, T.; Honda, Y.; Asano, Y.; Sato, S.; Akaike, A. Protective effect of bradykinin against glutamate neurotoxicity in cultured rat retinal neurons. *Investig. Ophtalmol. Vis. Sci.* **2000**, *41*, 2273–2278.

187. Mortari, M.R.; Cunha, A.O.S.; Carolino, R.O.G.; Coutinho-Netto, J.C.; Tomaz, N.P.; Coimbra, N.C.; Santos, W.F. Inhibition of acute nociceptive responses in rats after i.c.v. injection of Thr6-bradykinin, isolated from the venom of the social wasp, Polybia occidentalis. *BJP* **2007**, *151*, 860–869. [CrossRef] [PubMed]

188. Pellegrini, M.; Mierke, D.F. Threonine6-bradykinin: Molecular dynamics simulations in a biphasic membrane mimetic. *J. Med. Chem.* **1997**, *40*, 99–104. [CrossRef] [PubMed]

189. Strømgaard, K.; Andersen, K.; Krogsgaard-Larsen, P.; Jaroszewski, J.W. Recent Advances in the Medicinal Chemistry of Polyamine Toxins. *Mini Rev. Med. Chem.* **2001**, *1*, 317–338. [CrossRef] [PubMed]

190. Andersen, T.F.; Tikhonov, D.B.; Bølcho, U.; Bolshakov, K.; Nelson, J.K.; Pluteanu, F.; Mellor, R.; Egebjerg, J.; Strømgaard, K. Uncompetitive Antagonism of AMPA Receptors: Mechanistic Insights from Studies of Polyamine Toxin Derivatives. *J. Med. Chem.* **2006**, *49*, 5414–5423. [CrossRef] [PubMed]

191. Eldefrawi, A.T.; Eldefrawi, M.E.; Konno, K.; Mansour, N.A.; Nakanishi, K.; Oltz, E.; Usherwood, P.N.R. Structure and synthesis of a potent glutamate receptor antagonist in wasp venom. *Proc. Natl. Acad. Sci. USA* **1988**, *85*, 4910–4913. [CrossRef] [PubMed]

192. Mellor, I.R.; Usherwood, P.N.R. Targeting ionotropic receptors with polyamine-containing toxins. *Toxicon* **2004**, *43*, 493–508. [CrossRef] [PubMed]

193. Strømgaard, K.; Jensen, L.S.; Vogensen, S.B. Polyamine toxins: Development of selective ligands for ionotropic receptors. *Toxicon* **2005**, *45*, 249–254. [CrossRef] [PubMed]

194. Strømgaard, K.; Mellor, I. AMPA Receptor Ligands: Synthetic and Pharmacological Studies of Polyamines and Polyamine Toxins. *Med. Res. Rev.* **2004**, *24*, 589–620. [CrossRef] [PubMed]

195. Traynelis, S.F.; Wollmuth, L.P.; McBain, C.J.; Menniti, F.S.; Vance, K.M.; Ogden, K.K.; Kasper, B.; Yuan, H.H.; Myers, S.J.; Dingledine, R. Glutamate Receptor Ion Channels: Structure, Regulation, and Function. *Pharmacol. Rev.* **2010**, *62*, 405–496. [CrossRef] [PubMed]

196. Lemoine, D.; Jiang, A.; Taly, A.; Chataigneau, R.; Specht, A.; Grutter, T. Ligand-Gated Ion Channels: New Insights into Neurological Disorders and Ligand Recognition. *Chem. Rev.* **2012**, *112*, 6285–6318. [CrossRef] [PubMed]

197. Poulsen, M.H.; Simon, L.; Strømgaard, K.; Kristensen, A.S. Inhibition of AMPA Receptors by Polyamine Toxins is Regulated by Agonist Efficacy and Stargazin. *Neurochem. Res.* **2014**, *39*, 1906–1913. [CrossRef] [PubMed]

198. Lipton, S.A. Pathologically activated therapeutics for neuroprotection. *Nat. Rev. Neurosci.* **2007**, *8*, 803–808. [CrossRef] [PubMed]

199. Johnson, J.W.; Glasgow, N.G.; Povysheva, N.V. Recent insights into the mode of action of memantine and ketamine. *Curr. Opin. Pharmacol.* **2015**, *20*, 54–63. [CrossRef] [PubMed]

200. Tikhonov, D.B. Ion channels of glutamate receptors: Structural modeling. *Mol. Membr. Biol.* **2007**, *24*, 135–147. [CrossRef] [PubMed]

201. Nørager, N.G.; Jensen, C.B.; Rathje, M.; Andersen, J.; Madsen, K.L.; Kristensen, A.S.; Strømgaard, K. Development of potent fluorescent polyamine toxins and application in labeling of ionotropic glutamate receptors in hippocampal neurons. *ACS Chem. Biol.* **2013**, *8*, 2033–2041. [CrossRef] [PubMed]

© 2015 by the authors; licensee MDPI, Basel, Switzerland. This article is an open access article distributed under the terms and conditions of the Creative Commons Attribution (CC-BY) license (http://creativecommons.org/licenses/by/4.0/).

toxins

MDPI

Review

Therapeutic Effects of Bee Venom on Immunological and Neurological Diseases

Deok-Sang Hwang [1,†], Sun Kwang Kim [2,†] and Hyunsu Bae [2,*]

[1] Department of Korean Medicine Obstetrics and Gynecology, College of Korean Medicine, Kyung Hee University, 26 Kyungheedae-ro, Dongdaemun-gu, Seoul 130-701, Korea; soulhus@gmail.com
[2] Department of Physiology, College of Korean Medicine, Kyung Hee University, 26 Kyungheedae-ro, Dongdaemun-gu, Seoul 130-701, Korea; skkim77@khu.ac.kr
* Correspondence: hbae@khu.ac.kr; Tel.: +82-2-961-9316; Fax: +82-70-4194-9316
† These authors contributed equally to this work.

Academic Editor: Lai Ren
Received: 15 May 2015; Accepted: 24 June 2015; Published: 29 June 2015

Abstract: Bee Venom (BV) has long been used in Korea to relieve pain symptoms and to treat inflammatory diseases, such as rheumatoid arthritis. The underlying mechanisms of the anti-inflammatory and analgesic actions of BV have been proved to some extent. Additionally, recent clinical and experimental studies have demonstrated that BV and BV-derived active components are applicable to a wide range of immunological and neurodegenerative diseases, including autoimmune diseases and Parkinson's disease. These effects of BV are known to be mediated by modulating immune cells in the periphery, and glial cells and neurons in the central nervous system. This review will introduce the scientific evidence of the therapeutic effects of BV and its components on several immunological and neurological diseases, and describe their detailed mechanisms involved in regulating various immune responses and pathological changes in glia and neurons.

Keywords: Bee Venom; immunological diseases; neurological diseases

1. Introduction

Bee Venom (BV) therapy is a form of medicine originated from the ancient Greece and China. Several scientific reports suggesting the anti-rheumatic and anti-inflammatory effects of BV have been published for a hundred years [1,2]. In Korea, BV has long been used to relieve pain and to treat various diseases, such as arthritis, rheumatism, herniation nucleus pulpous, cancer, asthma, and skin diseases [3–5]. It is administered systemically or in the form of chemical stimulation of acupoints, so called "BV acupuncture" or "apipuncture". BV is known to contain many active components, including peptides (e.g., melittin and apamin), enzymes (e.g., phospholipase A_2 (PLA_2)), and small molecules (e.g., histamine). Recent studies suggested further that BV and BV-derived active components might have potent therapeutic effects on refractory immunological and neurodegenerative diseases including allergic disorders, autoimmune diseases, amyotrophic lateral sclerosis (ALS), and Parkinson's disease (PD) [3,6–9], however well-controlled, randomized clinical studies are still insufficient.

In this review, the underlying mechanisms of BV-induced regulation of immune responses as well as of neuronal and glial pathology in refractory immunological and neurological diseases will be discussed, based mainly on the articles that have been published in the last decade. In addition, the therapeutic effects and mechanisms of BV-derived active components, especially focusing on PLA_2, melittin and apamin will be introduced. Finally, we will comment on the future perspectives in the research area of BV therapy.

2. Therapeutic Effects of Bee Venom on Immunological Diseases

2.1. Effects on Allergic Disorders

The initial event responsible for the development of allergic disorders, such as asthma, allergic rhinoconjunctivitis, and atopic eczema, is the generation of allergen-specific CD4$^+$ T cells [10]. In a general view, allergy is a T helper 2 (Th2) cells-mediated disease that involves the hyperproduction of specific immunoglobulin E (IgE) antibodies to which interleukin-4 (IL-4) and IL-13, the key Th2-specific cytokines, mainly contribute [11].

BV therapy is a kind of allergen-specific immunotherapy (SIT) that has been carried out for a long time. Although the mechanism of SIT remains poorly understood, hitherto several features, including modifications of antigen presenting cells (APCs), T cells, and B cells, as well as both the number and the function of effector cells that mediate the allergic response have been clarified [12]. In clinical trials, it was reported that SIT increases the production of IL-10 by APCs, including B cells, monocytes, and macrophages [12]. The efficacy of SIT has been emphasized in insect venom allergy and respiratory allergies. BV immunotherapy has early and late influences on major cells of allergic inflammation [10]. Venom immunotherapy induces a monocyte activation characterized by a delayed overproduction of IL-12 and tumor necrosis factor alpha (TNF-α), which are cytokines related to the inhibition of Th2 cells [13]. BV immunotherapy is known to generate IL-10 and transforming growth factor beta (TGF-β), which potently suppresses IgE production and increases IgG4 and IgA, simultaneously [14]. Our previous study demonstrated that BV induces Th1 lineage development from CD4$^+$ T cells without affecting Th2 cells, by increasing the expression of a Th1-specific cytokine, interferon gamma (INF-γ), via an upregulation of Th1-specific transcription factor, T-bet [15].

CD4$^+$CD25$^+$Foxp3$^+$ regulatory T cells (Tregs) play a pivotal role in the maintenance of tolerance in the immune system and are involved in the control of transplantation tolerance, tumor immunity, allergy, and infection [16,17]. Tregs could regulate allergic disorders through several inhibitory pathways, including suppression of Th2 immune responses, of Th17 cells, and of T cell migration to tissues [11]. It has been suggested that an essential step in successful BV immunotherapy is associated with the presence of Tregs, which secrete IL-10, consequently inhibiting the secretion of cytokines IL-4, IL-5, and IL-13 from Th2 cells, in turn impeding specific IgE production [18]. For example, our recent study showed that BV treatment increased Treg populations, augmented the production of IL-10, and suppressed the production of Th1, Th2, and Th17-related cytokines, resulting in bronchial inflammation with a reduction in the degranulation of mast cells and eosinophils in an OVA-induced allergic asthma murine model [3]. Taken together, the immunological mechanisms of BV immunotherapy include a shift toward Th1 cytokines, an increase in the number of peripheral Tregs, and an upregulation of different markers expressed on CD4$^+$ T cells [10,19].

The most important allergen in BV is PLA$_2$, which induces rapid leukotriene C4 production from purified human basophils within 5 min, while IL-4 expression and production is induced at later time-points without histamine release [20]. Direct injection of the BV-derived PLA$_2$ (bvPLA$_2$) into inguinal lymph nodes enhanced allergen-specific IgG and T-cell responses and stimulated the production of the Th1-dependent subclass IgG2a [21]. Melittin, a major peptide component of BV, is reported to trigger lysis of a mast cells, which can lead to the release of histamine and other intracellular components into surrounding tissues [18]. In our unpublished data [22], bvPLA$_2$ increased Treg population *in vitro* and *in vivo* more potently than BV, resulting in prevention of ovalbumin-induced allergic asthma in mice with suppression of various effector cells, such as eosinophils, lymphocytes, and macrophages, and of Th2 cytokines and serum IgE.

2.2. Effects on Autoimmune and Inflammatory Diseases

Autoimmune diseases, such as rheumatoid arthritis, systemic lupus erythematosus, and multiple sclerosis, have been understood to be Th1-dominant diseases, however, the important roles of Th17 cells and Tregs in autoimmune diseases have recently emerged [23]. Rheumatoid arthritis is a common autoimmune disease, yet current conventional therapies are not always successful [24]. BV has been traditionally used to treat chronic inflammatory diseases, including rheumatoid arthritis [4]. Especially, the anti-rheumatic and anti-inflammatory effects of BV have been understood as of one hundred years ago [1,2]. Previous study has demonstrated that BV injection into the Zusanli acupoint has both anti-inflammatory and anti-nociceptive effects on Freund's adjuvant-induced arthritis in rats [25]. Effect of combined application of bee-venom therapy and medication is superior to the simple use of medication in relieving rheumatoid arthritis and might reduce the commonly-taken doses of Western medicines [26]. These anti-arthritis effects have been reported in several arthritis models, and these effects of BV might be associated with melittin, a major peptide component of BV, which has anti-inflammatory and anti-arthritis properties, and inhibitory activity on nuclear factor kappaB (NF-κB) [5].

We previously examined the effects of BV on the nitric oxide (NO) generation by lipopolysaccharide (LPS) or sodium nitroprusside (SNP) in RAW264.7 macrophages, and the expression of inducible nitric oxide synthase (iNOS), cyclooxygenase 2 (COX-2), NF-κB and mitogen-activated protein kinase (MAPK) with RT-PCR in LPS stimulated RAW 264.7 cells. The results showed that BV suppressed NO production and decreased the level of iNOS and COX-2 expression, possibly through suppressing NF-κB and MAPK [27]. We also performed microarray analysis to evaluate the global gene expression profiles of macrophage cell treated with BV. We found that BV decreased the expression of various genes, including mitogen-activated protein kinase kinase kinase 8 (MAP3K8), TNF, suppressor of cytokine signaling 3 (SOCS3), TNF-receptor-associated factor 1 (TRAF1), JUN, and CREB binding protein (CBP), related to the inflammatory effects, which occur in LPS-treated RAW264.7 cells [28]. Other studies support these observations. For example, BV and melittin prevent LPS- or SNP-induced NO and prostaglandin E2 production via c-Jun N-terminal kinase (JNK) pathway dependent inhibition of NF-κB [29]. BV also suppressed adjuvant-induced arthritis in rats by targeting TNF-α and NF-κB activation [30]. These findings indicate that BV may have anti-inflammatory effects in rheumatoid arthritis.

Lupus nephritis, a serious complication of systemic lupus erythematosus, is mediated by the glomerular inflammation involving the production of autoantibodies against the nucleus and of cytokines/chemokines, which ultimately results in irreversible renal damage [31,32]. New Zealand Black/White F1 female mice age-dependently develop autoimmune disease, which is characterized by glomerulonephritis, proteinuria, and renal dysfunction [33]. Using this animal model, we showed that BV treatment significantly delayed the development of proteinuria, prevented renal inflammation, reduced tubal damage, and decreased immune deposits in the glomeruli, and these results are closely associated with a BV-induced increase in splenic Tregs and decrease in renal proinflammatory cytokines, TNF-α and IL-6 [8]. These results suggest that BV therapy has the potential to modulate autoimmune response in lupus nephritis, possibly by enhancing Tregs and suppressing renal inflammation.

Multiple sclerosis is a chronic inflammatory disease of the central nervous system (CNS) that affects more than one million people worldwide. Its clinical symptoms include ataxia, loss of coordination, sensory impairment, cognitive dysfunction, and fatigue [34]. The pathogenesis of multiple sclerosis is known to be an autoimmune T cell responses, in which Th1 and Th17 cells are critically involved [35]. An animal model of experimental autoimmune encephalomyelitis (EAE) has been widely used for the study of multiple sclerosis, because clinical and pathological features of EAE are very similar to those of multiple sclerosis. We previously demonstrated that BV treatment has a neuroprotective effect against immune cell infiltration and Th1/Th17 differentiation via increasing Tregs in EAE mouse model [36]. Very recently, another research group has also reported that BV acupuncture attenuates the development and progression of EAE in rats by upregulating Tregs and suppressing Th1/Th17 cell responses [37]. These results suggest that BV has the potential to become a therapeutic agent for multiple sclerosis, and warrants further investigation of BV and bvPLA$_2$ as potent modulators of autoimmune T cell responses in the CNS.

3. Therapeutic Effects of Bee Venom on Neurological Diseases

Parkinson's disease (PD) is one of the most common progressive neurodegenerative disorders, which is characterized clinically by bradykinesia, resting tremor, rigidity, and disturbances in posture and gait resulting from the selective, irreversible loss of dopaminergic (DA) neurons in the substantia nigra (SN) and their terminals in the striatum [38,39]. Activated microglia, innate immune cells in the CNS, near the degenerating DA neurons is known to be a key mediator of neuroinflammation in PD [39,40]. BV acupuncture was reported to be anti-inflammatory and anti-neurodegenerative, and to improve motor symptoms in PD clinical trials [6]. In a 1-methyl-4-phenyl-1,2,3,6-tetrahydropyridine (MPTP)-induced mouse model of PD, BV improved the survival percentage of tyrosine hydroxylase$^+$ cells to 70% on day 1 and 78% on day 3 compared with normal mice, and reduced expression of the inflammation markers macrophage antigen complex-1 (MAC-1) and iNOS in the SN [41]. Our previous study also revealed that modulation of peripheral immune tolerance by Tregs may contribute to the neuroprotective effect of BV in the MPTP animal model of PD [42]. Recently, apamin, a specific component of BV, was also shown to have a protective effect in animal models of PD [43].

In an animal model of amyotrophic lateral sclerosis (ALS), mutant human superoxide dismutase 1 (hSOD1) transgenic mice, BV acupuncture inhibited microglia activation and phospho-p38 MAPK expression in the CNS, resulting in improvement of motor activity [9]. It also has been reported that melittin ameliorated the inflammation of lung and spleen in an ALS animal model [44]. BV might be helpful in reducing glutamatergic cell toxicity, which has been reported in many neurodegenerative diseases, including PD, Alzheimer's disease, and ALS, through the inhibition of MAP kinase activation (e.g., JNK, ERK, and p38) following exposure to glutamate [45]. Although scientific evidence is still limited in this research area, we strongly believe that the aforementioned results may lead to future advanced studies, elucidating the therapeutic effects of BV and its active components on various neurodegenerative diseases and its underlying mechanisms. Indeed, in several preliminary results, we found that BV and BV-derived active components ameliorate Alzheimer's disease, Parkinson's disease, and chronic neuropathic pain by modulating peripheral immune and inflammatory responses, as well as modulating central glial activation.

4. Conclusions and Perspectives

In this review, we introduced the therapeutic effects of BV and its major components on immunological and neurological diseases, and discussed its underlying mechanisms. We propose that BV is a strong immune modulator that may subsequently affect the CNS glia and neurons. BV also seems to play a role in maintaining homeostasis in our body's immune system and nervous system, because BV therapy can regulate two immunologically opposite conditions, *i.e.*, allergic disorders (Th2 dominant) and autoimmune diseases (Th1 dominant). It remains to be understood how the same treatments of BV or BV-derived active components could modulate both conflicting diseases. Thus, other T cell populations, such as Th17 cells and Tregs, have emerged as a key players in BV-induced modulation of immune and nervous system. Th17 cells are known to play an important role in the pathogenesis of autoimmune, as well as allergic, diseases [46,47]. In contrast, Tregs inhibits activation of both Th1 and Th2 cells, and of Th17 cells, thereby suppressing autoimmune and allergic diseases [11,35]. Indeed, several recent studies reported that BV or bvPLA$_2$ could upregulate peripheral Tregs and/or suppress Th17 responses in various animal models of both diseases [3,8,36,37,48]. Further studies on this issue might shed light on our understandings of such homeostatic therapeutic effects of BV.

In addition, it should be noted that BV is called a "double-edged sword" having nociceptive and anti-nociceptive effects [49], and BV itself could act as a strong allergen. BV induces the release of either of histamine or leukotriene C4 in skin of beekeepers [50], and bvPLA$_2$, the major allergen of BV components [51,52], induces a PLA$_2$-specific IgE immune responses in mice [53], although BV- and bvPLA$_2$-induced Th2 cell immunity and specific IgE production might be protective [54,55]. We also observed that a high dose of BV (2.5 mg/kg, s.c.) treatment could exacerbate oxaliplatin-induced neuropathic pain in rats, whereas low doses of BV (0.25 and 1.0 mg/kg, s.c.) strongly alleviate pain [56].

Thus, the optimal dose and treatment method without side effects should be determined in each disease conditions. Future studies including experimental elucidation of detailed cellular/molecular mechanisms, and well-controlled, randomized clinical trials will lead to a potential therapeutic alternative for treating refractory immunological and neurological diseases.

Acknowledgments: This work was supported by a grant of the Korea Health Technology R&D Project through the Korea Health Industry Development Institute (KHIDI), funded by the Ministry of Health and Welfare, Republic of Korea (grant number: HI14 C0738) and by a grant of the National Research Foundation of Korea (NRF) grant funded by the Korea government [MEST] (No. 2012-0005755).

Author Contributions: H. Bae conceived the idea for the manuscript. D.-S. Hwang and S. K. Kim wrote the manuscript. All three authors revised the manuscript and approved the final manuscript.

Conflicts of Interest: The authors declare no conflict of interest.

References

1. Billingham, M.E.; Morley, J.; Hanson, J.M.; Shipolini, R.A.; Vernon, C.A. Letter: An anti-inflammatory peptide from bee venom. *Nature* **1973**, *245*, 163–164. [CrossRef] [PubMed]
2. Walker, E.W. Bees' stings and rheumatism. *Br. Med. J.* **1908**, *2*, 1056–1060. [CrossRef] [PubMed]
3. Choi, M.S.; Park, S.; Choi, T.; Lee, G.; Haam, K.K.; Hong, M.C.; Min, B.I.; Bae, H. Bee venom ameliorates ovalbumin induced allergic asthma via modulating CD4⁺CD25⁺ regulatory T cells in mice. *Cytokine* **2013**, *61*, 256–265. [CrossRef] [PubMed]
4. Lee, J.-D.; Park, H.-J.; Chae, Y.; Lim, S. An overview of bee venom acupuncture in the treatment of arthritis. *Evid. Based Complement. Altern. Med.* **2005**, *2*, 79–84. [CrossRef] [PubMed]
5. Son, D.J.; Lee, J.W.; Lee, Y.H.; Song, H.S.; Lee, C.K.; Hong, J.T. Therapeutic application of anti-arthritis, pain-releasing, and anti-cancer effects of bee venom and its constituent compounds. *Pharmacol. Ther.* **2007**, *115*, 246–270. [CrossRef] [PubMed]
6. Cho, S.-Y.; Shim, S.-R.; Rhee, H.Y.; Park, H.-J.; Jung, W.-S.; Moon, S.-K.; Park, J.-M.; Ko, C.-N.; Cho, K.-H.; Park, S.-U. Effectiveness of acupuncture and bee venom acupuncture in idiopathic parkinson's disease. *Park. Relat. Disord.* **2012**, *18*, 948–952. [CrossRef] [PubMed]
7. Kang, S.Y.; Roh, D.H.; Yoon, S.Y.; Moon, J.Y.; Kim, H.W.; Lee, H.J.; Beitz, A.J.; Lee, J.H. Repetitive treatment with diluted bee venom reduces neuropathic pain via potentiation of locus coeruleus noradrenergic neuronal activity and modulation of spinal nr1 phosphorylation in rats. *J. Pain* **2012**, *13*, 155–166. [CrossRef] [PubMed]
8. Lee, H.; Lee, E.J.; Kim, H.; Lee, G.; Um, E.J.; Kim, Y.; Lee, B.Y.; Bae, H. Bee venom-associated Th1/Th2 immunoglobulin class switching results in immune tolerance of NZB/W F1 murine lupus nephritis. *Am. J. Nephrol.* **2011**, *34*, 163–172. [CrossRef] [PubMed]
9. Yang, E.J.; Jiang, J.H.; Lee, S.M.; Yang, S.C.; Hwang, H.S.; Lee, M.S.; Choi, S.-M. Bee venom attenuates neuroinflammatory events and extends survival in amyotrophic lateral sclerosis models. *J. Neuroinflamm.* **2010**, *7*, 69. [CrossRef] [PubMed]
10. Ozdemir, C.; Kucuksezer, U.; Akdis, M.; Akdis, C. Mechanisms of immunotherapy to wasp and bee venom. *Clin. Exp. Allergy* **2011**, *41*, 1226–1234. [CrossRef] [PubMed]
11. Palomares, O.; Yaman, G.; Azkur, A.K.; Akkoc, T.; Akdis, M.; Akdis, C.A. Role of treg in immune regulation of allergic diseases. *Eur. J. Immunol.* **2010**, *40*, 1232–1240. [CrossRef] [PubMed]
12. Larché, M.; Akdis, C.A.; Valenta, R. Immunological mechanisms of allergen-specific immunotherapy. *Nat. Rev. Immunol.* **2006**, *6*, 761–771. [CrossRef] [PubMed]
13. Magnan, A.; Marin, V.; Mely, L.; Birnbaum, J.; Romanet, S.; Bongrand, P.; Vervloet, D. Venom immunotherapy induces monocyte activation. *Clin. Exp. Allergy* **2001**, *31*, 1303–1309. [CrossRef] [PubMed]
14. Akdis, M.; Schmidt-Weber, C.; Jutel, M.; Akdis, C.A.; Blaser, K. Mechanisms of allergen immunotherapy. *Clin. Exp. Allergy Rev.* **2004**, *4*, 56–60. [CrossRef]
15. Nam, S.; Ko, E.; Park, S.K.; Ko, S.; Jun, C.Y.; Shin, M.K.; Hong, M.C.; Bae, H. Bee venom modulates murine Th1/Th2 lineage development. *Int. Immunopharmacol.* **2005**, *5*, 1406–1414. [CrossRef] [PubMed]
16. Sakaguchi, S.; Ono, M.; Setoguchi, R.; Yagi, H.; Hori, S.; Fehervari, Z.; Shimizu, J.; Takahashi, T.; Nomura, T. Foxp3⁺CD25⁺CD4⁺ natural regulatory T cells in dominant self-tolerance and autoimmune disease. *Immunol. Rev.* **2006**, *212*, 8–27. [CrossRef] [PubMed]

17. Vignali, D.A.; Collison, L.W.; Workman, C.J. How regulatory T cells work. *Nat. Rev. Immunol.* **2008**, *8*, 523–532. [CrossRef] [PubMed]

18. Bilò, M.B.; Antonicelli, L.; Bonifazi, F. Honeybee venom immunotherapy: Certainties and pitfalls. *Immunotherapy* **2012**, *4*, 1153–1166. [CrossRef] [PubMed]

19. Cabrera, C.M.; Urra, J.M.; Alfaya, T.; de la Roca, F.; Feo-Brito, F. Expression of Th1, Th2, lymphocyte trafficking and activation markers on CD4$^+$ T cells of hymenoptera allergic subjects and after venom immunotherapy. *Mol. Immunol.* **2014**, *62*, 178–185. [CrossRef] [PubMed]

20. Mustafa, F.; Ng, F.; Nguyen, T.H.; Lim, L. Honeybee venom secretory phospholipase A2 induces leukotriene production but not histamine release from human basophils. *Clin. Exp. Immunol.* **2008**, *151*, 94–100. [CrossRef] [PubMed]

21. Martínez-Gómez, J.M.; Johansen, P.; Erdmann, I.; Senti, G.; Crameri, R.; Kündig, T.M. Intralymphatic injections as a new administration route for allergen-specific immunotherapy. *Int. Arch. Allergy Immunol.* **2008**, *150*, 59–65.

22. Lee, G.; Bae, H. Phospholipase A2 from bee venom, a novel Foxp3$^+$ regulatory T cell inducer, suppresses immune disorders. *J. Immunol.* Under Revision. **2015**.

23. Alunno, A.; Manetti, M.; Caterbi, S.; Ibba-Manneschi, L.; Bistoni, O.; Bartoloni, E.; Valentini, V.; Terenzi, R.; Gerli, R. Altered immunoregulation in rheumatoid arthritis: The role of regulatory T cells and proinflammatory Th17 cells and therapeutic implications. *Med. Inflamm.* **2015**. [CrossRef] [PubMed]

24. McInnes, I.B.; Schett, G. The pathogenesis of rheumatoid arthritis. *N. Engl. J. Med.* **2011**, *365*, 2205–2219. [CrossRef] [PubMed]

25. Kwon, Y.-B.; Lee, J.-D.; Lee, H.-J.; Han, H.-J.; Mar, W.-C.; Kang, S.-K.; Beitz, A.J.; Lee, J.-H. Bee venom injection into an acupuncture point reduces arthritis associated edema and nociceptive responses. *Pain* **2001**, *90*, 271–280. [CrossRef]

26. Liu, X.; Zhang, J.; Zheng, H.; Liu, F.; Chen, Y. Clinical randomized study of bee-sting therapy for rheumatoid arthritis. *Zhen Ci Yan Jiu* **2008**, *33*, 197–200. [PubMed]

27. Jang, H.S.; Kim, S.K.; Han, J.B.; Ahn, H.J.; Bae, H.; Min, B.I. Effects of bee venom on the pro-inflammatory responses in RAW264.7 macrophage cell line. *J. Ethnopharmacol.* **2005**, *99*, 157–160. [CrossRef] [PubMed]

28. Jang, H.-S.; Chung, H.-S.; Ko, E.; Shin, J.-S.; Shin, M.-K.; Hong, M.-C.; Kim, Y.; Min, B.-I.; Bae, H. Microarray analysis of gene expression profiles in response to treatment with bee venom in lipopolysaccharide activated RAW264.7 cells. *J. Ethnopharmacol.* **2009**, *121*, 213–220. [CrossRef] [PubMed]

29. Park, H.J.; Lee, H.J.; Choi, M.S.; Son, D.J.; Song, H.S.; Song, M.J.; Lee, J.M.; Han, S.B.; Kim, Y.; Hong, J.T. JNK pathway is involved in the inhibition of inflammatory target gene expression and NF-κB activation by melittin. *J. Inflamm.* **2008**. [CrossRef]

30. Darwish, S.F.; El-Bakly, W.M.; Arafa, H.M.; El-Demerdash, E. Targeting TNF-α and NF-κB activation by bee venom: Role in suppressing adjuvant induced arthritis and methotrexate hepatotoxicity in rats. *PLoS ONE* **2013**, *8*, e79284. [CrossRef] [PubMed]

31. Agrawal, N.; Chiang, L.K.; Rifkin, I.R. Lupus nephritis. *Seminars Nephrol.* **2006**, *26*, 95–104. [CrossRef] [PubMed]

32. Mason, L.J.; Berden, J.H. Pathogenic factors for the development of lupus nephritis. *Lupus* **2008**, *17*, 251–255. [CrossRef] [PubMed]

33. Foster, M.H. Relevance of systemic lupus erythematosus nephritis animal models to human disease. *Seminars Nephrol.* **1999**, *19*, 12–24.

34. McFarland, H.F.; Martin, R. Multiple sclerosis: A complicated picture of autoimmunity. *Nat. Immunol.* **2007**, *8*, 913–919. [CrossRef] [PubMed]

35. Goverman, J. Autoimmune T cell responses in the central nervous system. *Nat. Rev. Immunol.* **2009**, *9*, 393–407. [CrossRef] [PubMed]

36. Lee, G.; Lee, H.; Park, S.; Jang, H.; Bae, H. Bee venom attenuates experimental autoimmune encephalomyelitis through direct effects on CD4$^+$CD25$^+$Foxp3$^+$ T cells. *Eur. J. Inflamm.* **2013**, *11*, 111–121.

37. Lee, M.J.; Jang, M.; Choi, J.; Lee, G.; Min, H.J.; Chung, W.S.; Kim, J.I.; Jee, Y.; Chae, Y.; Kim, S.H.; *et al.* Bee venom acupuncture alleviates experimental autoimmune encephalomyelitis by upregulating regulatory T cells and suppressing Th1 and Th17 responses. *Mol. Neurobiol.* **2015**. [CrossRef]

38. Paulus, W.; Jellinger, K. The neuropathologic basis of different clinical subgroups of parkinson's disease. *J. Neuropathol. Exp. Neurol.* **1991**, *50*, 743–755. [CrossRef] [PubMed]

39. Hirsch, E.C.; Breidert, T.; Rousselet, E.; Hunot, S.; Hartmann, A.; Michel, P.P. The role of glial reaction and inflammation in parkinson's disease. *Ann. N. Y. Acad. Sci.* **2003**, *991*, 214–228. [CrossRef] [PubMed]

40. Block, M.L.; Hong, J.S. Microglia and inflammation-mediated neurodegeneration: Multiple triggers with a common mechanism. *Prog. Neurobiol.* **2005**, *76*, 77–98. [CrossRef] [PubMed]

41. Kim, J.-I.; Yang, E.J.; Lee, M.S.; Kim, Y.-S.; Huh, Y.; Cho, I.-H.; Kang, S.; Koh, H.-K. Bee venom reduces neuroinflammation in the mptp-induced model of parkinson's disease. *Int. J. Neurosci.* **2011**, *121*, 209–217. [CrossRef] [PubMed]

42. Chung, E.S.; Kim, H.; Lee, G.; Park, S.; Kim, H.; Bae, H. Neuro-protective effects of bee venom by suppression of neuroinflammatory responses in a mouse model of parkinson's disease: Role of regulatory T cells. *Brain Behav. Immun.* **2012**, *26*, 1322–1330. [CrossRef] [PubMed]

43. Alvarez-Fischer, D.; Noelker, C.; Vulinović, F.; Grünewald, A.; Chevarin, C.; Klein, C.; Oertel, W.H.; Hirsch, E.C.; Michel, P.P.; Hartmann, A. Bee venom and its component apamin as neuroprotective agents in a parkinson disease mouse model. *PLoS ONE* **2013**, *8*, e61700. [CrossRef] [PubMed]

44. Lee, S.-H.; Choi, S.-M.; Yang, E.J. Melittin ameliorates the inflammation of organs in an amyotrophic lateral sclerosis animal model. *Exp. Neurobiol.* **2014**, *23*, 86–92. [CrossRef] [PubMed]

45. Lee, S.M.; Yang, E.J.; Choi, S.-M.; Kim, S.H.; Baek, M.G.; Jiang, J.H. Effects of bee venom on glutamate-induced toxicity in neuronal and glial cells. *Evid. Based Complement. Altern. Med.* **2011**, *2012*. [CrossRef] [PubMed]

46. Kim, S.K.; Bae, H. Acupuncture and immune modulation. *Auton. Neurosci.* **2010**, *157*, 38–41. [CrossRef] [PubMed]

47. Maddur, M.S.; Miossec, P.; Kaveri, S.V.; Bayry, J. Th17 cells: Biology, pathogenesis of autoimmune and inflammatory diseases, and therapeutic strategies. *Am. J. Pathol.* **2012**, *181*, 8–18. [CrossRef] [PubMed]

48. Kim, H.; Lee, G.; Park, S.; Chung, H.S.; Lee, H.; Kim, J.Y.; Nam, S.; Kim, S.K.; Bae, H. Bee venom mitigates cisplatin-induced nephrotoxicity by regulating CD4$^+$CD25$^+$Foxp3$^+$ regulatory T-cells in mice. *Evid. Based Complement. Altern. Med.* **2013**. [CrossRef] [PubMed]

49. Chen, J.; Lariviere, W.R. The nociceptive and anti-nociceptive effects of bee venom injection and therapy: A double-edged sword. *Prog. Neurobiol.* **2010**, *92*, 151–183. [CrossRef] [PubMed]

50. Annila, I.; Saarinen, J.V.; Nieminen, M.M.; Moilanen, E.; Hahtola, P.; Harvima, I.T. Bee venom induces high histamine or high leukotriene c4 release in skin of sensitized beekeepers. *J. Investig. Allergol. Clin. Immunol.* **2000**, *10*, 223–228. [PubMed]

51. Annila, I. Bee venom allergy. *Clin. Exp. Allergy* **2000**, *30*, 1682–1687. [CrossRef] [PubMed]

52. Sobotka, A.K.; Franklin, R.M.; Adkinson, N.F.; Valentine, M.; Baer, H.; Lichtenstein, L.M. Allergy to insect stings: II. Phospholipase A: The major allergen in honeybee venom. *J. Allergy Clin. Immunol.* **1976**, *57*, 29–40. [CrossRef]

53. Dudler, T.; Machado, D.C.; Kolbe, L.; Annand, R.R.; Rhodes, N.; Gelb, M.H.; Koelsch, K.; Suter, M.; Helm, B.A. A link between catalytic activity, IgE-independent mast cell activation, and allergenicity of bee venom phospholipase A2. *J. Immunol.* **1995**, *155*, 2605–2613. [PubMed]

54. Marichal, T.; Starkl, P.; Reber, L.L.; Kalesnikoff, J.; Oettgen, H.C.; Tsai, M.; Metz, M.; Galli, S.J. A beneficial role for immunoglobulin E in host defense against honeybee venom. *Immunity* **2013**, *39*, 963–975. [CrossRef] [PubMed]

55. Palm, N.W.; Rosenstein, R.K.; Yu, S.; Schenten, D.D.; Florsheim, E.; Medzhitov, R. Bee venom phospholipase A2 induces a primary type 2 response that is dependent on the receptor ST2 and confers protective immunity. *Immunity* **2013**, *39*, 976–985. [CrossRef] [PubMed]

56. Lim, B.S.; Moon, H.J.; Li, D.X.; Gil, M.; Min, J.K.; Lee, G.; Bae, H.; Kim, S.K.; Min, B.I. Effect of bee venom acupuncture on oxaliplatin-induced cold allodynia in rats. *Evid. Based Complement. Altern. Med.* **2013**. [CrossRef] [PubMed]

© 2015 by the authors; licensee MDPI, Basel, Switzerland. This article is an open access article distributed under the terms and conditions of the Creative Commons Attribution (CC-BY) license (http://creativecommons.org/licenses/by/4.0/).

toxins

MDPI

Review

Venom Proteins from Parasitoid Wasps and Their Biological Functions

Sébastien J. M. Moreau [1],* and Sassan Asgari [2],*

[1] Institut de Recherche sur la Biologie de l'Insecte, Centre National de la Recherche Scientifique Unité Mixte de Recherche 7261, Université François-Rabelais, Unité de Formation et de Recherche Sciences et Techniques, Parc Grandmont, 37200 Tours, France

[2] School of Biological Sciences, the University of Queensland, Brisbane, QLD 4067, Australia

* Correspondence: sebastien.moreau@univ-tours.fr (S.J.M.M.); s.asgari@uq.edu.au (S.A.);
  Tel.: +33-2-47-36-74-55 (S.J.M.M.); +61-7-3365-2043 (S.A.)

Academic Editor: R. Manjunatha Kini

Received: 20 May 2015; Accepted: 16 June 2015; Published: 26 June 2015

Abstract: Parasitoid wasps are valuable biological control agents that suppress their host populations. Factors introduced by the female wasp at parasitization play significant roles in facilitating successful development of the parasitoid larva either inside (endoparasitoid) or outside (ectoparasitoid) the host. Wasp venoms consist of a complex cocktail of proteinacious and non-proteinacious components that may offer agrichemicals as well as pharmaceutical components to improve pest management or health related disorders. Undesirably, the constituents of only a small number of wasp venoms are known. In this article, we review the latest research on venom from parasitoid wasps with an emphasis on their biological function, applications and new approaches used in venom studies.

Keywords: venom; endoparasitoid; ectoparasitoids; parasitism; host; wasp

---

1. Biological Functions of Parasitoid Wasp Venoms

Parasitoid wasps belong to the order Hymenoptera and are valuable insects in suppressing host populations either through natural or augmented biological control. Typically, the female wasp deposits its egg inside (endoparasitoid) or outside (ectoparasitoid) the host (mostly arthropods) where the emerged parasitoid larva continues to feed. Eventually, the host dies due to parasitism, although there seem to be examples in which the host may survive and continue to reproduce (e.g., [1]). As a consequence of the differing lifestyle, the physiological requirements and impacts on the host by endoparasitoids and ectoparasitoids may vary [2]. Components injected into the host at the time of parasitization play vital roles in facilitating successful parasitism, including venom and ovarian/calyx fluid. These may or may not contain symbiotic viruses or virus-like particles that contribute to host manipulation, in particular in endoparasitoids.

Similar to venom found in most venomous animals, venom fluid from parasitoid wasps consists of a cocktail of proteinacious and non-proteinacious compounds. While various studies that have focused on determining the venom profile of ecto- and endoparasitoid venoms (see below) have revealed the presence of several conserved proteins between the two parasitic wasp groups, venom appears to serve different purposes in the two groups. In general, venom from ectoparasitoids is largely involved in the host paralysis (short or long-term) to secure feeding of the ectoparasitic larva outside the host, whereas endoparasitoids' venom rarely causes paralysis but facilitates parasitization by interfering with the host immune system, development or synergizing the effects of other maternal factors introduced into the host (e.g., polydnaviruses, PDVs). In this review, we will discuss the major biological functions of parasitoid wasp venoms, latest approaches used for venom studies and some of the potential applications of venom proteins from those insects.

1.1. Ectoparasitoids

The primary function of venom in most ectoparasitoids (in particular when the host is at the active stage, e.g., larval/nymphal stage) is induction of short- to long-term paralysis/lethargy in the host and developmental arrest. However, venom may play other roles in facilitating parasitization, such as suppressing the host immunity (e.g., [3,4]) or interrupting development (e.g., [5,6]). Despite thousands of ectoparasitoids species known, there are only very limited number of venom components identified from a small number of ectoparasitoids.

The parasitoid *Ampulex compressa* injects a cocktail of neurotoxins into the central nervous system of its cockroach prey. This involves two consecutive stings, one in the thorax, which leads to transient paralysis of the front legs due to post-synaptic blockage of central cholinergic synaptic transmission, and a second one by injection of venom specifically inside the sub-esophageal ganglion of its cockroach prey, which induces a 30 min intense grooming in the prey (induced by dopamine) followed by a long-lasting lethargic effect [7,8]. The latter effect is most likely caused by venom affecting the opoid system [9] or octopaminergic receptor [10]. The venom from *A. compressa* contains GABA (inhibitory neurotransmitter) and ß-alanine (GABA receptor agonist), and taurine (impairs the re-update of GABA from the synaptic cleft) [11] (Table 1). It has been suggested that these three main components have both pre- and post-synaptic effects on GABA-gated chloride channels.

Venom from the digger wasp *Philanthus triangulum* contains philanthotoxins which affect both the central and the peripheral nervous system of the prey by presynaptic as well as a postsynaptic blockage of neuromuscular transmission [12,13]. Specifically, the toxins inhibit the re102lease of glutamate and block the post-synaptic glutamate receptors. In addition, δ-philanthotoxin inhibits the nicotinic acetylcholine receptors in the central nervous system [14]. *Bracon hebetor* is another ectoparasitoid with a potent venom causing host paralysis [15]. Three proteins with molecular masses of about 73 kDa were found in the venom, two of them (Brh-I and -II) being insecticidal when injected into lepidopteran larvae. Of the two, Brh-I was found to be more toxic against the larvae of the cotton bollworm, *Heliothis virescens*. *Liris niger*, which hunts, paralyses and parasitizes the mole cricket, injects venom into the nervous system leading to blockage of voltage-gated sodium inward currents, and synaptic transmission [16]. The constituents of the venom, which comprise of proteins from 3.4–200 kDa have not been well characterized [17]. The ectoparasitoid *Eupelmus orientalis* venom causes permanent host paralysis and developmental arrest, the two effects found to be independent of each other [18]. In the venom, hyaluronidase and phospholipase activities were detected.

Table 1. Major biological functions of venom from parasitoid wasps.

Biological Functions	Wasp	Parasitism	Host	Reference
Paralysis				
pimplin	*Pimpla hypochondriaca*	Endo	*Lacanobia oleracea*	[19]
philanthotoxins	*Philanthus triangulum*	Ecto	*Schistocerca gregaria*	[12]
Brh-I & -II	*Bracon hebetor*	Ecto	*Diaprepes abbreviatus*	[20]
GABA, β-alanine, taurine	*Ampulex compressa*	Ecto	*Periplaneta americana*	[11]
Hemocyte inactivation				
VPr1	*Pimpla hypochondriaca*	Endo	*L. oleracea*	[21]
VPr3	*Pimpla hypochondriaca*	Endo	*L. oleracea*	[22]
Vn.11	*Pteromalus puparum*	Endo	*Pieris rapae*	[23]
VP P4, RhoGAP	*Leptopilina boulardi*	Endo	*Drosophila melanogaster*	[24]
calreticulin	*Cotesia rubecula*	Endo	*P. rapae*	[25]
	Pteromalus puparum	Endo	*P. rapae*	[26]
SERCA *	*Ganaspis* sp.1	Endo	*D. melanogaster*	[27]

Table 1. *Cont.*

Biological Functions	Wasp	Parasitism	Host	Reference
Inhibition of melanization				
LbSPNy	*Leptopilina boulardi*	Endo	*D. melanogaster*	[28]
Vn50	*Cotesia rubecula*	Endo	*P. rapae*	[29]
Interrupting development				
Reprolysin	*Eulophus pennicornis*	Ecto	*L. oleracea*	[6]
Enhancing PDVs				
Vn1.5	*Cotesia rubecula*	Endo	*P. rapae*	[30]
Castration				
γ-glutamyl transpeptidase	*Aphidius ervi*	Endo	*Acyrthosiphon pisum*	[31]
Anti-microbial				
PP13, PP102, PP113	*Pteromalus puparum*	Endo	*P. rapae*	[32]

* sarco/endoplasmic reticulum calcium ATPase.

Nasonia vitripennis is a model ectoparasitoid wasp with its genome completely sequenced. It parasitizes the pupal stage of a number of fly species as its host. The wasp's venom inflicts a variety of effects on the host including developmental arrest and decrease in metabolism and immunity [33–36]. Using a suppression subtractive hybridization method, it was shown that the venom from *N. vitripennis* caused differential gene expression in the hemocytes of the host pupae *Musca domestica* [37]. At 1 h after venom application, 133 expressed sequence tags (ESTs) showed decrease in transcript levels and 111 ESTs were found to be upregulated. The altered genes were mostly related to various biological functions such as immunity, apoptosis, stress response, metabolism and regulation of transcription/translation. The outcome shows a profound impact of venom injection on the host hemocytes. In another study, the global effects of *N. vitripennis* on an alternative fly host, *Sarcophaga bullata*, were studied using high throughput RNA sequencing (RNA-seq) of the whole host body following venom treatment [38]. Overall, about 147 host genes were significantly differentially expressed due to envenomation with the percentage of differentially expressed genes increasing as the parasitization progressed. The genes were mostly related to chitin metabolism, cell death, immunity and metabolism. In a similar study, it was found that parasitization of *Sarcophaga crassipalpis* pupa by *N. vitripennis* led to differential expression of only one gene at 3 h after parasitization but 128 genes at 25 h post-parasitization [39]. Similarly, these genes were involved in metabolism, development, immune responses and apoptosis. While various proteins have been identified in *N. vitripennis* venom such as serpins, laccases, metalloproteases, calreticulin, chitinase and serine proteases, their functions in alterations of host physiology have mainly been implied rather than experimentally tested (reviewed in [36]).

In a different pupal ectoparasitoid, *Scleroderma guani*, the transcriptome of the host *Tenebrio molitor* was shown to change following envenomation and the differentially expressed genes were related to similar categories changed in *N. vitripennis* hosts [40]. This indicates that parasitoids manipulate similar genes and pathways that facilitate parasitization.

1.2. Endoparasitoids

In endoparasitoids, venom usually does not have a paralytic effect on the host, except in a few examples in which transient paralysis has been recorded [41–44]. The host normally recovers in a few minutes or within one hour after parasitization. By adopting a koinobiont life style, which allows further development of the host, and living inside the host, endoparasitoids do not require inducing a long-term paralysis in the host. However, the presence of toxin-like peptides in their venom strengthens the assumption that they shared a common ancestor with ectoparasitoids [45].

Deposition of the endoparasitoid egg inside the host exposes the developing parasitoid to host immune responses, mostly encapsulation, which is engulfing the egg/larva with multi-layers of hemocytes. This response is often accompanied by melanization, a cascade of proteolytic reactions leading to the deposition of melanin and production of phenolic intermediates [2]. In addition, as most endoparasitoids allow further development of their host while their juvenile stage is feeding inside the host, regulation of the host development and metabolism is essential. Components introduced into the host at parasitization play the main part in conditioning the host physiology to facilitate endoparasitoid development. While all endoparasitoids inject venom at parasitization, it may not be sufficient to subvert the host physiology. In a large number of parasitoid-host systems injection of supplementary proteins produced in the calyx region of the ovaries or viruses that replicate in the ovaries or venom glands are essential to guarantee successful parasitism.

In many instances, venom is the sole maternal factor that accompanies the endoparasitoid egg, which is sufficient to facilitate parasitization. A well-studied example is *Pimpla hypochondriaca*. The venom from *P. hypochondriaca* consists of several enzymes, protease inhibitors, neurotoxin-like factors and anti-hemocyte aggregation compounds (Table 1). While the function of most of these compounds remains unexplored, evidence suggests that they could be involved in venom homeostasis [46], transient paralysis [19], cytotoxicity [47], and inactivation of hemocytes [22,47]. In *Leptopilina boulardi*, that parasitizes *Drosophila* species, venom is essential to suppress the encapsulation response [24]. The major protein involved is a RhoGAP (Rac GTPase Activating protein) that affects the spreading and aggregation of lamellocytes rendering them incapable of forming a capsule [48]. This might be achieved by targeting two *Drosophila* Rho GTPases, Rac1 and Rac2, essential for encapsulation of parasitoid eggs, after entering the host hemocytes [49]. Venom from *Pteromalus puparum*, and in particular a 24.1 kDa protein (Vn.11), affects the host hemocytes causing reduction in total hemocyte count and their ability to encapsulate foreign objects [23,50]. Sequencing of forward subtractive libraries of *Pieris rapae* hemocytes and fat body after *P. puparum* venom injection showed that the expression levels of a large number of genes were significantly altered (113 in hemocytes and 221 genes in fat body down-regulated) [51]. Many of the identified genes were immune related, as well some that were in non-immune categories. Consistently, a C-type lectin was found down-regulated affecting the activation of the host immune system [52]. In addition to immune suppression, venom from *P. puparum* also affects host development by inducing endocrine changes in the host [53]. Accordingly, juvenile hormone (JH) titers were significantly higher, and JH esterase and ecdysteroid titers significantly lower in parasitized or venom-injected *P. rapae* larvae as compared to control non-treated larvae. These changes would ensure that the larvae have a prolonged larval period. Venom from *Aphidius ervi* causes cell death in the ovarial tissues of the host *Acyrthosiphon pisum* leading to host castration [54]. The apoptotic effect is presumably caused by a γ-glutamyl transpeptidase in the venom fluid [31].

In endoparasitoids that produce viruses or virus-like particles (VLPs), venom functions vary in different host-parasitoid systems ranging from no effect to having overlapping functions with genes expressed from the encapsidated genes within the VLPs or synergise their function. Venom from *Tranosema rostrale* had no effect on host alterations (reduction in total hemocyte count and inhibition of melanization) observed in natural parasitization or when calyx fluid alone was injected [55]. This implied that venom might not play a significant role in parasitization. Similarly, despite having a complex mixture of proteins, venom from *Hyposoter didymator* was found not required for successful parasitism [56]. In addition, in a number of other ichneumonid wasps with PDVs, venom has been found not essential for successful parasitism [57–60].

On the other hand, in a number of host-parasitoid systems venom is essential for proper function of PDVs. PDVs are virus-like particles that are produced in the calyx region of a large number of wasps from Ichneumonidae and Braconidae [61]. They are defective viruses in that they are not able to replicate independent of the parasitoid since the replication machinery (related to nudiviruses) is integrated into the wasp genome [62,63]. For this, they only replicate in the wasp ovaries and not in the parasitoid's host following parasitization. The genes encapsidated in the particles, which appear to be mostly of insect origin, are expressed in the host interfering with the host physiology, in particular suppressing the immune system [64].

Venom has been found to synergize the effect of PDVs. For instance, venom enhances the effects of *Microplitis demolitor* PDVs on host hemocytes (inhibition of cell spreading) in a dose-dependent manner [65] and in delaying development [66]. In *Cotesia melanoscela*, venom is required for entry and unpackaging of PDVs [67], and in *Cotesia rubecula*, in the absence of venom, PDV genes were not expressed in hemocytes [30]. A 1.5 kDa venom peptide (Vn1.5) was found to facilitate expression of CrPDV genes. In *Cotesia nigriceps* both venom and calyx fluid were needed to cause cessation of growth in *Heliothis virescens* larvae [68]. In addition to enhancing PDV functions, venom proteins from endoparasitoids may affect host immunity as well as development. For example, a number of venom proteins interfere with the proper function of host hemocytes. In *C. rubecula*, a calreticulin was shown to inhibit *P. rapae* hemocyte spreading debilitating them from the encapsulation response [25]. Calreticulin from *P. puparum* venom was also found to inhibit *P. rapae* hemocyte spreading and encapsulation response [26]. Calreticulin has been reported from the venom of other endoparasitoids (e.g., [69]) as well as ectoparasitoids (e.g., [70]). A sarco/endoplasmic reticulum calcium ATPase (SERCA) pump protein from a less known parasitoid of *D. melanogaster*, *Ganaspis* sp. 1, was shown to inhibit the activation of plasmatocytes by suppressing calcium burst required for their activation [27]. As a consequence, hemocytes failed to carry out encapsulation. In another parasitoid of *D. melanogaster*, *Asobara japonica*, venom suppressed hemocyte functions but had no effect on humoral responses [71].

Apart from the effects of venom on cellular immunity, the humoral (non-cellular) arm of the host immune system could be a target of venom proteins. For instance, inhibition of host hemolymph melanization is usually a consequence of parasitization in which venom proteins may play a role. A 50 kDa protein (Vn50) from *C. rubecula* was found to inhibit the activation of prophenoloxidase (proPO) to phenoloxidase (PO), a key enzyme in the melanization pathway [72]. This is due to structural resemblance of Vn50 to serine protease homologs (SPHs) [73], which normally facilitate activation of the enzyme by proPO activating protein (PAP) [74], and competitive binding to proPO and PAP [75]. Venom from *P. puparum* reduced transcription of antimicrobial peptides such as cecropin, lysozyme, attacin, lebocin, proline-rich AMP, *etc.* in hemocytes and fat body of *P. rapae* larvae [51]. In addition, transcript levels of genes involved in proPO activation cascade, such as PAPs, were down-regulated.

In addition to suppressing the host immune system, interfering with host development could be another function of venom from some endoparasitoids. A 66 kDa venom protein from *Cardiochiles nigriceps* in combination with calyx fluid was found responsible for delaying larval development and inhibit pupation in *H. virescens* larvae [76]. Calyx fluid alone was not able to induce the same effect in the host. This effect appears to be due to degradation of the prothoracic glands [68,77].

2. New Approaches in Venom Studies

2.1. RNAi

RNA interference is an ancient and conserved response to the presence of double stranded RNA (dsRNA) in eukaryotic cells [78]. The source of dsRNA might be exogenous or endogenous. Exogenous dsRNA could be viral RNA genome, viral replicative intermediates, overlapping viral transcripts produced during viral replication (mostly in DNA viruses), or *in vitro* synthesized dsRNA. Once the presence of dsRNA is sensed in the cell, a ribonuclease enzyme called Dicer, cleaves the dsRNA into short interfering RNAs (siRNAs). siRNAs induce formation of the RNA Induced Silencing Complex (RISC) in which an argonaute (Ago) protein plays a major role. siRNA-loaded RISC complex facilitates binding of the siRNAs to their complementary target sequences and their subsequent cleavage. Transfection of dsRNA/siRNAs into cells or whole organisms are routinely used to knockdown transcript levels of target genes. In insects, the level of success in application of dsRNA for gene silencing studies has been variable; working really well in some insects and not in others [79]. Given knocking out genes is not possible in many non-model insects, gene silencing by RNAi using long dsRNA or siRNA has been quite useful.

Application of RNAi could also be useful in studying the function of specific venom proteins in host-parasitoid interaction. In a recent study, Colinet *et al.* utilized RNAi to successfully silence the *RhoGap* gene abundant in *L. boulardi* venom [80]. Silencing was achieved by microinjection of dsRNA specific to the gene into the parasitoid pupae. The results showed near complete silencing of the gene and lack of the protein detection in the venom reservoir of the wasps emerged from gene-specific dsRNA injected pupae. Interestingly, the silencing effect remained stable throughout the entire wasps' lifetime. This initial step towards demonstration of successful silencing of a gene coding for a venom protein provided a new experimental tool in investigating the specific role of the proteins in host-parasitoid interactions and their importance in the success of parasitism.

2.2. High Throughput Methods: Transcriptomic, Proteomics, Peptidomics

Major advances and cost reductions in high throughput analyses of RNA and proteins have provided opportunities for researchers to identify and gain a better understanding of the diversity of venom proteins/peptides from various animals. In addition, these approaches could enable analysis of venom impacts on the host transcriptome in more depth. These studies in general show the presence of conserved proteins present in venom from endo- and ectoparasitoids but also some that are unique to each parasitoid. Further, a large number of proteins/peptides are being discovered with no significant similarity to other proteins with known functions.

In 2010, Zhu *et al.* performed a proteomic analysis of the venom from the endoparasitoid *P. puparum* which allowed identifying 12 out of 56 soluble proteins extracted from a venom apparatus homogenate. While a number of proteins highly similar to venom proteins identified in other hymenopteran species were found (e.g., venom acid phosphatase, calreticulin), the method used did not allow identification of other possible venom components among proteins of cellular origins with certainty.

In the same year, the first exhaustive identification of venom components of a parasitoid wasp was carried out using a combination of sequencing ESTs from a venom gland library and nano-LC-MS/MS analysis of peptides from pure venom isolated from the venom reservoir of the egg-larval parasitoid *Chelonus inanitus* [81]. The main venom components were a number of enzymes (chitinase, esterase, metalloprotease-like, C1A protease, serine protease), mucin-like peritrophins, lectin-like proteins and yellow-e3 like venom protein similar to *Apis mellifera* protein produced in the head and hypopharyngeal gland of honeybee workers. A number of proteins were also found to be unique to the parasitoid. In a complementary work, it was shown that the venom proteins enhance PDV entry into the host cells and facilitate placement of the parasitoid egg in the host embryo's hemocoel [82].

Using a combined transcriptomic and proteomic approach, Colinet *et al.* found 16 putative venom proteins from *A. ervi* [83]. The most abundant proteins were three γ-glutamyl transpeptidases (γ-GTs), two of which are likely coded by alleles of the same gene and the third one unrelated to the other two and most likely inactive due to a mutation in the active site. The product of one of the two alleles was previously shown to cause castration in the host aphid *A. pisum* by causing apoptosis and subsequent tissue degradation in the ovaries [31,54]. The study also resulted in the identification of proteins present in other parasitoid venoms such as SPHs, neprilysin-like and cysteine-rich toxins; and some being unique to the parasitoid; such as endoplasmin [83].

In another transcriptomic study, a large number of unigenes were identified from the venom of *Leptopilina heterotoma*, an endoparasitoid of *D. melanogaster* [84]. Similar to other such studies, several venom proteins were identified that were conserved among endo- and ectoparasitoids, as well as several unique genes. The components of venom from *L. heterotoma*, including VLPs, are responsible for suppression of the host immune system and delay in the host larval development [85,86]. The exact impact of venom *versus* VLPs in the effects observed in the host is not clear. In a different study, the venom composition of *L. heterotoma* and *L. boulardi*, two parasitoids of *D. melanogaster* with different parasitism strategies were explored using a combination of RNA-seq and proteomics approaches. *L. boulardi* is specialized on *D. melanogaster* and inhibits cellular immunity by inhibiting hemocytes from encapsulation, while *L. heterotoma* parasitizes different species of *Drosophila* and causes destruction of the host hemocytes [87]. This study led to the identification of 129 and 176 proteins in *L. boulardi* and *L. heterotoma* venoms, respectively. A large number of proteins were found in venom from both species but also some that were unique to each which presumably may contribute towards their different strategies in parasitism [69].

Following the availability of the complete genome of the ectoparasitoid wasp *Nasonia vitripennis*, de Graaf *et al.* used a combination of bioinformatics and proteomic analyses to determine the profile of proteins present in the wasp's venom reservoir [70]. The bioinformatics approach was based on digging into the genome sequences using similarity with other known venom proteins. The proteomics approach was based on using two mass spectrometry approaches: off-line 2D liquid chromatography matrix-assisted laser desorption/ionization time-of-flight (2D-LC-MALDI-TOF) MS and a 2D liquid chromatography electrospray ionization Founer transform ion cyclotron resonance (2D-LC-ESI-FT-ICR). The outcome was identification of 79 proteins among which half of them were proteins that were either unknown or not yet associated to insect venoms. The major groups were proteases/peptidases, protease inhibitors, enzymes involved in carbohydrate, DNA and glutathione metabolism, esterases, recognition proteins, and immune related. Serine proteases and protease inhibitors were overrepresented in the venom fluid [70].

Similarly, a combination of transcriptome sequencing and proteomics was used to identify the protein profile of venom proteins in *M. demolitor* [64]. This study demonstrated the presence of several venom proteins found in other parasitoids (e.g., a reprolysin-like metalloprotease and Ci-48a), but also some unique hypothetical proteins. This study presented further evidence of recruitment of insect proteins into venom by gene duplication and modification. Further, comparison of *M. demolitor* PDV gene products and venom proteins showed no overlap suggesting separate functions of the products.

3. Venom Protein Evolution and Diversity

3.1. Venom Diversity within the Hymenoptera: Who Are the Outliers?

Hymenoptera of the suborder Apocrita would gather more than 300,000 species, representing 10% to 20% of all insect species currently living on earth [2,88,89]. The suborder Apocrita is divided into two major groups, Parasitica, that possess an ovipositor (terebra or drill) functioning as a dual egg-laying and venom injecting organ, and Aculeata, in which the ancestral ovipositor has been fully modified for injection of venom (aculeus or sting) [90,91]. All modern Apocrita share a common ancestral parasitic origin [92] from which they have successfully evolved to display diversified lifestyles and nutritional behaviours, including parasitism of plants or arthropods, predation, phytophagy and omnivory. Depending on the species, venoms are being used as defensive agents against predators, competitors and pathogens, hunting weapons, manipulators of host physiology, repellents and trail, alarm, sex, recognition, aggregation and attractant-recruitment pheromones [93,94]. Available data on the composition of Apocrita venoms are highly heterogeneous depending on the considered superfamilies, and the extent of the complexity and diversity of these arsenals is still difficult to estimate with precision.

During the past 50 years, proteins and peptides have focused most of the attention of investigators interested in the composition of Hymenoptera venoms. Around 70 Hymenoptera species (out of 300,000 venom-producing species!) were studied to this aim. In the Vespoidea superfamily, which gathers ants and solitary and social wasps, an overall of 138 different proteins and peptides out of 43 species investigated were characterized. In Chalcidoidea, Apoidea and Ichneumonoidea, a lower number of species have been studied (3, 6 and 11 respectively) with lead species deeply investigated through venomic approaches (e.g., *N. vitripennis*, *A. mellifera*, *C. inanitus*, *M. demolitor*). In Cynipoidea, investigations were only performed on three parasitoids of *Drosophila* belonging to the genus *Leptopilina* or to the genus *Ganaspis*. In the seven remaining superfamilies of Hymenoptera, which represent an estimated number of more than 20,000 species [2], there is simply no data available. This underlines how far we are from grasping the richness and diversity of Hymenoptera venoms even for the most studied groups of species. It stresses too how extensive venomic studies can quantitatively and qualitatively improve our knowledge of this molecular diversity. An overview of main families of proteins and peptides characterized until now from Hymenoptera venoms can be found in Table 1.

Most parasitoid species have evolved under strong selective pressures and have adapted to a restricted range of hosts thank to specific strategies of virulence in which venoms can play a predominant role [95]. The very first analytical works led on parasitoid venoms from the late eighties intended to compare their composition to that of social hymenopteran species which were, by far, the best known at this period. These pioneer studies suggested that unlike venoms from social Aculeates, parasitoid venoms lacked small proteins and peptides and could be characterized instead by the presence of large venom proteins whose molecular masses frequently exceeded 100 kDa [90,96,97]. This statement was taken up during the following two decades and seemed to be confirmed for a while by the description of large proteins in venoms of parasitoids [15,18,19,72,98–102]. In parallel, however, an increasing number of peptides and small proteins of molecular masses lower than 15 kDa have also been discovered in the venoms of parasitoid wasps belonging to distant families such as Eupelmidae, Pteromalidae, Braconidae or Ichneumonidae [18,19,30,32,41,72,81,101,103,104]. In fact, there is such a diversity of functions represented among proteins that were identified to date in parasitoid venoms, that presence of proteins of high molecular masses is definitely of second importance and cannot reasonably be hold as a common distinctive feature of parasitoids' venoms.

Incidentally, if their functional diversity was more explored and taken into consideration, this could put an end to the temptation to see in the venom of social aculeate Hymenoptera a classical pattern for all hymenopteran venoms and to consider parasitoids as outliers. At most it may be considered now that venoms of social species, which only gather a fraction of the most recent species of the order and which have independently evolved from several parasitoid lineages [2,89], constitute rather an exception than a paradigm. Indeed, these venoms are particularly rich in neurotoxic, cytolytic and antimicrobial peptides that fulfill key roles in capture and conservation of preys, defense against competitors and prevention of brood nest contamination by microorganisms [105]. The presence and abundance of such compounds make sense only with respect to particular lifestyles (eusocial, subsocial or solitary life) and feeding modes (omnivory, phytophagy, predation) and have certainly played an important role in the diversification of these species and their ecological dominance [89]. But they have little in common with venom compounds used as virulence factors needed to finely adjust and manipulate the internal physiological balance of hosts, a constraint experienced by most parasitic wasps, and endoparasitoid species in particular.

This confusion mainly originates from a widespread anthropocentric view in addressing the issue of venom, that some extensive works such as "Venoms of the Hymenoptera" [90] and other recent papers [2,106,107] have not totally succeeded to clear up. According to this conception, the most important venoms to man would naturally constitute the most important venoms. One may understand that this outdated view has served as a primitive conceptual matrix to the pioneer comparative works of the eighties because they preceded the functional examination of parasitoid venoms. It is a bit more surprising how often this view is fostered in more recent comparative papers (see for instance [108,109]) aiming at underlying the potential of venomic approaches for pharmaceutical discovery, but in which the composition richness and subtlety of functions displayed by the venoms of parasitic wasps and other Hymenoptera are simply ignored to the benefit of a lapidary mention of the "predatory" and "defensive" roles of the venoms from "ants, wasps and bees". It does not only deny hymenopteran venom diversity, it also neglects venom variability and complexity, its inter- and intra-individual corollary dimensions. To go further and break with "an anthropocentric view of toxicity", Fry *et al.* (2009) [110] have proposed a global definition regarding "(…) venom as a secretion, produced in a specialized gland in one animal, and delivered to a target animal through the infliction of a wound regardless of how tiny it could be, which contains molecules that disrupt normal physiological or biochemical processes so as to facilitate feeding or defense by the producing animal". With a little effort this interesting definition could have been useful, but it excludes important cases, like the possibility for venom to be injected into a host plant and not just into an animal. The case is frequent for instance in Cynipidae which develop as parasites of wild roses or oaks and which inject venom in host plant tissues during oviposition [111]. The oak gall wasp *Biorhiza pallida* even possesses one of the largest venom apparatuses found in a hymenopteran in proportion of the body size [112]. The exact functions of *B. pallida*'s venom are not known to date as those of hundreds of thousands of other parasitic wasps associated to animal or plant hosts. It may thus be also hazardous to define what is or not venom by reference to a restrictive set of functions, such as feeding and defense, because we still largely ignore all what venoms can achieve. The proposed definition also discards the cases in which venoms can act on another organism in the absence of wounding, for instance through venom spraying for prophylactic or defensive purposes and venom deposition in order to serve as a pheromone [105].

It is noteworthy that little is known, in Hymenoptera, on small size venom components belonging to other biochemical classes than proteins and peptides. Knowledge acquired in this domain only comes from the study of some social species belonging to Apoidea and Vespoidea superfamilies. A small set of biologically active amines either acting as smooth muscle agonists, pain-inducing or paralytic factors have been described in venoms of solitary or social Aculeates [11,90]. They notably include histamine, acetylcholine, 5-hydroxytryptamine, bradykinin, GABA, taurine, β-alanine, serotonine, tyramine, dopamine, noradrenaline and adrenaline. Formic acid is the most famous ant venom component and to date, the single organic acid known from Hymenoptera venoms. Some ant venoms may also contain a diversified range of alkaloids, monoterpene hydrocarbons, aromatic nitrogen-containing compounds and lactones [93]. New available techniques for metabolomics analyses could be useful to investigate the presence of such molecules in venoms of other Hymenoptera superfamilies and may even reveal other unexplored classes of active venom metabolites (e.g., free amino acids, lipids, polysaccharides, sterols, *etc.*).

In summary, by highlighting papers of particular interest that focus on the main venomous functions of species of importance, one may sometimes be at risk of simply missing the essential aspects of an issue.

3.2. Factors Shaping Venom Complexity in Parasitoid Species

A given parasitoid species is supposed to obtain several adaptive advantages from the production of a complex venom [99]: (1) the combined actions of different venom components allow targeting of several host functions either simultaneously or sequentially; (2) the effects of the venom components may be complementary or even cumulative; (3) the likelihood that hosts simultaneously develop resistances against multiple venom components is low. On the other hand, the production of venom, which may start before adult emergence [97], is reputed to be costly. Biochemical, proteomic and transcriptomic analyses on parasitoid venoms and venom glands have shown a long time ago that these tissues generally express a small number of highly abundant proteins and peptides and a large number of low abundance products [27,64,70,80,81,96,113]. This raises at least two questions: First, why do investigators continue to expend public funds into costly high-throughput transcriptomic methods that will generate millions of redundant sequence reads to identify only a few dozen major venom proteins? Second, what allows a secreted product to be selected among these few dozen key components? There are a whole bunch of politically incorrect answers to the first question which might be easily found elsewhere, suffice to say that some like to be exhaustive at someone else's expense, and that deep sequencing methods are appropriate tools to address the issue of inter- and intraspecific venom variability [114].

Concerning the second point, Fry *et al.* (2009) [110,115] have noticed a high proportion of convergently recruited protein families among secretions of a wide range of venomous organisms, suggesting that similar structural and/or functional constraints could influence toxins recruitment across the animal kingdom. In parasitoids, venom complexity is the result of a balance between benefits and costs that seems to have favored the selection and production in abundance of a restricted number of venom proteins that are congruent with strict requirements of safety towards the producing organism and efficiency towards targets. As in other venomous animals, potentialities for recruitment and evolution of parasitoid venom proteins greatly depend on individual, populational and ecological factors.

3.2.1. Individual Factors

Individual factors gather physiological features that may affect venom complexity. They may include specific biochemical properties such as the acidic nature of the venomous secretions, the histological organization of venom glands in a simple glandular epithelium and the occasional presence of structures like an internal chitin layer in the secretory duct and the reservoir of the venom apparatus [116]. The tissue organization could have greatly influenced the recruitment of compatible compounds, selected under the double necessity to be devoid of any autotoxic effect and to remain (or to only become) bioactive upon injection. The recruitment of new venom proteins is supposedly mediated, in parasitic wasps, by gene duplication events eventually followed by changes in protein addressing and processing steps [41,114]. Prevention of autotoxicity and conservation of bioactivity may be achieved by additional changes in substrate specificity or catalytic sites of the venom enzymes by comparison to their cellular homologs [72,80,101,114,117]. It can also rely on the presence in the venom of specific enzyme inhibitors [118] and molecular chaperones [119] or secretion of venom enzymes as inactive precursors [41].

In most parasitoid species studied to date, venoms were either reported to constitute the main predominant factors of virulence or to facilitate the action of other factors [30,82,116,120–123]. It is worthy to underline that in the former case, numerous studies have focused on the identification of prime venom components while paying little attention to other secondary venom molecules capable of potentiating or regulating their action. When venoms act synergistically with other factors of virulence (*i.e.*, symbiotic PDVs, VLPs, ovarian fluids, larval secretions, *etc.*), the molecular basis underlying facilitating processes remain largely unknown, with few exceptions [30]. Even more intriguing are parasitoid venoms that were acknowledged to be non-essential for the survival of the parasitoid wasps' progeny. This is notably the case for the venoms of the braconid *Cotesia congregata* [124] and the ichneumonid *C. sonorensis* [60], *Tranosema rostrale* [125] and *H. didymator* [56]. Subtle interactions are hence likely to take place (1) between venom components; (2) between venom and other factors of virulence; and (3) between venom and various targets. These interactions probably weighted significantly on venom complexity. Remarkably, functional redundancy seems not to be a widespread rule in parasitoid wasp venoms as suggested by the presence, in several species, of venom proteins that arose from gene duplications but whose key functional residues are often mutated [83,114]. This is in sharp contrast with PDVs whose genomes frequently contain gene sets forming large multigenic families and which may be co-expressed in host tissues [126]. The fact that the former are produced by the parasitic wasp itself and the latter at the expense of host insects may explain why diversification operated differently on both factors of virulence. Definitely, some like to be exhaustive at someone else's expense.

PDV- and VLP-associated wasps are apparently undergoing a process of subfunctionalization, or functional partitioning, of their venom that probably started with the integrations of the ancestors of actual bracoviruses (BV), Ichnoviruses (IV) (the two subgroups of PDVs) and producers of VLPs in the genomes of different organisms belonging to Braconidae, Ichneumonidae and Figitidae [127]. In BV-associated parasitoids such as *C. inanitus* or *M. demolitor*, we observe that almost no overlap exists between venom proteins and PDV gene products [64,81] and, for *M. demolitor*, between venom and teratocytes [64]. According to Burke and Strand (2014) [64], this functional partitioning would provide "wasps the means to deliver and express effector molecules in hosts for protracted periods".

Consequently, the presence of the PDV allows relaxing the selective pressure exerted by a functional constraint identified by Fry *et al.* [110]: the need to produce a rapid effect in order to be effective and successful. Therefore it opens the possibility for these venoms to evolve in new directions, like the recruitment of venom proteins fulfilling structural functions or promoting cell growth and/or tissue differentiation. It seems to be the case for the venom of *C. inanitus* which contains an Imaginal disc Growth Factor (IDGF)-like protein (Ci-48b) and two putative mucin-like peritrophins (Ci-23c and Ci-220) [81].

Dorémus *et al.* (2013) [56] have suggested that several IV-associated wasps, such as *H. didymator*, have been further in the subfunctionalization process in producing venoms which reveal to be unnecessary to the success of the parasitoid although a number of venom proteins are still abundantly produced. Loss of regulatory functions may have followed viral acquisition or alternately, acquisition of these functions was only performed by the symbiotic virus. In parasitic wasps whose venom glands produce VLPs such as *L. herotoma* and *L. boulardi*, the most abundant venom proteins are in fact constitutive of the VLP capsid [69,114]. Since VLPs are devoid of nucleic acids and cannot externalize the production of regulatory proteins in parasitized host, venom gland plays here the role of "VLP factory".

These examples highlight how, in parasitoids, venom composition and functional diversity are interdependently linked to the evolution of other factors of virulence. Here probably reside the main origins of their singularity regarding other animal venoms. Structural and functional constraints only explain a part of parasitoids' venom complexity. Existence of inter-individual variability is another important parameter to understand how venom complexity and diversity have arisen.

3.2.2. Populational and Ecological Factors

Studies led on BV-associated parasitoid wasps of the Microgastrinae complex have recently suggested that the highly diverse gene content of BV genomes could drive adaptation or specialization of parasitoid wasps to particular hosts [128]. For instance, former cross-protection experiments with species of the genus *Microplitis* have shown that BV-mediated immunosuppression was one important determinant of host range along with other factors [129]. Investigating similarly whether venom composition and effects could influence major life traits of parasitoid Hymenoptera, and could in turn be influenced by ecological constraints, necessitates the study of particular biological models. It can be achieved for example through the study of species devoid of symbiotic viruses and offering intraspecific variations of their venomous properties and successful parasitism rates (SPRs) toward a reference host model.

In the case of the evolution of the genus *Asobara* (Braconidae: Alysiinae), the cross-influences of the levels of resistance displayed by local species or strains of *Drosophila* hosts and of levels of virulence exhibited by sympatric parasitoid species are quite well documented [130]. The richness of these interactions has led to a surprising diversification of the composition and functional properties of the venomous secretions in *Asobara* parasitoids with direct and major consequences on the strategies of virulence of these species [116,131,132].

The SPR of the solitary endoparasitoid *Asobara tabida* towards *D. melanogaster* has been correlated with geographic localization [133]: the strain called A1 originates from the south of France and develops more successfully on *D. melanogaster* than the WOPV strain originating from the Netherlands [134]. The SPR of both strains in controlled conditions has been shown to be of 74% ± 2.6% and 18.8% ± 8.8%, respectively [42]. In this biological system, the parasitoid female lays a single egg into a young *Drosophila* larva, which may escape encapsulation if its chorion strongly adheres to the basal lamina surrounding the internal tissues of the host [135] (Moreau S., unpublished data). If an *A. tabida* egg is not able to bury itself between the host's organs, it is rapidly encapsulated by circulating hemocytes, unless it has been oviposited into a host deprived of encapsulation abilities [133,136]. Whatever the outcome of the parasitic relationship, parasitized *D. melanogaster* larvae retain substantially their ability to mount effective immune reactions but experience a transient paralysis, an altered weight gain and a significant increase in the time required before the onset of pupariation [42,137]. These observations suggested that factors of parasitic origins, and notably venoms, had a precocious effect on activity and development of parasitized hosts, but almost not on their immunity, and that crucial difference could take place between the two strains studied. Interestingly, while parasitism by both strains induced equivalent mortality rates before the parasitoid's emergence (approximately 20%), experimental injection of venom proteins from the WOPV strain significantly increased the mortality rate of *D. melanogaster* larvae. At the highest tested dose of venom (equivalent to a tenth of the venom produced during the first five days after emergence), 95% of *D. melanogaster* larvae died before reaching the pupal stage and none completed its development. In comparison, when the same quantity of venom proteins from the A1 strain was injected into *D. melanogaster* larvae, the observed mortality rate was only 35%. Venoms of both strains also exhibited variations in their ability to induce transient paralysis, the venom of the WOPV strain having the strongest effect [42]. Finally, electrophoretic profiles of venom extracts showed minor band differences (Moreau S., unpublished data). The abilities of *A. tabida*'s venom to induce host paralysis, to be lethal at high doses and most probably to delay development in parasitized hosts, are reminiscent of its ectoparasitic origins [138], even though its lifestyle is now undoubtedly endoparasitic. On the basis of their ancestral functional legacy and despite the loss (or the non-acquisition) of an immune-suppressive venom, geographically distant populations of *A. tabida* have thus evolved at least two strategies to adapt to endoparasitism. Eggs of the A1 strain take benefit from the increased stickiness of their chorion and from the lesser toxicity of their mother's venom to successfully parasitize immune-reactive host larvae. Conversely, the eggs of the WOPV strain need to develop into immunocompromised hosts and the higher toxicity of the female's venom could serve here to weaken or even eliminate some potential competitors (e.g., eggs of *L. boulardi* already present in the host) in order to allow a kleptoparasitic development of *A. tabida* eggs [133].

Countless other inter-individual variations in venom composition have occurred within hymenopteran parasitoids over the past millions of years of evolution and many of them have probably been selected under constraints imposed by interacting species within local communities. This question is now the subject of renewed interest [119,139] and should benefit from the availability of high-throughput sequencing methods that provide access to the inter-individual variability of venom gland gene expression within natural populations.

Beyond understanding the evolution of particular parasitoids-hosts relationships, investigation of the venom gland content may also directly inform us about the evolution of the order Hymenoptera. The venom composition of *C. inanitus* hence appeared as a mixture of conserved venom components and of recent proteins potentially specific of the lineage [81]. The phylogeny of several conserved venom proteins has been reconstructed and the authors identified Allergen 5 proteins, a group of major allergen components of ants and wasps venoms, as one of the most ancient family among insect venom proteins: the ancestral *Allergen 5* gene was likely already expressed by the venom glands of the common ancestor to Ichneumonoidea and Aculeata, 155 to 185 million years ago and has apparently been lost in Apoidea. This study has also confirmed that genes coding for honeybee's Major Royal Jelly Proteins derived from a progenitor gene (*yellow-e3*) which probably possessed an ancestral venomous function, as previously suggested [140,141]. These examples show that aside from quantitative benefits expected from the achievement of extensive inventories, combined genomic, transcriptomic and proteomic approaches constitute appropriate tools to explore evolutionary purposes. Such approaches have notably been used with success to unravel the origins of PDVs produced by parasitoid wasps [62]; reconstructing the composition of some "paleovenoms" should thus become an achievable objective with exciting perspectives. Transcriptomic data can also help us understand the functioning of venom glands cells through the identification of gene products directly involved in production, delivery and activation of venom toxins of a broad range of venomous animals, such as the dipeptidyl peptidase IV (DPPIV) family enzymes [56].

Toxins **2015**, *7*, 174–196

4. Pharmaceutical and Biological Potential of Parasitoid Wasp Venoms

With a complexity generally comprised between 10 to 100 different venom proteins and peptides per species, the 250,000 known hymenopteran parasitoid species represent a source of millions of bioactive molecules which remain almost totally unexplored. The applied perspectives expected from their study are just as vast. In medical areas, they range for instance from the prevention and treatment of venom hypersensitivity to the discovery of innovative drug candidates. Given that parasitoid venoms also attract a growing attention as a rich source of bioactive substances for the control of insect pests [142], an advantage could be taken from the acquisition of a better knowledge on venom composition in a greater number of species, to optimize new strategies of integrated pest management.

4.1. Pharmaceutical Perspectives

The therapeutic value of venom immunotherapy to improve the quality of life of patients which are hypersensitive to the venom of social Hymenoptera is acknowledged since more than eighty years [143,144]. Recently, venomic approaches have allowed the discovery of new venom constituents which were proven to be of immunological significance and have opened the way to optimization of immunotherapeutic strategies through the use of cocktails of recombinant allergens [70]. Interestingly, the toxicity, allergenicity and algogenicity (potential for pain induction) of solitary predatory or parasitoid wasps towards human and domestic animals have almost never been assessed in laboratory. Although rare, accidental envenomation events have nevertheless been documented [105]. The continued increase in human populations' densities could expose a greater number of people to these non-intentional contacts.

On the other hand, a number of venom serine proteases from wasps and bees were shown to exert a potent anticoagulant effect, inhibiting platelet aggregation and degrading the β-chain of fibrinogen [145,146]. Enzymes with similar functions have been reported from the venom of the ectoparasitoid *N. vitripennis* [70] and from those of the endoparasitoids *P. hypochondriaca* [99], *P. puparum* [113] and *A. japonica* [71]. Their study would raise potential applications for the treatment of thrombotic disorders.

Venoms of Hymenoptera also often contain antimicrobial peptides (AMPs) [32,147–149] or may stimulate the antimicrobial immune defenses of the targeted organism [105]. Such molecules may serve as templates to inspire new antibiotic agents, an invaluable resource considering the worrying current increase in the number of multi-drug resistant pathogens and the expected changes in the distribution of terrestrial ectotherms and epidemiology of infectious diseases which are likely to be induced by the ongoing global warming [150,151]. The antinociceptive effects of other venom peptides involved in the blockage of ionic channels can also represent potential sources for drug development to treat pain. Several examples of neurotoxic peptides from solitary and parasitoid wasps are already known [12,13,152].

Finally, one of the most interesting properties of venom components and venom cocktails are probably their natural stability as injectable solutes, their effectiveness in reaching targeted tissues and their ability to synergize their actions, shaped by millions years of "R&D". Their features may inspire the design of recombinant hydrolases and innovative strategies of enzyme replacement therapy for patients suffering from rare lysosomal storage disorders [95,153]. Other venom molecules with protease inhibiting, pro-apoptotic or cytotoxic properties and described from several endoparasitoid species [94,95] may also worth to be considered for the development of anti-tumor or anti-viral agents.

4.2. Biological Control: Development, Reproduction and Immune Modulators

Twenty-one years ago, Zeneca Ltd. patented a synthetic DNA capable of expressing a venom peptide from *Conus* marine snails in pest insects via a baculoviral vector [154]. The year after, the same company patented a chimeric toxin resulting from the fusion of part of an endotoxin from *Bacillus thuringiensis* and an insecticidal venom toxin from the scorpion *Androctonus australis* Hector [155]. The chimeric toxin was thought to be applied directly to crop plants or to be produced by transgenic plants and delivered to pest insects through ingestion. In 1996, several patent applications concerning the potential use of *B. hebetor* venom neurotoxins as insecticidal toxins have been simultaneously deposited by Sandoz Ltd. [156], Zeneca Ltd. [157] and NPS Pharmaceuticals, Inc. [158,159]. The U.S. Patent by Quistad *et al.* (1996) [156] was notably directed to "toxins active against insects which are isolated from the parasitic wasp *B. hebetor*, the nucleic acids which encode the toxins, cloning of the toxins, use of the toxins to control insects, and genetically engineered virus vectors carrying the toxin gene". Many have seen in these patent applications the opening of a new era in the use of parasitoid wasps in biological control, not only as living organisms, but also as sources of genes and molecules of interest to control pests [95,160,161]. However, because sequence information and experimental results were kept confidential, these patents have seriously hindered the diffusion of a useful knowledge. They have resulted in delaying both fundamental works about venom diversity and evolution, and the effective use of interesting molecules in the field. Beyond the fact that these "inventions" constitute regrettable attempts to enclose and usurp the genetic patrimony of wild species and data constitutive of the common knowledge of the human kind, they fuelled the public mistrust towards biotechnologies and the potential use of venom compounds from parasitoids for plant protection. In 1999, the British registered charity ActionAid was already wondering about the possible impacts on environment of genetically modified (GM) crops or baculoviruses expressing the *B. hebetor*'s venom toxins [162]: what would happen if the venom toxins were ingested by non-targeted organisms or if their genes were horizontally transferred to microorganisms or to wild plants? Are they allergenic or toxic to humans? Will these genetically modified organisms contribute to improving or degrading the situation of farmers in developing countries? It seems that answers to these important questions will also remain confidential for a while.

A very different approach has been followed during the next decade by a group working on the venom of the widely used biological control agent *Aphidius ervi* (Braconidae: Aphidiinae). The venom of *A. ervi* induces the castration of the pea aphid *Acyrthosiphon pisum* via the specific degeneration of the germaria and of the young apical embryos [54]. The sequence of the bioactive venom component inducing castration was published in 2007 [31]. It corresponds to a dimeric γ-glutamyl transpeptidase (γ-GT) which acts by specifically triggering apoptosis in germarial cells and cells of the ovariole sheath of the parasitized aphid. Recently this γ-GT has also been found in egg extracts of *A. ervi* which suggests that the expression of the venom enzyme would not be restricted to the venom gland of the parasitoid [163]. In addition these authors have precisely dosed the quantity of venom γ-GT injected in the aphid host during oviposition (approximately 4 ng) thanks to an elegant and transferable experimental setup relying on chitosan beads. Finally, a third group of investigators has recently performed a combined transcriptomic and proteomic approach on the venom of *A. ervi* [80]. Surprisingly, their work has revealed the presence of two additional γ-GTs in the venom, from which one would not be functional while the other would represent the product of an allelic variant of the original γ-GT gene. Interestingly, a reconstruction of the phylogenetic relationships between known hymenopteran γ-GTs suggests that an independent and converging duplication event would be at the origin of the presence of two other γ-GTs in the venom of the ectoparasitoid *N. vitripennis* [70,80]. Finally, the presence of an endoplasmin has also been detected in the venom of *A. ervi*. This protein, which belongs to a family of molecular chaperones could play a role in the transport and stabilization of the other venom proteins including the γ-GTs [80]. If it was confirmed, this finding could have important implications for future applications that would aim at using γ-GTs or other venom proteins to efficiently control aphid populations. This example illustrates, if we needed reminding, how open collaboration more than mercantile enclosures, encourages the acquisition of useful knowledge and makes progress possible.

The most advanced project to date concerns the selection of venom proteins from the endoparasitoid *P. hypochondriaca* able to help the control in the field of two pest insects, *Lacanobia oleracea* and *Mamestra brassicae* [164,165]. The originality of the envisaged strategy of control resides here in the use of the immunosuppressive properties of two venom proteins (VPr3 and VPr1) to increase sensitivity of pest insects to biological control agents (BCA) such as *Beauvaria bassiana* and *B. thuringiensis* (Richards and Dani, 2008). Injections of the recombinant rVPr1 suppressed the ability of *L. oleracea* and *M. brassicae* to mount hemocyte-mediated immune responses [165]. Two modes of delivery of rVPr1 to the targeted pest insects are studied in view of future practical applications: either by directly spraying rVPr1 onto plants attacked by the pests (which would require protecting the protein from degradation and inactivation) or via the expression of rVPr1 by the BCA itself [164]. The latest option would necessitate a careful development to avoid a reckless widening of the biological spectrum of BCAs, notably towards non-targeted species of Lepidoptera.

5. Conclusions

Despite the large diversity of parasitoid wasp species, there are only a small number of venom proteins that have been described from the wasps. There is a wealth of unexplored biomolecules present in parasitoid venoms that are of value in basic evolutionary studies, venom biology, host-parasite interactions, evolution of life strategies, and may potentially contain components that could be used in agriculture and pharmacology. The available state-of-the-art approaches in proteomics and transcriptomics provide us with valuable tools and unique opportunities to explore these diverse biomolecules. By characterizing parasitoids' venoms at the functional level, we can gain a better understanding of neglected interactions to achieve knowledge that can enable us to utilize them for improved appreciation of life and diversity, pest management and health.

Author Contributions: S.J.M.M. and S.A. wrote the review manuscript.

Conflicts of Interest: The authors declare no conflict of interest.

References

1. Dheilly, N.; Maure, F.; Ravallec, M.; Galinier, R.; Doyon, J.; Duval, D.; Leger, L.; Volkoff, A.; Missé, D.; Nidelet, S.; *et al.* Who is the puppet master? Replication of a parasitic wasp-associated virus correlates with host behaviour manipulation. *Proc. Biol. Sci.* **2015**, *282*. [CrossRef] [PubMed]
2. Pennacchio, F.; Strand, M.R. Evolution of developmental strategies in parasitic Hymenoptera. *Annu. Rev. Entomol.* **2006**, *51*, 233–258. [CrossRef] [PubMed]
3. Tian, C.; Wang, L.; Ye, G.; Zhu, S. Inhibition of melanization by a *Nasonia* defensin-like peptide: Implications for host immune suppression. *J. Insect Physiol.* **2010**, *56*, 1857–1862. [CrossRef] [PubMed]
4. Kryukova, N.; Dubovskiy, I.; Chertkova, E.; Vorontsova, Y.; Slepneva, I.; Glupov, V. The effect of *Habrobracon hebetor* venom on the activity of the prophenoloxidase system, the generation of reactive oxygen species and encapsulation in the haemolymph of *Galleria mellonella* larvae. *J. Insect Physiol.* **2011**, *57*, 769–800. [CrossRef] [PubMed]
5. Edwards, J.P.; Bell, H.A.; Audsley, N.; Marris, G.C.; Kirkbride-Smith, A.; Bryning, G.; Frisco, C.; Cusson, M. The ectoparasitic wasp *Eldophus pennicornis* (Hymenoptera: Eulophiclae) uses instar-specific endocrine disruption strategies to suppress the development of its host *Lacanobia oleracea* (Lepidoptera: Noctuidae). *J. Insect Physiol.* **2006**, *52*, 1153–1162. [CrossRef] [PubMed]
6. Price, D.; Bell, H.; Hinchliffe, G.; Fitches, E.; Weaver, R.; Gatehouse, J. A venom metalloproteinase from the parasitic wasp *Eulophus pennicornis* is toxic towards its host, tomato moth (*Lacanobia oleracae*). *Insect Mol. Biol.* **2009**, *18*, 195–202. [CrossRef] [PubMed]
7. Libersat, F. Wasp uses venom cocktail to manipulate the behavior of its cockroach prey. *J. Comp. Physiol. A Neuroethol. Sens. Neural Behav. Physiol.* **2003**, *189*, 497–508. [CrossRef] [PubMed]
8. Gal, R.; Libersat, F. A wasp manipulates neuronal activity in the sub-esophageal ganglion to decrease the drive for walking in its cockroach prey. *PLoS ONE* **2010**, *5*, e10019. [CrossRef] [PubMed]
9. Gavra, T.; Libersat, F. Involvement of the opioid system in the hypokinetic state induced in cockroaches by a parasitoid wasp. *J. Comp. Physiol. A Neuroethol. Sens. Neural Behav. Physiol.* **2011**, *197*, 279–291. [CrossRef] [PubMed]

10. Libersat, F.; Gal, R. Wasp voodoo rituals, venom-cocktails, and the zombification of cockroach hosts. *Integr. Comp. Biol.* **2014**, *54*, 129–142. [CrossRef] [PubMed]

11. Moore, E.; Haspel, G.; Libersat, F.; Adams, M. Parasitoid wasp sting: A cocktail of GABA, taurine, and beta-alanine opens chloride channels for central synaptic block and transient paralysis of a cockroach host. *J. Neurobiol.* **2006**, *66*, 811–820. [CrossRef] [PubMed]

12. Piek, T. Delta-philanthotoxin, a semi-irreversible blocker of ion-channels. *Comp. Biochem. Physiol. C* **1982**, *72*, 311–315. [CrossRef]

13. Eldefrawi, A.; Eldefrawi, M.; Konno, K.; Mansour, N.; Nakanishi, K.; Oltz, E.; Usherwood, P. Structure and synthesis of a potent glutamate receptor antagonist in wasp venom. *Proc. Natl. Acad. Sci. USA* **1988**, *85*, 4910–4913. [CrossRef] [PubMed]

14. Rozental, R.; Scoble, G.; Albuquerque, E.; Idriss, M.; Sherby, S.; Sattelle, D.; Nakanishi, K.; Konno, K.; Eldefrawi, A.; Eldefrawi, M. Allosteric inhibition of nicotinic acetylcholine receptors of vertebrates and insects by philanthotoxin. *J. Pharmacol. Exp. Ther.* **1989**, *249*, 123–130. [PubMed]

15. Quistad, G.; Nguyen, Q.; Bernasconi, P.; Leisy, D. Purification and characterization of insecticidal toxins from venom glands of the parasitic wasp, *Bracon hebetor*. *Insect Biochem. Mol. Biol.* **1994**, *24*, 955–961. [CrossRef]

16. Ferber, M.; Horner, M.; Cepok, S.; Gnatzy, W. Digger wasp *versus* cricket: Mechanisms underlying the total paralysis caused by the predator's venom. *J. Neurobiol.* **2001**, *47*, 207–222. [CrossRef] [PubMed]

17. Gnatzy, W.; Volknandt, W. Venom gland of the digger wasp *Liris niger*: Morphology, ultrastructure, age-related changes and biochemical aspects. *Cell Tissue Res.* **2000**, *302*, 271–284. [CrossRef] [PubMed]

18. Periquet, G.; Bigot, Y.; Doury, G. Physiological and biochemical analysis of factors in the female venom gland and larval salivary secretions of the ectoparasitoid wasp *Eupelmus orientalis*. *J. Insect Physiol.* **1997**, *43*, 69–81. [PubMed]

19. Parkinson, N.; Smith, I.; Audsley, N.; Edwards, J.P. Purification of pimplin, a paralytic heterodimeric polypeptide from venom of the parasitoid wasp *Pimpla hypochondriaca*, and cloning of the cDNA encoding one of the subunits. *Insect Biochem. Mol. Biol.* **2002**, *32*, 1769–1773. [CrossRef]

20. Quintela, E.D.; McCoy, C.W. Synergistic effect of imidacloprid and two entomopathogenic fungi on the behavior and survival of larvae of *Diaprepes abbreviatus* (Coleoptera: Curculionidae) in soil. *J. Econ. Entomol.* **1998**, *91*, 110–122. [CrossRef]

21. Dani, M.P.; Richards, E.H. Cloning and expression of the gene for an insect haemocye anti-aggregation protein (VPr3), from the venom of the endoparasitic wasp, *Pimpla hypochondriaca*. *Arch. Insect Biochem. Physiol.* **2009**, *71*, 191–204. [CrossRef] [PubMed]

22. Richards, E.H.; Dani, M.P. Biochemical isolation of an insect haemocyte anti-aggregation protein from the venom of the endoparasitic wasp, *Pimpla hypochondriaca*, and identification of its gene. *J. Insect Physiol.* **2008**, *54*, 1041–1049. [CrossRef] [PubMed]

23. Wu, M.-L.; Ye, G.-Y.; Zhu, J.Y.; Chen, X.-X.; Hu, C. Isolation and characterization of an immunosuppressive protein from venom of the pupa-specific endoparasitoid *Pteromalus puparum*. *J. Invertebr. Pathol.* **2008**, *99*, 186–191. [CrossRef] [PubMed]

24. Labrosse, C.; Eslin, P.; Doury, G.; Drezen, J.M.; Poirie, M. Haemocyte changes in *D. melanogaster* in response to long gland components of the parasitoid wasp *Leptopilina boulardi*: A Rho-GAP protein as an important factor. *J. Insect Physiol.* **2005**, *51*, 161–170. [CrossRef] [PubMed]

25. Zhang, G.; Schmidt, O.; Asgari, S. A calreticulin-like protein from endoparasitoid venom fluid is involved in host hemocyte inactivation. *Dev. Comp. Immunol.* **2006**, *30*, 756–764. [CrossRef] [PubMed]

26. Wang, L.; Fang, Q.; Qian, C.; Wang, F.; Yu, X.; Ye, G. Inhibition of host cell encapsulation through inhibiting immune gene expression by the parasitic wasp venom calreticulin. *Insect Biochem. Mol. Biol.* **2013**, *43*, 936–946. [CrossRef] [PubMed]

27. Mortimer, N.; Goecks, J.; Kacsoh, B.; Mobley, J.; Bowersock, G.; Taylor, J.; Schlenke, T. Parasitoid wasp venom SERCA regulates *Drosophila* calcium levels and inhibits cellular immunity. *Proc. Natl. Acad. Sci. USA* **2013**, *110*, 9427–9432. [CrossRef] [PubMed]

28. Colinet, D.; Dubuffet, A.; Cazes, D.; Moreau, S.; Drezen, J.M.; Poirié, M. A serpin from the parasitoid wasp *Leptopilina boulardi* targets the *Drosophila* phenoloxidase cascade. *Dev. Comp. Immunol.* **2009**, *33*, 681–689. [CrossRef] [PubMed]

29. Asgari, S.; Zareie, R.; Zhang, G.; Schmidt, O. Isolation and characterization of a novel venom protein from an endoparasitoid, *Cotesia rubecula* (Hym: Braconidae). *Arch. Insect Biochem. Physiol.* **2003**, *53*, 92–100. [CrossRef] [PubMed]

30. Zhang, G.; Schmidt, O.; Asgari, S. A novel venom peptide from an endoparasitoid wasp is required for expression of polydnavirus genes in host hemocytes. *J. Biol. Chem.* **2004**, *279*, 41580–41585. [CrossRef] [PubMed]

31. Falabella, P.; Riviello, L.; Caccialupi, P.; Rossodivita, T.; Valente, M.T.; de Stradis, M.L.; Tranfaglia, A.; Varricchio, P.; Gigliotti, S.; Graziani, F.; *et al.* A γ-glutamyl transpeptidase of *Aphidius ervi* venom induces apoptosis in the ovaries of host aphids. *Insect Biochem. Mol. Biol.* **2007**, *37*, 453–465. [CrossRef] [PubMed]

32. Shen, X.; Ye, G.; Cheng, X.; Yu, C.; Yao, H.; Hu, C. Novel antimicrobial peptides identified from an endoparasitic wasp cDNA library. *J. Peptide Sci.* **2009**, *16*, 58–64. [CrossRef] [PubMed]

33. Rivers, D.B.; Ruggiero, L.; Hayes, M. The ectoparasitic wasp *Nasonia vitripennis* (Walker) (Hymenoptera: Pteromalidae) differentially affects cells mediating the immune response of its flesh fly host, *Sarcophaga bullata* Parker (Diptera: Sarcophagidae). *J. Insect Physiol.* **2002**, *48*, 1053–1064. [CrossRef]

34. Rivers, D.B.; Denlinger, D.L. Developmental fate of the flesh fly, *Sarcophaga Bullata*, envenomated by the pupal ectoparasitoid, *Nasonia Vitripennis*. *J. Insect Physiol.* **1994**, *40*, 121–127. [CrossRef]

35. Rivers, D.B.; Denlinger, D.L. Venom-induced alterations in fly lipid metabolism and Its impact on larval development of the ectoparasitoid *Nasonia vitripennis* (Walker) (Hymenoptera, Pteromalidae). *J. Invertebr. Pathol.* **1995**, *66*, 104–110. [CrossRef]

36. Danneels, E.L.; Rivers, D.B.; de Graaf, D.C. Venom proteins of the parasitoid wasp *Nasonia vitirpennis*: Recent discovery of an untapped pharmacopee. *Toxins* **2010**, *2*, 494–516.

37. Qian, C.; Liu, Y.; Fang, Q.; Min-Li, Y.; Liu, S.; Ye, G.; Li, Y. Venom of the ectoparasitoid, *Nasonia vitripennis*, influences gene expression in *Musca domestica* hemocytes. *Arch. Insect Biochem. Physiol.* **2013**, *83*, 211–231. [CrossRef] [PubMed]

38. Martinson, E.; Wheeler, D.; Wright, J.; Alini, M.; Siebert, A.; Werren, J. *Nasonia vitripennis* venom causes targeted gene expression changes in its fly host. *Mol. Ecol.* **2014**, *23*, 5918–5930. [CrossRef] [PubMed]

39. Danneels, E.; Formesyn, E.; Hahn, D.; Denlinger, D.; Cardoen, D.; Wenseleers, T.; Schoofs, L.; de Graaf, D. Early changes in the pupal transcriptome of the flesh fly *Sarcophagha crassipalps* to parasitization by the ectoparasitic wasp, *Nasonia vitripennis*. *Insect Biochem. Mol. Biol.* **2013**, *43*, 1189–1200. [CrossRef] [PubMed]

40. Zhu, J.; Wu, G.; Ze, S.; Stanley, D.; Yang, B. Parasitization by *Scleroderma guani* influences protein expression in *Tenebrio molitor* pupae. *J. Insect Physiol.* **2014**, *66*, 37–44. [CrossRef] [PubMed]

41. Moreau, S.J.M.; Cherqui, A.; Doury, G.; Dubois, F.; Fourdrain, Y.; Sabatier, L.; Bulet, P.; Saarela, J.; Prevost, G.; Giordanengo, P. Identification of an aspartylglucosaminidase-like protein in the venom of the parasitic wasp *Asobara tabida* (Hymenoptera: Braconidae). *Insect Biochem. Mol. Biol.* **2004**, *34*, 485–492. [CrossRef] [PubMed]

42. Moreau, S.J.M.; Dingremont, A.; Doury, G.; Giordanengo, P. Effects of parasitism by *Asobara tabida* (Hymenoptera: Braconidae) on the development, survival and activity of *Drosophila melanogaster* larvae. *J. Insect Physiol.* **2002**, *48*, 337–347. [CrossRef]

43. Desneux, N.; Barta, R.J.; Delebecque, C.J.; Heimpel, G.E. Transient host paralysis as a means of reducing self-superparasitism in koinobiont endoparasitoids. *J. Insect Physiol.* **2009**, *55*, 321–327. [CrossRef] [PubMed]

44. Ergin, E.; Uckan, F.; Rivers, D.B.; Sak, O. *In vivo* and *in vitro* activity of venom from the endoparasitic wasp *Pimpla turionellae* (L.) (Hymenoptera: Ichneumonidae). *Arch. Insect Biochem. Physiol.* **2006**, *61*, 87–97. [CrossRef] [PubMed]

45. Dowton, M.; Austin, A.D. Molecular phylogeny of the insect order hymenoptera: Apocritan relationships. *Proc. Natl. Acad. Sci. USA* **1994**, *91*, 9911–9915. [CrossRef] [PubMed]

46. Parkinson, N.M.; Conyers, C.; Keen, J.; MacNicoll, A.; Smith, I.; Audsley, N.; Weaver, R.J. Towards a comprehensive view of the primary structure of venom proteins from the parasitoid wasp *Pimpla hypochondriaca*. *Insect Biochem. Mol. Biol.* **2004**, *34*, 565–571. [CrossRef] [PubMed]

47. Richards, E.H.; Parkinson, N.M. Venom from the endoparasitic wasp *Pimpla hypochondriaca* adversely affects the morphology, viability, and immune function of hemocytes from larvae of the tomato moth, *Lacanobia oleracea*. *J. Invertebr. Pathol.* **2000**, *76*, 33–42. [CrossRef] [PubMed]

48. Labrosse, C.; Stasiak, K.; Lesobre, J.; Grangeia, A.; Huguet, E.; Drezen, J.M.; Poirie, M. A RhoGAP protein as a main immune suppressive factor in the *Leptopilina boulardi* (Hymenoptera, Figitidae)-*Drosophila melanogaster* interaction. *Insect Biochem. Mol. Biol.* **2005**, *35*, 93–103. [CrossRef] [PubMed]

49. Colinet, D.; Schmitz, A.; Depoix, D.; Crochard, D.; Poirié, M. Convergent use of RhoGAP toxins by eukaryotic parasites and bacterial pathogens. *PLoS Pathog.* **2007**, *3*, e203. [CrossRef] [PubMed]

50. Cai, J.; Ye, G.-Y.; Hu, C. Parasitism of *Pieris rapae* (Lepidoptera: Pieridae) by a pupal endoparasitoid, *Pteromalus puparum* (Hymenoptera: Pteromalidae): Effects of parasitization and venom on host hemocytes. *J. Insect Physiol.* **2004**, *50*, 315–322. [CrossRef] [PubMed]

51. Fang, Q.; Wang, L.; Zhu, J.; Li, Y.; Song, Q.; Stanley, D.; Akhtar, Z.; Ye, G. Expression of immune-response genes in lepidopteran host is suppressed by venom from an endoparasitoid, *Pteromalus puparum*. *BMC Genomics* **2010**, *11*, 484. [CrossRef] [PubMed]

52. Fang, Q.; Wang, F.; Gatehouse, J.; Gatehouse, A.; Chen, X.; Hu, C.; Ye, G. Venom of parasitoid, *Pteromalus puparum*, suppresses host, *Pieris rapae*, immune promotion by decreasing host C-type lectin gene expression. *PLoS ONE* **2011**, *6*, e26888. [CrossRef] [PubMed]

53. Zhu, J.; Ye, G.; Dong, S.; Fang, Q.; Hu, C. Venom of *Pteromalus puparum* (Hymenoptera: Pteromalidae) induced endocrine changes in the hemolymph of its host, *Pieris rapae* (Lepidoptera: Pieridae). *Arch. Insect Biochem. Physiol.* **2009**, *71*, 45–53. [CrossRef] [PubMed]

54. Digilio, M.C.; Isidoro, N.; Tremblay, E.; Pennacchio, F. Host castration by *Aphidius ervi* venom proteins. *J. Insect Physiol.* **2000**, *46*, 1041–1050. [CrossRef]

55. Doucet, D.; Cusson, M. Role of calyx fluid in alterations of immunity in *Choristoneura fumiferana* larvae parasitized by *Tranosema rostrale*. *Comp. Biochem. Physiol.* **1996**, *114*, 311–317. [CrossRef]

56. Dorémus, T.; Urbach, S.; Jouan, V.; Cousserans, F.; Ravallec, M.; Demettre, E.; Wajnberg, E.; Poulain, J.; Azéma-Dossat, C.; Darboux, I.; *et al.* Venom gland extract is not required for successful parasitism in the polydnavirus-associated endoparasitoid *Hyposoter didymator* (Hym. Ichneumonidae) despite the presence of numerous novel and conserved venom proteins. *Insect Biochem. Mol. Biol.* **2013**, *43*, 292–307. [CrossRef] [PubMed]

57. Davies, D.H.; Strand, M.R.; Vinson, S.B. Changes in differential haemocyte count and *in vitro* behaviour of plasmatocytes from host *Heliothis virescens* caused by *Campolethis sonorensis* polydnavirus. *J. Insect Physiol.* **1987**, *33*, 143–153. [CrossRef]

58. Dover, B.A.; Davies, D.H.; Strand, M.R.; Gray, R.S.; Keeley, L.L.; Vinson, S.B. Ecdysteroid-titre reduction and developmental arrest of last instar *Heliothis virescens* larvae by calyx fluid from the parasitoid *Campoletis sonorensis*. *J. Insect Physiol.* **1987**, *33*, 333–338. [CrossRef]

59. Guzo, D.; Stoltz, D.B. Observation on cellular immunity and parasitism in the tussock moth. *J. Insect Physiol.* **1987**, *33*, 19–31. [CrossRef]

60. Webb, B.A.; Luckhart, S. Evidence for an early immunosuppressive role for related *Campoletis sonorensis* venom and ovarian proteins in *Heliothis virescens*. *Arch. Insect Biochem. Physiol.* **1994**, *26*, 147–163. [CrossRef] [PubMed]

61. Stoltz, D.B.; Vinson, S.B. Viruses and parasitism in insects. *Adv. Virus Res.* **1979**, *24*, 125–171. [PubMed]

62. Bézier, A.; Annaheim, M.; Herbinière, J.; Wetterwald, C.; Gyapay, G.; Bernard-Samain, S.; Wincker, P.; Roditi, I.; Heller, M.; Belghazi, M.; *et al.* Polydnaviruses of braconid wasps derive from an ancestral nudivirus. *Science* **2009**, *323*, 926–930. [CrossRef] [PubMed]

63. Burke, G.; Thomas, S.; Eum, J.; Strand, M. Mutualistic polydnaviruses share essential replication gene functions with pathogenic ancestors. *PLoS Pathog.* **2013**, *9*, e1003348. [CrossRef] [PubMed]

64. Burke, G.; Strand, M.R. Systematic analysis of a wasp parasitism arsenal. *Mol. Ecol.* **2014**, *23*, 890–901. [CrossRef] [PubMed]

65. Strand, M.R.; Noda, T. Alterations in the haemocytes of *Pseudoplusia includens* after parasitism by *Microplitis demolitor*. *J. Insect Physiol.* **1991**, *37*, 839–850. [CrossRef]

66. Strand, M.R.; Dover, B.A. Developmental disruption of *Pseudoplusia includens* and *Heliothis virescens* larvae by calyx fluid and venom of *Microplitis demolitor*. *Arch. Insect Biochem. Physiol.* **1991**, *18*, 131–145. [CrossRef] [PubMed]

67. Stoltz, D.B.; Guzo, D.; Belland, E.R.; Lucarotti, C.J.; MacKinnon, E.A. Venom promotes uncoating *in vitro* and persistence *in vivo* of DNA from a braconid polydnavirus. *J. Gen. Virol.* **1988**, *69*, 903–907. [CrossRef]

68. Pennacchio, F.; Falabella, P.; Vinson, S.B. Regulation of *Heliothis virescens* prothoracic glands by *Cardiochiles nigriceps* polydnavirus. *Arch. Insect Biochem. Physiol.* **1998**, *38*, 1–10. [CrossRef]

69. Goecks, J.; Mortimer, N.; Mobley, J.; Bowersock, G.; Taylor, J.; Schlenke, T. Integrative approach reveals composition of endoparasitoid wasp venoms. *PLoS ONE* **2013**, *8*, e64125. [CrossRef] [PubMed]

70. De Graaf, D.; Aerts, M.; Brunain, M.; Desjardins, C.; Jacobs, F.; Werren, J.; Devreese, B. Insights into the venom composition of the ectoparasitoid wasp *Nasonia vitripennis* from bioinformatic and proteomic studies. *Insect Mol. Biol.* **2010**, *19* (Suppl. 1), 11–26. [CrossRef] [PubMed]

71. Furihata, S.; Matsumoto, H.; Kimura, M.; Hayakawa, Y. Venom components of *Asobara japonica* impair cellular immune responses of host *Drosophila melanogaster*. *Arch. Insect Biochem. Physiol.* **2013**, *83*, 86–100. [CrossRef] [PubMed]

72. Asgari, S.; Zhang, G.; Zareie, R.; Schmidt, O. A serine proteinase homolog venom protein from an endoparasitoid wasp inhibits melanization of the host hemolymph. *Insect Biochem. Mol. Biol.* **2003**, *33*, 1017–1024. [CrossRef] [PubMed]

73. Thomas, P.; Asgari, S. Inhibition of melanization by a parasitoid serine protease homolog venom protein requires both the clip and the non-catalytic protease-like domains. *Insects* **2011**, *2*, 509–514. [CrossRef]

74. Jiang, H.; Kanost, M.R. The clip-domain family of serine proteinases in arthropods. *Insect Biochem. Mol. Biol.* **2000**, *30*, 95–105. [CrossRef]

75. Zhang, G.; Lu, Z.-Q.; Jiang, H.; Asgari, S. Negative regulation of prophenoloxidase (proPO) activation by a clip-domain serine proteinase homolog (SPH) from endoparasitoid venom. *Insect Biochem. Mol. Biol.* **2004**, *34*, 477–483. [CrossRef] [PubMed]

76. Tanaka, T.; Vinson, S.B. Interaction between venom and calyx fluids of three parasitoids, *Cardiochiles nigriceps*, *Microplitis croceipes* (Hymenoptera:Braconidae), and *Campoletis sonorensis* (Hymenoptera: Ichneumonidae) in affecting a delay in the pupation of *Heliothis virescens* (Lepidoptera: Noctuidae). *Ann. Entomol. Soc. Am.* **1991**, *84*, 87–92.

77. Pennacchio, F.; Flabella, P.; Sordetti, R.; Varricchio, P.; Malva, C.; Vinson, S.B. Prothoracic gland inactivation in *Heliothis virescens* (F.) (Lepidoptera: Noctuidae) larvae parasitized by *Cardiochiles nigriceps* Viereck (Hymenoptera: Braconidae). *J. Insect Physiol.* **1998**, *44*, 845–857. [CrossRef]

78. Xu, J.; Cherry, S. Viruses and antiviral immunity in *Drosophila*. *Dev. Comp. Immunol.* **2014**, *42*, 67–84. [CrossRef] [PubMed]

79. Kolliopoulou, A.; Swevers, L. Recent progress in RNAi research in Lepidoptera: Intracellular machinery, antiviral immune response and prospects for insect pest control. *Curr. Opin. Insect Sci.* **2015**, *6*, 28–34. [CrossRef]

80. Colinet, D.; Kremmer, L.; Lemauf, S.; Rebuf, C.; Gatti, J.; Poirié, M. Development of RNAi in a *Drosophila* endoparasitoid wasp and demonstration of its efficiency in impairing venom protein production. *J. Insect Physiol.* **2014**, *63*, 56–61. [CrossRef] [PubMed]

81. Vincent, B.; Kaeslin, M.; Roth, T.; Heller, M.; Poulain, J.; Cousserans, F.; Schaller, J.; Poirié, M.; Lanzrein, B.; Drezen, J.; *et al.* The venom composition of the parasitic wasp *Chelonus inanitus* resolved by combined expressed sequence tags analysis and proteomic approach. *BMC Genomics* **2010**, *11*, 693. [CrossRef] [PubMed]

82. Kaeslin, M.; Reinhard, M.; Bühler, D.; Roth, T.; Pfister-Wilhelm, R.; Lanzrein, B. Venom of the egg-larval parasitoid *Chelonus inanitus* is a complex mixture and has multiple biological effects. *J. Insect Physiol.* **2010**, *56*, 686–694. [CrossRef] [PubMed]

83. Colinet, D.; Anselme, C.; Deleury, E.; Mancini, D.; Poulain, J.; Azéma-Dossat, C.; Belghazi, M.; Tares, S.; Pennacchio, F.; Poirié, M.; *et al.* Identification of the main venom protein components of *Aphidius ervi*, a parasitoid wasp of the aphid model *Acyrthosiphon pisum*. *BMC Genomics* **2014**, *15*, 342. [CrossRef] [PubMed]

84. Heavner, M.; Gueguen, G.; Rajwani, R.; Pagan, P.; Small, C.; Govind, S. Partial venom gland transcriptome of a *Drosophila* parasitoid wasp, *Leptopilina heterotoma*, reveals novel and shared bioactive profiles with stinging Hymenoptera. *Gene* **2013**, *526*, 195–204. [CrossRef] [PubMed]

85. Dupas, S.; Brehelin, M.; Frey, F.; Carton, Y. Immune suppressive virus-like particles in a *Drosophila* parasitoid: Significance of their intraspecific morphological variations. *Parasitology* **1996**, *113*, 207–212. [CrossRef] [PubMed]

86. Labrosse, C.; Carton, Y.; Dubuffet, A.; Drezen, J.M.; Poirie, M. Active suppression of *D. melanogaster* immune response by long gland products of the parasitic wasp *Leptopilina boulardi*. *J. Insect Physiol.* **2003**, *49*, 513–522. [CrossRef]

87. Schlenke, T.A.; Morales, J.; Govind, S.; Clark, A.G. Contrasting infection strategies in generalist and specialist wasp parasitoids of *Drosophila melanogaster*. *PLoS Pathog.* **2007**, *3*, e158. [CrossRef] [PubMed]

88. Dowton, M.; Austin, A.D. Simultaneous analysis of 16S, 28S, CO1 and morphology in the Hymenoptera: Apocrita-evolutionary transitions among parasitic wasps. *Biol. J. Linn. Soc.* **2001**, *74*, 87–111.

89. Whitfield, J.B. Phylogenetic insights into the evolution of parasitism in Hymenoptera. *Adv. Parasitol.* **2003**, *54*, 69–100. [PubMed]

90. Piek, T. *Venoms of the Hymenoptera. Biochemical, pharmacological and Behavioural Aspects*; Academic Press: London, UK, 1986.

91. Quicke, D.L.J. *Parasitic Wasps*; Chapman and Hall: London, UK, 1997.

92. Grimaldi, D.; Engel, M. *Evolution of the Insects*; Cambridge University Press: New York, NY, USA, 2005.

93. Schmidt, J. Chemistry, pharmacology and chemical ecology of ant venoms. In *Venoms of the Hymenoptera. Biochemical, Pharmacological and Behavioural Aspects*; Piek, T., Ed.; Academic Press: London, UK, 1986; pp. 425–508.

94. Asgari, S. Venoms from endoparasitoids. In *Parasitoid Viruses, Symbionts and Pathogens*; Beckage, N., Drezen, J.-M., Eds.; Academic Press: London, UK, 2012; pp. 217–231.

95. Moreau, S.J.M.; Guillot, S. Advances and prospects on biosynthesis, structures and functions of venom proteins from parasitic wasps. *Insect Biochem. Mol. Biol.* **2005**, *35*, 1209–1223. [CrossRef] [PubMed]

96. Leluk, J.; Schmidt, J.; Jones, D. Comparative studies on the protein composition of hymenopteran venom reservoirs. *Toxicon* **1989**, *27*, 105–114. [CrossRef] [PubMed]

97. Jones, D.; Wozniak, M. Regulatory mediators in the venom of *Chelonus* sp.: Their biosynthesis and subsequent processing in homologous and heterologous systems. *Biochem. Biophys. Res. Commun.* **1991**, *178*, 213–220. [CrossRef]

98. Rappuoli, R.; Montecucco, C. *Guidebook to Protein Toxins and Their Use in Cell Biology*; Oxford University Press: Oxford, UK, 1997.

99. Parkinson, N.; Richards, E.H.; Conyers, C.; Smith, I.; Edwards, J.P. Analysis of venom constituents from the parasitoid wasp *Pimpla hypochondriaca* and cloning of a cDNA encoding a venom protein. *Insect Biochem. Mol. Biol.* **2002**, *32*, 729–735. [CrossRef]

100. Parkinson, N.M.; Conyers, C.; Keen, J.N.; MacNicoll, A.D.; Weaver, I.S.R. cDNAs encoding large venom proteins from the parasitoid wasp *Pimpla hypochondriaca* identified by random sequence analysis. *Comp. Biochem. Physiol. C* **2003**, *134*, 513–520. [CrossRef]

101. Dani, M.P.; Richards, E.H.; Isaac, R.E.; Edwards, J.P. Antibacterial and proteolytic activity in venom from the endoparasitic wasp *Pimpla hypochondriaca* (Hymenoptera: Ichneumonidae). *J. Insect Physiol.* **2003**, *49*, 945–954. [CrossRef]

102. Nakamatsu, Y.; Tanaka, T. Venom of ectoparasitoid, *Euplectrus* sp. near plathypenae (Hymenoptera: Eulophidae) regulates the physiological state of *Pseudaletia separata* (Lepidoptera: Noctuidae). *J. Insect Physiol.* **2003**, *49*, 149–159. [CrossRef]

103. Crawford, A.M.; Brauning, R.; Smolenski, G.; Ferguson, C.; Barton, D.; Wheeler, T.T.; Mcculloch, A. The constituents of *Microtonus* sp. parasitoid venoms. *Insect Mol. Biol.* **2008**, *17*, 313–324. [CrossRef] [PubMed]

104. Ye, J.; Zhao, H.; Wang, H.; Bian, J.; Zheng, R. A defensin antimicrobial peptide from the venoms of *Nasonia vitripennis*. *Toxicon* **2010**, *56*, 101–106. [CrossRef] [PubMed]

105. Moreau, S. "It stings a bit but it cleans well": Venoms of Hymenoptera and their antimicrobial potential. *J. Insect Physiol.* **2013**, *59*, 186–204. [CrossRef] [PubMed]

106. Asgari, S.; Rivers, D.B. Venom proteins from endoparasitoid wasps and their role in host-parasite interactions. *Annu. Rev. Entomol.* **2011**, *56*, 313–335. [CrossRef] [PubMed]

107. Formesyn, E.M.; Danneels, E.L.; de Graaf, D.C. Proteomics of the venom of the parasitoid *Nasonia vitripennis*. In *Parasitoid Viruses, Symbionts and Pathogens*; Beckage, N.E., Drezen, J.-M., Eds.; Academic Press: London, UK, 2012; pp. 233–246.

108. Escoubas, P.; Quinton, L.; Nicholson, G.M. Venomics: Unravelling the complexity of animal venoms with mass spectrometry. *J. Mass Spectrom.* **2008**, *43*, 279–295. [CrossRef] [PubMed]

109. Casewell, N.; Wüster, W.; Vonk, F.; Harrison, R.; Fry, B. Complex cocktails: The evolutionary novelty of venoms. *Trends Ecol. Evol.* **2013**, *28*, 219–229. [CrossRef] [PubMed]

110. Fry, B.G.; Roelants, K.; Champagne, D.E.; Scheib, H.; Tyndall, J.D.A.; King, G.F.; Nevalainen, T.J.; Norman, J.A.; Lewis, R.J.; Norton, R.S.; *et al.* The toxicogenomic multiverse: Convergent recruitment of proteins into animal venoms. *Annu. Rev. Genomics Hum. Genet.* **2009**, *10*, 483–511. [CrossRef] [PubMed]

111. Bronner, R. Anatomy of the ovipositor and oviposition behavior of the gall wasp *Diplolepis rosae* (Hymenoptera: Cynipidae). *Can. Entomol.* **1985**, *117*, 849–858. [CrossRef]

112. Vårdal, H. Venom gland and reservoir morphology in cynipoid wasps. *Arthropod Struct. Dev.* **2006**, *35*, 127–136. [CrossRef] [PubMed]

113. Zhu, J.; Fang, Q.; Wang, L.; Hu, C.; Ye, G. Proteomic analysis of the venom from the endoparasitoid wasp *Pteromalus puparum* (Hymenoptera: Pteromalidae). *Arch. Insect Biochem. Physiol.* **2010**, *75*, 28–44. [CrossRef] [PubMed]

114. Colinet, D.; Deleury, E.; Anselme, C.; Cazes, D.; Poulain, J.; Azema-Dossat, C.; Belghazi, M.; Gatti, J.-L.; Poirié, M. Extensive inter- and intraspecific venom variation in closely related parasites targeting the same host: The case of *Leptopilina* parasitoids of *Drosophila*. *Insect Biochem. Mol. Biol.* **2013**, *43*, 601–611. [CrossRef] [PubMed]

115. Fry, B.G.; Roelants, J.; Norman, J.A. Tentacles of venom: Toxic protein convergence in the Kingdom Animalia. *J. Mol. Evol.* **2009**, *68*, 311–321. [CrossRef] [PubMed]

116. Moreau, S.J.M.; Vinchon, S.; Cherqui, A.; Prévost, G. Components of *Asobara* venoms and their effects on hosts. *Adv. Parasitol.* **2009**, *70*, 217–232. [PubMed]

117. Krishnan, A.; Nair, P.N.; Jones, D. Isolation, cloning and characterization of new chitinase stored in active form in chitin-lined venom reservoir. *J. Biol. Chem.* **1994**, *269*, 20971–20976. [PubMed]

118. Parkinson, N.M.; Weaver, R.J. Noxious components of venom from the pupa-specific parasitoid *Pimpla hypochondriaca*. *J. Invertebr. Pathol.* **1999**, *73*, 74–83. [CrossRef] [PubMed]

119. Colinet, D.; Mathé-Hubert, H.; Allemand, R.; Gatti, J.-L.; Poirié, M. Variability of venom components in immune suppressive parasitoid wasps: From a phylogenetic to a population approach. *J. Insect Physiol.* **2013**, *59*, 205–212. [CrossRef] [PubMed]

120. Kitano, H. The role of *Apanteles glomeratus* venom in the defensive response of its host, *Pieris rapae crucivora*. *J. Insect Physiol.* **1986**, *32*, 369–375. [CrossRef]

121. Wago, H.; Tanaka, T. Synergistic effects of calyx fluid and venom of *Apanteles kariyai* Watanabe (Hymenoptera: Braconidae) on the granular cells of *Pseudaletia separata* Walker (Lepidoptera: Noctuidae). *Zool. Sci.* **1989**, *6*, 691–696.

122. Lanzrein, B.; Pfister-Wilhelm, R.; Kaeslin, M.; Wespi, G.; Roth, T. The orchestrated manipulation of the host by *Chelonus inanitus* and its polydnavirus. In *Parasitoid Viruses: Symbionts and Pathogens*; Beckage, N., Drezen, J.-M., Eds.; Academic Press: London, UK, 2012; pp. 169–178.

123. Prévost, G.; Eslin, P.; Cherqui, A.; Moreau, S.; Doury, G. When parasitoids lack polydnaviruses, can venoms subdue the hosts? The case study of *Asobara* species. In *Parasitoid Viruses: Symbionts and Pathogens*; Beckage, N., Drezen, J.-M., Eds.; Academic Press: London, UK, 2012; pp. 255–266.

124. Beckage, N.E.; Tan, F.F.; Schleifer, K.W.; Lane, R.D.; Cherubin, L.L. Characterization and biological effects of *Cotesia congregata* polydnavirus on host larvae of the tobacco hornworm, *Manduca sexta*. *Arch. Insect Physiol. Biochem.* **1994**, *26*, 165–195. [CrossRef]

125. Doucet, D.; Cusson, M. Alteration of developmental rate and growth of *Choristoneura fumiferana* parasitized by *Tranosema rostrale*—Role of the calyx fluid. *Entomol. Exp. Appl.* **1996**, *81*, 21–30. [CrossRef]

126. Moreau, S.; Huguet, E.; Drezen, J.-M. Polydnaviruses as tools to deliver wasp virulence factors to impair lepidopteran host immunity. In *Insect Infection and Immunity: Evolution, Ecology and Mechanisms*; Reynolds, S.E., Ed.; Oxford University Press: Oxford, UK, 2009; pp. 137–158.

127. Volkoff, A.; Jouan, V.; Urbach, S.; Samain, S.; Bergoin, M.; Wincker, P.; Demettre, E.; Cousserans, F.; Provost, B.; Coulibaly, F.; *et al.* Analysis of virion structural components reveals vestiges of the ancestral ichnovirus genome. *PLoS Pathog.* **2010**, *6*, e1000923. [CrossRef] [PubMed]

128. Herniou, E.A.; Huguet, E.; Thézé, J.; Bézier, A.; Periquet, G.; Drezen, J.-M. When parasitic wasps hijacked viruses: Genomic and functional evolution of polydnaviruses. *Philos. Trans. R. Soc. Lond. B Biol. Sci.* **2013**, *368*, 20130051. [CrossRef] [PubMed]

129. Kadash, K.; Harvey, J.A.; Strand, M.R. Cross-protection experiments with parasitoids in the genus *Microplitis* (Hymenoptera: Braconidae) suggest a high level of specificity in their associated bracoviruses. *J. Insect Physiol.* **2003**, *49*, 473–482. [CrossRef]

130. Fellowes, M.D.; Godfray, H.C. The evolutionary ecology of resistance to parasitoids by *Drosophila*. *Heredity* **2000**, *84*, 1–8. [CrossRef] [PubMed]

131. Prévost, G.; Doury, G.; Mabiala-Moundoungou, A.; Cherqui, A.; Eslin, P. Strategies of avoidance of host immune defenses in *Asobara* species. *Adv. Parasitol.* **2009**, *70*, 235–255. [PubMed]

132. Mabiala-Moundoungou, A.D.N.; Doury, G.; Eslin, P.; Cherqui, A.; Prevost, G. Deadly venom of *Asobara japonica* parasitoid needs ovarian antidote to regulate host physiology. *J. Insect Physiol.* **2010**, *56*, 35–41. [CrossRef] [PubMed]

133. Kraaijeveld, A.R. Kleptoparasitism as an explanation for paradoxical oviposition decisions of the parasitoid *Asobara tabida*. *J. Evol. Biol.* **1999**, *12*, 129–133. [CrossRef]

134. Eslin, P.; Prévost, G. Racing against host's immunity defenses: A likely strategy for passive evasion of encapsulation in *Asobara tabida* parasitoids. *J. Insect Physiol.* **2000**, *46*, 1161–1167. [CrossRef]

135. Eslin, P.; Giordanengo, P.; Fourdrain, Y.; Prévost, G. Avoidance of encapsulation in the absence of VLP by a braconid parasitoid of *Drosophila* larvae: An ultrastructural study. *Can. J. Zool.* **1996**, *74*, 2193–2198. [CrossRef]

136. Havard, S.; Doury, G.; Ravallec, M.; Brehélin, M.; Prévost, G.; Eslin, P. Structural and functional characterization of pseudopodocyte, a shaggy immune cell produced by two *Drosophila* species of the obscura group. *Dev. Comp. Immunol.* **2012**, *36*, 323–331. [CrossRef] [PubMed]

137. Moreau, S.J.M.; Doury, G.; Giordanengo, P. Intraspecific variation in the effects of parasitism by *Asobara tabida* on phenoloxidase activity of *Drosophila melanogater* larvae. *J. Invertebr. Pathol.* **2000**, *76*, 151–153. [CrossRef] [PubMed]

138. Dowton, M.; Austin, A.D.; Antolin, M.F. Evolutionary relationships among the Braconidae (Hymenoptera: Ichneumonoidea) inferred from partial 16S rDNA gene sequences. *Insect Mol. Biol.* **1998**, *7*, 129–150. [CrossRef] [PubMed]

139. Mathé-Hubert, H.; Gatti, J.-L.; Colinet, D.; Poirié, M.; Malausa, T. Statistical analysis of the individual variability of 1D protein profiles as a tool in ecology: An application to parasitoid venom. *Mol. Ecol. Res.* **2015**, in press. [CrossRef] [PubMed]

140. Peiren, N.; Vanrobaeys, F.; de Graaf, D.; Devreese, B.; van Beeumen, J.; Jacobs, F. The protein composition of honeybee venom reconsidered by a proteomic approach. *Biochem. Biophys. Acta* **2005**, *1752*, 1–5. [CrossRef] [PubMed]

141. Drapeau, M.; Albert, S.; Kucharski, R.; Prusko, C.; Maleszka, R. Evolution of the Yellow/Major Royal Jelly Protein family and the emergence of social behavior in honey bees. *Genome Res.* **2006**, *16*, 1385–1394. [CrossRef] [PubMed]

142. Beckage, N.E.; Gelman, D.B. Wasp parasitoid disruption of host development: Implications for new biologically based strategies for insect control. *Annu. Rev. Entomol.* **2004**, *49*, 299–330. [CrossRef] [PubMed]

143. Braun, L. Notes on desensitization of a patient hypersensitive to bee stings. *S. Afr. Med. Res.* **1925**, *23*, 408–409.

144. Biló, B.; Rueff, F.; Mosbech, H.; Bonifazi, F.; Oude-Elberink, J. The EAACI interest group on insect venom hypersensitivity: Diagnosis of Hymenoptera venom allergy. *Allergy* **2005**, *60*, 1339–1349. [CrossRef] [PubMed]

145. Czaikoski, P.; Menaldo, D.; Marcussi, S.; Baseggio, A.; Fuly, A.; Paula, R.; Quadros, A.; Romão, P.; Buschini, M.; Cunha, F.; *et al.* Anticoagulant and fibrinogenolytic properties of the venom of *Polybia occidentalis* social wasp. *Blood Coagul. Fibrinolysis* **2010**, *21*, 653–659. [CrossRef] [PubMed]

146. Choo, Y.; Lee, K.; Yoon, H.; Kim, B.; Sohn, M.; Roh, J.; Je, Y.; Kim, N.; Kim, I.; Woo, S.; *et al.* Dual function of a bee venom serine protease: Prophenoloxidase-activating factor in arthropods and fibrin(ogen)olytic enzyme in mammals. *PLoS ONE* **2010**, *5*, e10393. [CrossRef] [PubMed]

147. Kuhn-Nentwig, L. Antimicrobial and cytolytic peptides of venomous arthropods. *Cell. Mol. Life Sci.* **2003**, *60*, 2651–2668. [CrossRef] [PubMed]

148. Konno, K.; Rangel, M.; Oliveira, J.; Dos Santos Cabrera, M.; Fontana, R.; Hirata, I.; Hide, I.; Nakata, Y.; Mori, K.; Kawano, M.; *et al.* Decoralin, a novel linear cationic alpha-helical peptide from the venom of the solitary eumenine wasp *Oreumenes decoratus*. *Peptides* **2007**, *28*, 2320–2327. [CrossRef] [PubMed]

149. Zhu, J.-Y.; Ye, G.-Y.; Hu, C. Venom of the endoparasitoid wasp *Pteromalus puparum*: An overview. *Psyche* **2011**, *2011*, 520926. [CrossRef]

150. Khasnis, A.; Nettleman, M. Global warming and infectious disease. *Arch. Med. Res.* **2005**, *36*, 689–696. [CrossRef] [PubMed]

151. Deutsch, C.; Tewksbury, J.; Huey, R.; Sheldon, K.; Ghalambor, C.; Haak, D.; Martin, P. Impacts of climate warming on terrestrial ectotherms across latitude. *Proc. Natl. Acad. Sci. USA* **2008**, *105*, 6668–6672. [CrossRef] [PubMed]

152. Konno, K.; Hisada, M.; Naoki, H.; Itagaki, Y.; Yasuhara, T.; Juliano, M.; Juliano, L.; Palma, M.; Yamane, T.; Nakajima, T. Isolation and sequence determination of peptides in the venom of the spider wasp (*Cyphononyx dorsalis*) guided by matrix-assisted laser desorption/ionization time of flight (MALDI-TOF) mass spectrometry. *Toxicon* **2001**, *39*, 1257–1260. [CrossRef]

153. Rohrbach, M.; Clarke, J. Treatment of lysosomal storage disorders: Progress with enzyme replacement therapy. *Drugs* **2007**, *67*, 2697–2716. [CrossRef] [PubMed]

154. Windass, J.D.; Suner, M.-M.; Earley, F.G.P.; Guest, P.J. Biological control agents containing mollusc toxins. International Patent WO/1994/023047, 13 October 1994.

155. Ely, S. Insecticidal Proteins. International Patent WO 1995/011305 A2, 27 April 1995.

156. Quistad, G.B.; Leisy, D.J. Insecticidal toxins from the parasitic wasp, *Bracon hebetor*. Patent WO 1993018145 A1, 16 September 1993.

157. Windass, J.D.; Duncan, R.E.; Christian, P.D.; Baule, V.J. Agents biologiques antiparasites contenant des toxines de mollusques. International Patent WO 1994023047 A1, 13 October 1994.

158. Johnson, J.H.; Kral, R.M., Jr.; Krapcho, K. Insecticidal toxins from *Bracon hebetor*. International Patent WO/1996/025429 A1, 22 August 1996.

159. Johnson, J.H.; Kral, R.M., Jr.; Krapcho, K. Insecticidal toxins from *Bracon hebetor* nucleic acid encoding said toxin and methods of use. US Patent No. 5,874,298, 23 February 1999.

160. Pennachio, F.; Tranfaglia, A.; Malva, C. Host-parasitoid antagonism in insects: New opportunities for pest control? *Agro FOOD Industry Hi Tech.* **2003**, *14*, 53–56.

161. Manzoor, A.; Zain-ul-Abdin; Arshad, M.; Gogi, M.D.; Shaina, H.; Mubarik, E.; Abbas, S.K.; Khan, M.A. Biological activity of the toxic peptides from venom of *Bracon hebetor* (Say.) (Hymenoptera: Braconidae). *Pak. Entomol.* **2011**, *33*, 125–130.

162. ActionAid AstraZeneca and Its Genetic Research: Feeding the world or fuelling hunger? Available online: http://www.actionaid.org.uk/sites/default/files/doc_lib/astrazeneca.pdf (accessed on 24 June 2015).

163. Nguyen, T.; Magnoli, I.; Cloutier, C.; Michaud, D.; Muratori, F.; Hance, T. Early presence of an enolase in the oviposition injecta of the aphid parasitoid *Aphidius ervi* analyzed with chitosan beads as artificial hosts. *J. Insect Physiol.* **2013**, *59*, 11–18. [CrossRef] [PubMed]

164. Richards, E.; Bradish, H.; Dani, M.; Pietravalle, S.; Lawson, A. Recombinant immunosuppressive protein from *Pimpla hypochondrica* venom (rVPr1) increases the susceptibility of *Mamestra brassicae* larvae to the fungal biological control agent, *Beauveria bassiana*. *Arch. Insect Biochem. Physiol.* **2011**, *78*, 119–131. [CrossRef] [PubMed]

165. Richards, E.; Dani, M.; Bradish, H. Immunosuppressive properties of a protein (rVPr1) from the venom of the endoparasitic wasp, *Pimpla hypochondriaca*: Mechanism of action and potential use for improving biological control strategies. *J. Insect Physiol.* **2013**, *59*, 213–222. [CrossRef] [PubMed]

© 2015 by the authors; licensee MDPI, Basel, Switzerland. This article is an open access article distributed under the terms and conditions of the Creative Commons Attribution (CC-BY) license (http://creativecommons.org/licenses/by/4.0/).

toxins

MDPI

Review

Three Valuable Peptides from Bee and Wasp Venoms for Therapeutic and Biotechnological Use: Melittin, Apamin and Mastoparan

Miguel Moreno * and Ernest Giralt *

Chemistry and Molecular Pharmacology, Institute for Research in Biomedicine (IRB Barcelona), Baldiri i Reixac, 10, Barcelona 08028, Spain

* Authors to whom correspondence should be addressed; miguel.irbbarcelona@gmail.com (M.M.); ernest.giralt@irbbarcelona.org (E.G.); Tel.: +65-8191-9601 (M.M.); +34-9340-37126 (E.G.).

Academic Editor: Sokcheon Pak

Received: 4 February 2015; Accepted: 25 March 2015; Published: 1 April 2015

Abstract: While knowledge of the composition and mode of action of bee and wasp venoms dates back 50 years, the therapeutic value of these toxins remains relatively unexploded. The properties of these venoms are now being studied with the aim to design and develop new therapeutic drugs. Far from evaluating the extensive number of monographs, journals and books related to bee and wasp venoms and the therapeutic effect of these toxins in numerous diseases, the following review focuses on the three most characterized peptides, namely melittin, apamin, and mastoparan. Here, we update information related to these compounds from the perspective of applied science and discuss their potential therapeutic and biotechnological applications in biomedicine.

Keywords: bee; wasp; venom; melittin; apamin; mastoparan

1. Introduction

The order of *Hymenoptera* divides into two suborders *Symphyta* and *Apocrita*. This latter represents the first evolutionary step in the development of the hymenopteran venom system [1]. Furthermore, the suborder *Apocrita* is traditionally divided into two groups, the *Aculeata* and Parasitica. At the same time, *Aculeata* contains several superfamilies, *Vespoidea* and *Apoidea* among others. Inside Vespoidea is the family *Vespidae*, which represents a large and diverse family of cosmopolitan wasps as does the family *Apidae*, inside the family *Apoidea*, which comprises many species of bee, among them the common honey bee. The sting of members of the *Aculeata* group is modified for injecting venom into prey or predators. The chemical composition of these insect venoms is complex, encompassing, a mixture of many kinds of compounds, proteins, peptides, enzymes, and other smaller molecules. This mixture of biologically active substances can exert toxic effects, contributing to certain clinical signs and symptoms of envenomation. Human responses to stings include pain, small edema, redness, extensive local swelling, anaphylaxis, and systemic toxic reaction [2]. However, several venom components have been widely used in Oriental medicine to relieve pain and to treat inflammatory diseases such as rheumatoid arthritis and tendonitis. Other potential venom-related treatments for immune-related diseases, infections, and tumor therapies are currently under investigation. In this review, we focus our attention on the most recent and innovative therapeutic and biological applications of three of the most widely known components of bee and wasp venom, namely melittin, apamin and mastoparan (see Table 1). Melittin and apamin are the only found in the genus *Apis*. However, mastoparan is found in more genera, such as *Vespa*, *Parapolybia*, *Protonectarina*, *Polistes* and *Protopolybia*.

Table 1. Protein primary structure of melittin, apamin and mastoparan.

Melittin	GIGAVLKVLTTGLPALISWIKRKRQQ
Apamin	$C_1NC_2KAPETALC_1ARRC_2QQH$ *
Mastoparan	INLKALAALAKKIL

* The cysteines' subscripts of Apamin sequence represent the disulfide bridges between Cys in positions 1 with 11, and Cys in positions 3 with 15.

Hymenoptera venom therapy, in particular that involving bee venom (apitoxin), was practiced in ancient Egypt, Greece, and China and, improved by modern studies of apitherapy during the 19th century. However, precise knowledge of the composition and mode of action of such venom dates back only 50 years. The advent of electrophoresis, chromatography and gel-filtration, together with pharmacological and biochemical techniques, brought about the identification of a number of components of bee and wasp venoms. Later on, improved and novel techniques of purification and sequence analysis by Edman degradation and the new analytical chemistry technique mass spectrometry (MS) allowed accurate characterization of the major components in venoms. On the other hand, the advance in transcriptomic and genomic analysis also have helped to identify genes expressed in venom glands. The amount of venom protein released in a sting varies between species, ranging between 50 and 140 micrograms for bees [3,4] and between 1.5 and 20 micrograms for wasps [3,5]. Proteins and peptides comprise the main components of the venoms of these insects (see Table 2). The venoms also contain volatile alarm pheromones (4%–8%), such as iso-pentyl acetate, 2-nonanol, and n-butyl acetate, which trigger defensive responses from nearby insects [6]. Bee and wasp venom share several biologically active proteins and neurotransmitters, such as phospholipases A_2 and B, hyaluronidase, serotonin, histamine, dopamine, noradrenaline and adrenaline. However, some peptides are exclusive to each insect, namely melittin, apamin and mast cell degranulating (MCD) peptide to bees, and mastoparan and bradykinin to wasps.

Table 2. Main proteins and peptides found in bee and wasp venom.

Bee venom	Wasp venom	Type and MW (Da)	% Compound *	Toxic **
Phospholipase A_2	Phospholipase A_2	Enzyme (~18 kDa)	10–12	Yes
Phospholipase B	Phospholipase B	Enzyme (~26 kDa)	1	Yes
Hyaluronidase	Hyaluronidase	Enzyme (~54 kDa)	1.5–2	Yes
Phosphatase	Phosphatase	Enzyme (~60 kDa)	1	No
α-Glucosidase	α-Glucosidase	Enzyme (~170 kDa)	0.6	No
Melittin	-	Peptide (2847 Da)	40–50	Yes
Apamin	-	Peptide (2027 Da)	2–3	Yes
MCD peptide	-	Peptide (2593 Da)	2–3	Yes
-	Mastoparan	Peptide (1422 Da)	No data	Yes
-	Bradykinin	Peptide (1060 Da)	No data	No

* The percentages of compounds correspond to the venom itself and do not take into account the water content. Concentration can differ between bee and wasp species. ** This toxicity refers to the potential toxicity that each component could have. It is based on the cytotoxic and immunologic effect of each protein described in the text.

1.1. Enzymes

Focusing on enzymes related to toxicity, phospholipase and hyaluronidase are the two major enzymatic proteins present in hymenoptera venom. These enzymes can trigger an immune response, inducing IgE response in susceptible individuals [7].

Phospholipase A_2 (PLA_2) is a calcium-dependent enzyme that hydrolyzes the sn-2 ester of glycerophospholipids to produce a fatty acid and a lysophospholipid. It destroys phospholipids, disrupting the integrity of the lipid bilayers, thus making cells susceptible to further degradation. In fact, PLA_2 reaction products, such as lysophosphatidylcholine, lysophosphatidic acid and sphingosine 1-phosphate, can have cytotoxic or immunostimulatory effect on diverse cell types, causing inflammation and immune responses [8].

Phospholipase B (PLB), also known as lysophospholipase, is an enzyme found in very low concentrations in some venoms. With the capacity to cleave acyl chains from both sn-1 and sn-2 positions of a phospholipid, PLB shows a combination of PLA_1 and PLA_2 activities [9].

Hyaluronidase is commonly known as a "spreading factor" because it hydrolyzes the viscous polymer hyaluronic acid into non-viscous fragments. When extracellular matrix is destroyed by hyaluronidase, the gaps between cells facilitate the invasion of venom toxins. Therefore, venom penetrates tissues and enters blood vessels, thus catalysing systemic poisoning. Furthermore, hydrolyzed hyaluronan fragments are pro-inflammatory, pro-angiogenic and immunostimulatory, thus inducing faster systemic envenomation [10].

1.2. Peptides

Mast cell degranulating (MCD) peptide is a cationic peptide with 22 amino acid residues that has a similar structure to apamin, being cross-linked by two disulphide bonds [11]. This peptide is a potent anti-inflammatory agent; however, at low concentration it is a strong mediator of mast cell degranulation and histamine release from mast cells, which are present in the blood supply and in all tissues perfused by blood [12].

Bradykinin is a physiologically active peptide that belongs to the kinin group of proteins. Bradykinin and related kinins act on two receptors, designated as B1 and B2. The former is expressed only as a result of tissue injury and it is thought to play a role in chronic pain. In contrast, the B2 receptor is constitutively expressed, participating in vasodilatation via the release of prostacyclin, nitric oxide, and endothelium-derived hyperpolarizing factor, thus contributing to lowering blood pressure [13].

Adolapin is a peptide that was first isolated from bee venom in the 80s. It exerts a potent analgesic effect and anti-inflammatory activity in rats, blocking prostaglandin [14]. Tertiapin, also from bee venom, is a 21 amino acid peptide that blocks certain types of inwardly rectifying potassium channels [15]. The peptides Scapin, Scapin-1, and Scapin-2 are all 25 amino acid residues in length and share a similar secondary structure, with a disulfide bridge between Cys 9 and Cys 20. These peptides have been isolated from the venom of various species, such as Scapin from European *Apis mellifera* [16], Scapin-1 from Chinese *Apis mellifera* [17], and Scapin-2 from the Africanized honeybee. These compounds induce leukotriene-mediated hyperalgesia and edema [18]. Melittin F contains 19 amino acid residues and differs from melittin in that the first seven residues of the *N*-terminus are absent, therefore it resembles a fragment of melittin [19]. Cardiopep is a peptide isolated from whole bee venom that has beta adrenergic and anti-arrhythmic effects [20]. Antigen 5, one of the major allergens in all wasp venoms, has an unknown biological function [21]. Other recently isolated and featured peptides show antimicrobial activity, playing a key role in preventing potential infection by microorganism during prey consumption by insect larvae. Examples of such peptides include Anoplin from *Anoplius smariensis* [22], Crabrolin from *Vespa crabro* [23], Decoralin from *Oreumene decoratus* [24], Eumentin from *Eumenes rubronotatus* [25], Melectin from *Melecta albifrons* [26], and Protonectin from *Agelaia pallipe pallipes* [27].

1.3. Low Molecular Weight Compounds

Bee and wasp venoms also contain small molecules, such as minerals, amino acids, and physiologically active amines, such as catecholamines. Among this category, histamine is one of the major components. This organic nitrogenous compound participates in the inflammatory response by increasing the permeability of capillaries. In a similar manner, the catecholamines dopamine and nor-adrenaline increase heartbeat, thereby enhancing venom circulation and thus, its distribution [28]. However, like histamine, the effects of these two catecholamines are largely overshadowed by those of other components of venom. Serotonin can act as an irritant and can contribute to the pain caused by the venom. Finally, high levels of acetylcholine are detected only in wasp venom. Acetylcholine can increase perceived pain of a sting by stimulating pain receptors synergically with histamine effects.

2. Bee Venom (Apitoxin)

Since the first studies in apitherapy at the beginning of the 20th century, multiple therapeutic applications for bee venom have been developed for certain diseases. However, although we have a better understanding of the mechanisms of action of bee venom components, many questions remain unanswered. Given the anti-inflammatory properties of this venom, various forms of traditional bee venom therapy, including the administration of live stings, injection of venom, and venom acupuncture have been used to relieve pain and to treat chronic inflammatory diseases such as rheumatoid arthritis and multiple sclerosis [29,30]. This traditional medicine also has been used for other diseases like cancer [29], skin conditions [31], and recently even for Parkinson's disease [32]. In addition, Apitox®️ (Apimeds, Inc., Seongnam-si, Korea), purified bee venom from *Apis mellifera*, is an FDA-approved subcutaneous injectable product for relieving pain and swelling associated with rheumatoid arthritis, tendinitis, bursitis and multiple sclerosis [33]. From a scientific perspective, special mention is given to two particular reviews [34,35] that assessed the evidence of a non-systematic manner of designing, performing, and analyzing clinical studies of bee venom acupuncture for rheumatoid arthritis and musculoskeletal pain. After evaluating the safety and efficacy of these studies, the results showed evidence of effectiveness. However, the authors highlighted that not only was the total human sample size too small, but the quality of experimental design was varied and sometimes inadequate. Recently, two pilot studies addressing chronic pain of the neck and lower back have were designed under a rigorous randomized clinical trial with the aim to evaluate the true effect of bee venom acupuncture [36,37]. Therefore, new protocolized studies are in the pipeline to validate the efficiency of this novel therapy. Furthermore, purified and synthesized bee venom components and their derivatives have led to novel pharmaceutical agents. In the following sections, we address in more detail the applications of two of the most studied peptides obtained from bee venom, namely melittin and apamin.

2.1. Therapeutic and Biotechnological Applications of Melittin

Melittin, the main component of bee venom, hyaluronidase and PLA_2 are the three major causes of allergic reactions to this venom [38]. An amphiphilic peptide comprising 26 amino acid residues, and in which the amino-terminal region is predominantly hydrophobic and, the carboxyl-terminal region is hydrophilic. Melittin is the principal active component of apitoxin and is responsible for breaking up and killing cells. When several melittin peptides accumulate in the cell membrane, phospholipid packing is severely disrupted, thus leading to cell lysis [39]. Melittin triggers not only the lysis of a wide range of plasmatic membranes but also of intracellular ones such as those found in mitochondria. PLA_2 and melittin act synergistically, breaking up membranes of susceptible cells and enhancing their cytotoxic effect [40]. This cell damage, in turn, may lead to the release of other harmful compounds, such as lysosomal enzymes from leukocytes, serotonin from thrombocytes, and histamine from mast cells, which can all lead to pain.

Although melittin is the most studied and known bee venom peptide, its development for clinical applications remains mainly in preclinical phases. At the moment of writing, no products for human use are available on the market. Some patents and promising studies have focused on bacterial and viral infections, immunologic adjuvants, rheumatoid arthritis, arteriosclerosis, cancer, and endosomolytic properties for drug delivery.

2.1.1. Antimicrobial Properties of Melittin for Therapeutic Use

Antimicrobial peptides (AMPs) have been widely studied as an alternative to conventional antibiotics, especially for the treatment of drug-resistant infections [41]. Hundreds of AMPs have been isolated, and several thousand have been *de novo* designed and synthesized. Despite displaying extensive sequence heterogeneity, most of these peptides share two functionally important features, namely a net positive charge and the ability to adopt an amphipathic structure. Melittin is considered to show strong antimicrobial properties and it also has hemolytic activity and marked allergenic properties. Early studies using individual peptide analogs of melittin showed that the initial step of the mechanism underlying the hemolytic and antimicrobial activity of this venom peptide involves interactions with the lipid groups of the membrane [42]. The structural requirements for the action of melittin, its orientation, aggregation state, current view of pore formation, and also its various cellular actions are discussed in detail in an excellent review by Dempsey [43]. Bruce Merrifield performed pioneering work on improving the features of antimicrobial peptides, shortening their sequences and increasing their activity. In particular, a hybrid undecapeptide derived from the well-known cecropin A and melittin was found to be sufficient for antifungal and antibacterial activities, while displaying low cytotoxicity [44]. This hybrid version was later improved with retro and retroenantio analogs [45]. Indeed, a patent of several active D-peptides with antibiotic and antimalarial activity was even filed [46]. Despite the therapeutic efficacy of antimicrobial peptides, their use is limited due to poor *in vivo* bioavailability caused by instability, cytotoxicity, hydrophobicity, in addition, the cost production is an issue [47]. In parallel to antimicrobial peptides for therapeutic use in humans, these peptides can be applied to fight economically important plant pathogens, which are currently one of the major factors limiting crop production worldwide [48]. A library of linear undecapeptides derived from cecropin-melittin hybrids have been tested against phytopathogenic bacteria and patented for future use in phytosanitary compositions [49]. In this regard, a promising peptide called BP76 has been identified for this purpose [50].

2.1.2. Antimicrobial Properties of Melittin for Biotechnological Use

The idea of using antimicrobial peptides has also been translated to coatings for medical devices. Currently, a number of companies are turning their attention to the use of antimicrobial coatings of cationic peptides, such as melittin, for contact lenses in order to prevent the growth of undesirable microorganisms. Contact lenses made of materials comprising hydrogels and antimicrobial ceramics that contain at least one metal (selected from Ag, Cu and Zn) are available. However, although these polymeric compositions do have antimicrobial properties, they do not have all the properties desired for extended-wear contact lenses. Antimicrobial coatings containing covalently bound antimicrobial peptides exhibit diminished activity when compared that of the unbound corresponding antimicrobial peptides in solution. To overcome these drawbacks, Novartis has patented a method to produce contact lenses with an antimicrobial metal-containing layer-by-layer (LbL) [51]. In its LbL design, at least one layer has a negatively charged polyionic material, having -COO-Ag groups or silver nanoparticles.

2.1.3. Anti-Viral Properties of Melittin for Therapeutic Use

The antiviral activities described for melittin and its analogs are caused by specific intracellular events, with the selective reduction of the biosynthesis of some viral proteins, as reported for the melittin analog Hecate on herpes virus-1 [52], and for melittin itself on HIV-1-infected lymphoma cells [53]. In the 90s, active melittin was presented to provide an improved composition complementary to azidothymidine (AZT) to inhibit the reverse transcriptase and growth of HIV-infected cells [54]. Recently, a similar idea has been patented, whereby melittin is carried in a nanoparticle construct designed to be used as a topical vaginal virucide [55,56].

2.1.4. Vaccines

Related to fields of immunology and vaccinology, the 90s also witnessed great progress in therapeutic approaches based on vaccination against infectious pathogens. Despite these advances in the identification of new antigens and their immunological mechanisms, the immune response in most cases continues to be very weak. Therefore, to improve the response, effective adjuvants to enhance the immunogenicity of target antigens must be used. A few years ago, Rinaldo Zurbriggen presented a novel adjuvant system based on melittin and analogs capable of eliciting strong immune responses against target antigens, thus reducing the risk of toxic side effects associated with the use of adjuvants [57].

2.1.5. Inflammatory and Rheumatic Applications of Melittin

Uncontrolled inflammation can cause extensive tissue damage and is the hallmark of numerous diseases, including rheumatoid arthritis, which results in joint destruction and permanent disability. PLA_2 is the enzyme responsible for hydrolyzing arachidonic acid from phospholipids, and arachidonic acid is the precursor of eicosanoids, which are thought to mediate inflammation. Melittin and related peptides have been described as anti-inflammatory drugs as they have the capacity to inhibit PLA_2 [58,59]. However, in this field, melittin competes with a wide variety of non-steroidal drugs, methotrexate, and other biological disease-modifying antirheumatic drugs [60].

2.1.6. Atherosclerosis Applications of Melittin

Atherosclerosis is the major cause of morbidity and mortality worldwide. This specific form of arteriosclerosis is a chronic inflammatory disease of the arteries caused by the accumulation and interaction of white blood cells, remnants of dead cells, cholesterol, and triglycerides on the artery wall. This complex inflammatory process is characterized by the presence of monocytes/macrophages and T lymphocytes in the atheroma, where macrophages secrete pro-inflammatory cytokines, a main cellular component in the development of atherosclerotic plaques [61]. Several *in vitro* studies have shown positive effects of melittin for the treatment of atherosclerosis [62,63]. In addition, *in vivo* experiments have demonstrated the molecular mechanism of the anti-atherosclerotic effects of melittin in mouse models of this disease [64]. This has been the major finding regarding the capacity of melittin to prevent lipopolysaccharide (LPS)/high-fat-induced expression of inflammatory cytokines, proatherogenic proteins, and adhesion molecules.

2.1.7. Cancer Applications of Melittin

Many studies report that melittin inhibits tumor cell growth and induces apoptosis, thereby indicating a potential application of this venom peptide as an alternative or complementary medicine for the treatment of human cancers. A valuable review describing the mechanisms underlying the anticancer effects of melittin has been published [65]. Cells in several types of cancer, such as renal, lung, liver, prostate, bladder, breast, and leukemia, can be targeted by melittin. It is well-known that melittin is a natural detergent with the capacity to form tetramer aggregates on membranes, which lead to disorders in the structure of phospholipid bilayers, changes in membrane potential, aggregation of membrane proteins, as well as the induction of hormone secretion [66]. Furthermore, this membrane disruption directly or indirectly leads to alterations in enzymatic systems, such as G-protein [67], protein kinase C [68], adenylate cyclase [69], and phospholipase A [70]. Melittin can even inhibit calmodulin, a calcium-binding protein that plays a crucial role in cell proliferation [71]. Tumoral cells expose anionic phospholipids, mainly phosphatidylserine, on the external leaflet of the plasma membrane [72], and this feature can allow the preferential binding of cationic peptides, like melittin, relative to normal cells. Melittin studies with numerous types of cancer cells and *in vivo* animal models have demonstrated its antiproliferative activity [73,74]. Furthermore, recent studies have demonstrated that melittin has anti-angiogenesis properties [75–77].

However, when a therapeutic dose of melittin is injected *in vivo*, some side effects, such as liver injury and hemolysis, were observed. To minimize these emerging lesions in off-target tissues, the following three strategies have been designed: (1) conjugaton of melittin to an antibody or a targeting component; (2) development of shielded pro-cytolytic melittin systems; and (3) synthesis of melittin-transporting carriers.

With regard to the first approach, a melittin-based recombinant immunotoxin obtained by fusion of genes that encoded an antibody fragment derived from the murine monoclonal antibody

K121 with an oligonucleotide encoding melittin was tested successfully *in vitro* [78]. Another study was based on a recombinant immunotoxin of melittin fused to an anti-asialoglycoprotein receptor (ASGPR) single-chain variable fragment antibody (Ca) which conferred targeting and ASGPR-specific cytotoxicity to hepatocellular carcinoma cells [79]. Finally, a recent study characterized a CTLA-4-targeted scFv-melittin fusion protein as a potential immunosuppressive agent for organ transplant. In this regard, the selective cytotoxicity of the peptide construction was confirmed in preliminary biological activity assays [80].

Related to the pro-cytolytic melittin, by taking advantage of tumor matrix metalloproteinase 2 (MMP2) overexpressed on cancer cell membranes, an MMP2 cleavable melittin/avidin conjugate was built. Melittin coupled to avidin becomes inactive, but when released from the conjugate it induces immediate cell lysis [81]. A similar idea was published years later, this time using avadin, the latency-associated peptide (LAP) domain of the transforming growth factor beta (TGF-β). In this approach, LAP dimerization conferred latency to the MMP2-cleavable melittin-LAP fusion protein [82].

Regarding pro-cytotoxic melittin systems, a design was based on the mixture of melittin with the anionic detergent sodium dodecyl sulfate formulated into poly(D,L-lactide-co-glycolide acid) nanoparticles by an emulsion solvent diffusion method. The inhibitory *in vitro* effects of these 130 nm-diameter melittin-loaded nanoparticles on breast cancer MCF-7 cells were promising [83]. Another interesting carrier was a pegylated immunoliposome coupled to a humanized antihepatocarcinoma single-chain antibody variable region fragment and loaded with a bee venom peptide fraction [84]. A similar pegylated immunoliposome but using only melittin as cargo and the complete antibody trastuzumab as targeting component was designed to combat HER2-overexpressing human breast cancer cell lines [85]. The three aforementioned nanoparticles are not suitable for systemic administration because melittin can be released in blood vessels during transport, particularly in liposomes, which can be disrupted by the lytic peptide [86]. To overcome this drawback, Samuel A. Wickline's group developed a perfluorocarbon nanoemulsion vehicle incorporating melittin into its outer lipid monolayer [87]. This nanocarrier of approximately 270 nm in diameter presented favorable pharmacokinetics, accumulating melittin in murine tumors *in vivo* and causing a dramatic reduction in tumor growth without any apparent signs of toxicity [88,89]. Finally, the most recent ultra-small diameter melittin-nanoparticle (<40 nm) successfully tested *in vivo* with few side effects is the patented α-melittin-NP [90,91]. This nanoparticle comprise 1,2 dimyristoyl-sn-glycero-3-phosphatidylcholine (DMPC) decorated with the hybrid peptide formed by peptide D-4F and melittin via a GSG linker, the peptide D-4F being a peptide that mimics a high-density lipoprotein (HDL) [92].

2.1.8. Endosomolytic Properties of Melittin

The strategy of packing and carrying small interference RNA (siRNA) using a wide variety of systems for gene therapy has been increasingly followed in recent years. The efficiency mediated by these drug delivery systems is strongly dependent on their endosomal escape capability, otherwise the siRNA would be degraded in endolysosomes [93]. One mechanism designed for endosomal release is the use of fusogenic peptides, which are generally short amphipathic sequences between 20 and 30 amino acids in length and capable of disrupting biological membranes at endosomal pH [94,95]. One of the first highly innovative studies using melittin consisted of reversibly masking the membrane-active peptide using maleic anhydride derivative [96]. At neutral pH, the lysine residues of melittin were covalently acylated with anhydride, thereby inhibiting the membrane disruption activity of the peptide. Under acidic conditions such as those present within endosomes, the amide bond of the maleamate was cleaved, thus unmasking melittin. Similar studies performed by Ernest Wagner *et al.* showed that melittin analogs with high lytic activity at acid pH enhance the transfection of oligonucleotides in cell cultures and in *in vivo* mouse models [82,97–99]. Very recently, a derivative of melittin (p5RHH) was reported to successfully trigger siRNA release into the cellular cytoplasm [100,101]. The company Arrowhead Therapeutics is currently developing ARC-520 as a novel siRNA-based therapeutic to knock down the expression of viral RNAs of chronic hepatitis B virus.

They describe the use of a coinjection of a hepatocyte-targeted, *N*-acetylgalactosamine-conjugated melittin-like peptide (NAG-MLP) with a liver-tropic cholesterol-conjugate siRNA (chol-siRNA) targeting coagulation factor VII [102,103]. Preclinical studies with animals as well as Phase I assays have revealed that melittin promotes delivery without generating anti-melittin antibodies. In March 2014, Phase II trials of ARC-520 were started for patients with chronic hepatitis B virus [104].

2.2. Therapeutic Applications of Apamin

Apamin is a peptide neurotoxin comprising 18 amino acid residues that is tightly cross-linked by the presence of two disulphide bonds which connect position 1 with 11 and position 3 with 15 [105]. Apamin selectively blocks the small conductance of Ca^{2+}-dependent K^+ channels (SK channels) expressed in the central nervous system (CNS). This type of channel plays a crucial role in repetitive activities in neurons [106], blocking many hyperpolarising-inhibitory effects, including alpha-adrenergic, cholinergic, purinergic, and neurotensin-induced relaxations [107,108].

Since the Hahn and Leditschke's first descriptions in the 30s of mouse convulsions caused by apamin injection, other symptoms and properties have been ascribed to this rigid octadecapeptide [106]. After intraperitoneal injection in animals, apamin locates not only in the grey matter of the brain but also in the liver and the adrenal cortex [109,110]. Therefore, apamin can no longer be considered an exclusive neurotoxin. Unlike melittin, apamin is a peptide with a highly specific mode of action. It binds and occludes the pore of small conductance Ca^{2+}-triggered K^+ channels (SK), thus acting as an allosteric inhibitor [111] and depressing delayed cellular hyperpolarization. This binding specificity provides apamin electrical properties which have been exploited in biomedical research. Apamin acts mainly on the CNS, where SK channels are widely expressed [112]. SK channels are divided into the following three main classes on the basis of their conductance: (1) large conductance (BK or K1); (2) intermediate conductance (IK or K2); and (3) small conductance (SK or K3) [113]. These channels, which are activated solely by increases in intracellular Ca^{2+} contribute to regulating the excitability and function of many cell types, including neurons, epithelial cells, T-lymphocytes, and skeletal muscle cells [114]. SK channels are activated by submicromolar concentrations of Ca^{2+}, and this activation is mediated by calmodulin [115].

2.2.1. Learning Deficit

In excitable cells, the activation of SK channels generates a hyperpolarizing K^+ current which contributes to the afterhyperpolarisation (AHP) that follows an action potential [116]. This AHP modulates cell firing frequency and spike frequency adaptation, thereby influencing neuronal excitability. SK channels have been implicated in diverse physiological functions such as synaptic enhancement and long-term potentiation. Furthermore, early studies showed that systemic apamin administration facilitates learning and memory. The first such study, using appetitive learning paradigms, reported that systemic apamin injections accelerate acquisition of the bar-pressing response and also accelerate bar-pressing rates [117]. Several studies, listed in Table 3, underscore the relevance of SK channels in information processing and storage at the systems level. Such studies propose that SK channels would be appropriate targets for apamin as a therapeutic treatment for learning deficits.

Table 3. Evidences of enhanced learning in animals receiving apamin injections.

Animal studies with apamin	References
Apamin improved rat performance in the novel object recognition task, where habituation of exploratory activity was assessed	[118]
Apamin improved spatial navigation in medial septal-lesioned mice	[119]
Apamin dose-dependently alleviated deficits in spatial reference and working memory induced by partial electrolytic hippocampal lesion	[120]
Apamin attenuated the memory deficits caused by scopolamine, which affect hippocampal and cortical activity	[121]
Apamin-treated mice exhibited fater learning of the platform location during the initial trials in the Morris water maze	[122]
Apamin improved task acquisition in a learned extinction operant behavior protocol	[123]
Apamin enchanced working memory in a medical prefrontal cortex-dependent spatial delayed alternation task	[124]
Apamin facilitated the encoding of contextual fear memory	[125]
Apamin improved performance on the water task in mice with neurofibromatosis 1	[126]

2.2.2. Parkinson's Disease

Parkinson's disease (PD) is a neurodegenerative disorder characterized by the progressive loss of dopaminergic (DA) neurons in the subtantia nigra, leading to typical motor symptoms (akinesia, rigidity, rest tremor) [127]. Although therapy with L-Dopa, a precursor of dopamine, has provided benefit for years, the disease progresses slowly, resulting in disability [128]. Recent studies indicate that bee venom [129,130] and apamin [131–133] protect DA neurons from degeneration in experimental PD. The use of apamin was patented to overcome the drawbacks of drugs used in the treatment of PD, *i.e.*, L-Dopa [134]. According to this patent, treatment would consist of using from 1–10 micrograms of apamin in a single dose injection. In this approach, apamin would not only protect undamaged neurons but would restore the function of silent neurons. Another recent patent related to the previous one, claims that degenerative brain diseases can be treated with a pharmaceutical composition comprising apamin as an active ingredient and at least one other compound for the treatment of PD and related Parkinsonian disorders [135].

2.2.3. Preserving Red Blood Cells

Whole blood can be stored at 4 °C for three weeks using a CPD (citrate, phosphate, dextrose) anti-coagulant solution. Adding adenine, glucose and/or manitol can prolong blood storage time by two weeks or more. However, there is a necessity to reach a longer period of blood storage. There is a patent that provides methods, compositions, and kits for storing red blood cells for extended periods of time, preventing red blood cell storage lesions, retaining red blood cell deformability, and increasing survival of the cells following transfusion [136]. In some formulas, the composition comprises at least one K^+ channel blocker agent, including apamin among others.

2.2.4. Blood-Brain Barrier Shuttle

The blood-brain barrier (BBB) is a highly selective part of the neurovascular system that prevents the entry of many substances, including most therapeutics, into the CNS. Paracellular transport between endothelial cells is restricted by tight junctions and transendothelial transport is reduced, thus hindering the use of a high percentage of potential commercialized molecules intended for treatment of the CNS [137]. Several strategies have been implemented to deliver drugs across the BBB, some of which cause structural damage to the barrier by forcibly opening it to allow the uncontrolled passage of drugs. The ideal method for transporting drugs across the BBB should be controlable and should not damage the structure. While a wide range of nanoparticulate delivery systems have been studied with the aim to target therapeutics (low MW drugs, nucleic acids or proteins) to the brain [138], their success rate has been low. The specific distribution of apamin in the CNS and its capacity to cross the BBB [139] make the design of an apamin-based drug delivery system feasible. In fact, a recent study described the promising therapeutic effects of drug-loaded micelles targeted with apamin in reparing spinal cord injury (SCI) in mouse models [140]. However, apamin is neurotoxic at high concentrations, having a relatively low LD_{50} in mice (2.5 micromol/kg) [141]. Bearing this in mind, we assayed a non-toxic analog derived from apamin that bears two ornithine residues instead of arginines residues (ApOO) [142]. We compared and demonstrated the capacity of both peptides to cross the BBB in a cell-based model, thus revealing their potential as BBB carriers [142].

3. Wasp Venom

Wasp venom is more variable in composition among species. However, bee and wasp venoms have similar enzymatic composition (see Table 2). A significant difference in the peptide composition of wasp venom is the predominance of mastoparan and bradykinins. Although wasp venom has attracted much less attention than bee venom, extensive research over recent decades has shown the pharmacological properties [143]. In this section, we will focus on the therapeutic applications of the most studied peptide in wasp venom, namely mastoparan.

3.1. Therapeutic Applications of Mastoparan

Mastoparan is a membrane-active amphipathic peptide with 14 amino acid residues. It is rich in hydrophobic and basic residues that form amphipathic helical structures, the latter with the capacity to form pores in membranes. Mastoparan induces a potent mitochondrial permeability transition that affects cell viability [144]. The net effect of the mode of action of mastoparan depends on the cell type. In this regard, it causes the secretion of histamine from mast cells, serotonin from platelets, catecholamines from chromaffin cells, and prolactin from the anterior pituitary [145].

Mastoparan exhibits a wide variety of biological effects, including insertion into the membrane bilayer causing membrane destabilization with consequent lysis [146] or direct interaction with G proteins on the cytoplasmatic face, thus perturbing transmembrane signaling [145,147–149], stimulation of phospholipases, mobilization of Ca^{2+} from mitochondria and sarcoplasmic reticulum, and cell death by necrosis and/or apoptosis [150,151]. Thus, key biological activities have been described for this peptide, including antimicrobial activity, increased histamine release from mast cells, hemolytic activity [152,153], induction of a potent mitochondrial permeability transition, and tumor cell cytotoxicity. Related to its capacity to induce mitochondrial permeability transition [144], mastoparan has recently been reported to interact with the phospholipid phase of the mitochondrial membrane to induce permeabilization in cyclosporine A-sensitive and insensitive manners but does not interact with any specific receptors or enzymes [154].

3.1.1. Antimicrobial Properties of Mastoparan for Therapeutic Use

Mastoparan alone or in combination with other antibiotics could be a promising alternative for combating multiple-antibiotic resistant bacteria in clinical practice [155]. A number of strategies for optimizing the potency of mastoparan have been addressed, including structural stabilization and charge modification [156,157], achieving synthesis of derivatives and enantiomers [158,159], modulation of hydrophobicity [160,161], and selective acylation/alkylation [162]. However, these studies show that mastoparan activity is gained at the expense of impaired membrane selectivity or *vice versa*, with no distinction between bacterial and mammalian membranes. Thus, the development of new strategies to reduce the toxic side effects of mastoparan, thereby improving the feasibility of clinical applications, are required. However, three independent *in vivo* studies on sepsis, systematic inflammation caused by an infection where bacteria and LPS are potent activators of immune cells, have shown that an analog of mastoparan (mastoparan-1) protects mice from lethal challenge by live bacteria and LPS [163–165]. The effects of mastoparan-1 were associated with its bactericidal action and its capacity to neutralize LPS and attenuate inflammatory responses by macrophages.

3.1.2. Anti-Viral Properties of Mastoparan for Therapeutic Use

As for multidrug-resistant bacteria, there is a pressing need to identify novel and broad-spectrum antiviral agents that can be used as therapeutics. A recent study has demonstrated that a mastoparan derivative shows broad-spectrum antiviral activity *in vitro* against five families of enveloped viruses directly via disruption of their lipid envelope structure [166]. However, further studies are needed to demonstrate its therapeutic use.

3.1.3. Cancer Applications of Mastoparan and Mitoparan

As we mentioned previously, mastoparan targets the mitochondrial membrane and causes mitochondrial permeability transition to mediate its tumor cell cytotoxicity. Several studies have demonstrated the antitumor activity of mastoparan and analogs *in vitro* [145,167–169]. One potential way to deliver mastoparan and avoid side-effects was presented by Hiyedoshi Harashima *et al.* [170], whereby a transferrin (Tf)-modified liposomes decorated with endosomolytic GALA peptides and, in addition, encapsulating mastoparan, were designed to target the upregulated Tf receptor in tumor cells. Only one *in vivo* study has been carried out, in which mastoparan, administered in a peritumoral way, delayed the subcutaneous development of melanoma in a well-established subcutaneous murine melanoma model and increased survival [171]. It is noteworthy to highlight a potent mastoparan analog, called mitoparan. This peptide shows enhanced amphiphilicity, presenting two additional lysyl side chains in the cationic face and the replacement of a α-aminoisobutiric acid (Aib), a known helix promoter, by an Ala at position 10 [157]. A novel cell-penetrating mitochondriotoxic peptide, mitoparan was modified at its *N*-terminus by adding an RGD motif to confer capacity to selectively bind cell adhesion molecules [172] overexpressed in tumors, which play a significant role in cancer progression and metastasis. This modification would potentially improve the pharmacodynamic parameters of the chimeric mitoparan *in vivo* [168].

Rui Wang *et al.* later patented a new type of acid-activated mitoparan complex by conjugating normal mitoparan with its anionic binding partner via a disulfide linker, where the anionic partner mitoparan, which has three Lys residues replaced by three Glu residues, and two Lys residues by two His residues, shields the cytotoxic activity of mitoparan [173]. Although this chimeric mitoparan complex has been tested only *in vitro*, the designers are optimistic because the complex showed significant enzymatic stability compared with normal mitoparan, thus supporting its potential for *in vivo* application. Recently, we presented a peptide-polymer design strategy to obtain a pro-cytotoxic system based on mitoparan, as cytotoxic peptide, conjugated to a polyglutamic acid polymer through specific cleavage sites that are sensitive to overexpressed tumor proteases, such as MMP-2 and cathepsin B [174]. Our system was also decorated with a specific targeting peptide to HER2$^+$ breast tumor cells, thus allowing mitoparan to be released with exquisite spatiotemporal control. It should be noted that the mitoparan that we used for our experiments was the enantiomer form, because normal mitoparan was easily degraded by cellular proteases before its release from the polymer.

3.1.4. Cell-Penetrating Peptide Properties

Cell-penetrating peptides (CPPs) have potential pharmaceutical application in delivering macromolecules into cells. Hundreds of sequences now fall in the CPP classification, and several interesting reviews focusing on internalization mechanism, effect of the cargo, CPP modifications or extensions, protocols and significant effects on penetration capacity have been published [40,105,175–177]. Given the capacity of mastoparan and mitoparan to efficiently cross plasma membranes, some researchers have demonstrated the value of these peptides as CPPs [158]. Very recently, the influence of various CCPs on the transport of doxorubicin encapsulating Tf-liposomes across BBB, *in vitro* and *in vivo*, has been addressed [178]. Although mastoparan showed efficient translocation, the other CPPs, namely TAT peptide and penetratin, were more efficient. On the other hand, there is a chimeric galanin-mastoparan peptide called transportan, which contains the first 13 amino acids from the highly conserved amino-terminal part of galanin and the 14 amino acids sequence of mastoparan in the carboxyl terminus [179]. Transportan and some derivatives have the capacity to carry macromolecules [180–182].

4. Conclusions

Despite the many studies published on bee and wasp venoms, little has been reported on the practical applications of these substances. The main clinical uses of these toxins are based on meridian therapy, focusing on the application of bee venom acupuncture to relevant sites in function of a specific disease or to acupoints. Of note, the single most promising and advanced Phase II trial involving melittin deals with its use as an endosomolytic agent as an effective siRNA delivery system for hepatitis B virus infection [102]. The antimicrobial properties of melittin and mastoparan have been the most studied and developed among the components of wasp and bee venoms. Their mode of action on membranes has drawn attention to AMPs as a universal solution to the growing incidence of drug-resistant infections. However, expectations have not been fulfilled as quickly as initially imagined as a result of the uncompetitive costs of peptide production and also impaired membrane selectivity of peptides between bacteria and eukaryotic cells. Although a wide variety of compounds are currently available for the treatment of cancer, there is still hope to discover a cancer application for potent cytotoxic peptides derived from bees and wasps. Our targeted pro-cytotoxic system based on mitoparan, which transports this potent cytotoxic peptide to the tumor and allows its accumulation in a controlled manner, emerges as a plausible approach to overcome off-target side effects of current cancer treatment [174]. Another promising and feasible idea tested *in vivo* to combat cancer has been presented in the form of a patent, which discloses an ultra-small lipid nanoparticle carrying melittin with potential use in clinical practice [91]. We consider that the remaining potential applications described here are currently in earlier stages of investigation and their conversion into realistic therapeutic or biotechnological applications is still far away. The future therapeutic applications of wasp and bee venom components are further complicated by strong competition with millions of potential new molecules and systems that are coming to light.

Acknowledgments: This study was supported by grants from MICIN-FEDER (BIO2013-40716R) and the *Generalitat Catalunya* (XRB and 2014SGR-1251).

Author Contributions: Both authors wrote, read and approved the final manuscript.

Conflicts of Interest: The authors declare no conflict of interest.

References

1. Nakajima, T. Biochemistry of vespid venoms. In *Tu AT (Org). Insect Poisons, Allergens, and Other Invertebrate Venoms. Handbook of Natural Toxins*; Anthony, T., Ed.; Marcel Dekker: New York, NY, USA, 1984; Volume 2, pp. 109–133.
2. Hoffman, D.R. Hymenoptera venom allergens. *Clin. Rev. Allergy Immunol.* **2006**, *30*, 109–128. [CrossRef] [PubMed]
3. Hoffman, D.R.; Jacobson, R.S. Allergens in hymenoptera venom XII: How much protein is in a sting? *Ann. Allergy* **1984**, *52*, 276–278. [PubMed]
4. Schumacher, M.J.; Tveten, M.S.; Egen, N.B. Rate and quantity of delivery of venom from honeybee stings. *J. Allergy Clin. Immunol.* **1994**, *93*, 831–835. [CrossRef] [PubMed]
5. Dohtsu, K.; Okumura, K.; Hagiwara, K.; Palma, M.S.; Nakajima, T. Isolation and sequence analysis of peptides from the venom of *Protonectarina sylveirae* (Hymenoptera-Vespidae). *Nat. Toxins* **1993**, *1*, 271–276. [CrossRef] [PubMed]
6. Hider, R.C. Honeybee venom: A rich source of pharmacologically active peptides. *Endeavour* **1988**, *12*, 60–65. [CrossRef] [PubMed]
7. Cichocka-Jarosz, E. Hymenoptera venom allergy in humans. *Folia Med. Cracov.* **2012**, *52*, 43–60. [PubMed]
8. Graler, M.H.; Goetzl, E.J. Lysophospholipids and their G protein-coupled receptors in inflammation and immunity. *Biochim. Biophys. Acta* **2002**, *1582*, 168–174. [CrossRef] [PubMed]

9. Doery, H.M.; Pearson, J.E. Phospholipase B in snake venoms and bee venom. *Biochem. J.* **1964**, *92*, 599–602. [PubMed]

10. Girish, K.S.; Kemparaju, K. The magic glue hyaluronan and its eraser hyaluronidase: A biological overview. *Life Sci.* **2007**, *80*, 1921–1943. [CrossRef] [PubMed]

11. Dotimas, E.M.; Hamid, K.R.; Hider, R.C.; Ragnarsson, U. Isolation and structure analysis of bee venom mast cell degranulating peptide. *Biochim. Biophys. Acta* **1987**, *911*, 285–293. [CrossRef] [PubMed]

12. Ziai, M.R.; Russek, S.; Wang, H.C.; Beer, B.; Blume, A.J. Mast cell degranulating peptide: A multi-functional neurotoxin. *J. Pharm. Pharmacol.* **1990**, *42*, 457–461. [CrossRef] [PubMed]

13. Sharma, J.N. Basic and clinical aspects of bradykinin receptor antagonists. *Prog. Drug Res.* **2014**, *69*, 1–14. [PubMed]

14. Shkenderov, S.; Koburova, K. Adolapin—A newly isolated analgetic and anti-inflammatory polypeptide from bee venom. *Toxicon* **1982**, *20*, 317–321. [CrossRef] [PubMed]

15. Kitamura, H.; Yokoyama, M.; Akita, H.; Matsushita, K.; Kurachi, Y.; Yamada, M. Tertiapin potently and selectively blocks muscarinic K^+ channels in rabbit cardiac myocytes. *J. Pharmacol. Exp. Ther.* **2000**, *293*, 196–205. [PubMed]

16. Vlasak, R.; Kreil, G. Nucleotide sequence of cloned cDNAs coding for preprosecapin, a major product of queen-bee venom glands. *Eur. J. Biochem.* **1984**, *145*, 279–282. [CrossRef] [PubMed]

17. Meng, Y.; Yang, X.X.; Zhang, J.L.; Yu, D.Q. A novel peptide from Apis mellifera and solid-phase synthesis of its analogue. *Chin. Chem. Lett.* **2012**, *23*, 1161–1164. [CrossRef]

18. Mourelle, D.; Brigatte, P.; Bringanti, L.D.; de Souza, B.M.; Arcuri, H.A.; Gomes, P.C.; Baptista-Saidemberg, N.B.; Ruggiero Neto, J.; Palma, M.S. Hyperalgesic and edematogenic effects of Secapin-2, a peptide isolated from Africanized honeybee (*Apis mellifera*) venom. *Peptides* **2014**, *59*, 42–52. [CrossRef] [PubMed]

19. Gauldie, J.; Hanson, J.M.; Shipolini, R.A.; Vernon, C.A. The structures of some peptides from bee venom. *Eur. J. Biochem.* **1978**, *83*, 405–410. [CrossRef] [PubMed]

20. Vick, J.A.; Shipman, W.H.; Brooks, R., Jr. Beta adrenergic and anti-arrhythmic effects of cardiopep, a newly isolated substance from whole bee venom. *Toxicon* **1974**, *12*, 139–144. [CrossRef] [PubMed]

21. Monsalve, R.I.; Lu, G.; King, T.P. Expressions of recombinant venom allergen, antigen 5 of yellowjacket (*Vespula vulgaris*) and paper wasp (*Polistes annularis*), in bacteria or yeast. *Protein Expr. Purif.* **1999**, *16*, 410–416. [CrossRef] [PubMed]

22. Konno, K.; Hisada, M.; Fontana, R.; Lorenzi, C.C.; Naoki, H.; Itagaki, Y.; Miwa, A.; Kawai, N.; Nakata, Y.; Yasuhara, T.; *et al.* Anoplin, a novel antimicrobial peptide from the venom of the solitary wasp *Anoplius samariensis*. *Biochim. Biophys. Acta* **2001**, *1550*, 70–80. [CrossRef] [PubMed]

23. Krishnakumari, V.; Nagaraj, R. Antimicrobial and hemolytic activities of crabrolin, a 13-residue peptide from the venom of the European hornet, *Vespa crabro*, and its analogs. *J. Pept. Res.* **1997**, *50*, 88–93. [CrossRef] [PubMed]

24. Konno, K.; Rangel, M.; Oliveira, J.S.; Dos Santos Cabrera, M.P.; Fontana, R.; Hirata, I.Y.; Hide, I.; Nakata, Y.; Mori, K.; Kawano, M.; *et al.* Decoralin, a novel linear cationic alpha-helical peptide from the venom of the solitary eumenine wasp *Oreumenes decoratus*. *Peptides* **2007**, *28*, 2320–2327. [CrossRef] [PubMed]

25. Konno, K.; Hisada, M.; Naoki, H.; Itagaki, Y.; Fontana, R.; Rangel, M.; Oliveira, J.S.; Cabrera, M.P.; Neto, J.R.; Hide, I.; *et al.* Eumenitin, a novel antimicrobial peptide from the venom of the solitary eumenine wasp *Eumenes rubronotatus*. *Peptides* **2006**, *27*, 2624–2631. [CrossRef] [PubMed]

26. Cerovsky, V.; Hovorka, O.; Cvacka, J.; Voburka, Z.; Bednarova, L.; Borovickova, L.; Slaninova, J.; Fucik, V. Melectin: A novel antimicrobial peptide from the venom of the cleptoparasitic bee *Melecta albifrons*. *Chembiochem* **2008**, *9*, 2815–2821. [CrossRef] [PubMed]

27. Mendes, M.A.; de Souza, B.M.; Marques, M.R.; Palma, M.S. Structural and biological characterization of two novel peptides from the venom of the neotropical social wasp *Agelaia pallipes pallipes*. *Toxicon* **2004**, *44*, 67–74. [CrossRef] [PubMed]

28. Habermann, E. Bee and wasp venoms. *Science* **1972**, *177*, 314–322. [CrossRef] [PubMed]

29. Orsolic, N. Bee venom in cancer therapy. *Cancer Metastasis Rev.* **2012**, *31*, 173–194. [CrossRef] [PubMed]

30. Munstedt, J.; Hackethal, A.; Schmidt, K. Bee venom therapy, bee venom acupuncture of apiculture: What is the evidence behind the various health claims? *Am. Bee J.* **2005**, *145*, 665–668.

31. Han, S.M.; Lee, K.G.; Pak, S.C. Effects of cosmetics containing purified honeybee (*Apis mellifera* L.) venom on acne vulgaris. *J. Integr. Med.* **2013**, *11*, 320–326. [CrossRef] [PubMed]

32. Cho, S.Y.; Shim, S.R.; Rhee, H.Y.; Park, H.J.; Jung, W.S.; Moon, S.K.; Park, J.M.; Ko, C.N.; Cho, K.H.; Park, S.U. Effectiveness of acupuncture and bee venom acupuncture in idiopathic Parkinson's disease. *Parkinsonism Relat. Disord.* **2012**, *18*, 948–952. [CrossRef] [PubMed]

33. Alves, E.M.; Heneine, L.G.D.; Pesquero, J.L.; Albuquerque, M.L.D. Pharmaceutical Composition Containin an Apitoxin Fraction and Use Thereof. WO2011041865, 14 April 2011.

34. Lee, M.S.; Pittler, M.H.; Shin, B.C.; Kong, J.C.; Ernst, E. Bee venom acupuncture for musculoskeletal pain: A review. *J. Pain* **2008**, *9*, 289–297. [CrossRef] [PubMed]

35. Lee, J.A.; Son, M.J.; Choi, J.; Jun, J.H.; Kim, J.I.; Lee, M.S. Bee venom acupuncture for rheumatoid arthritis: A systematic review of randomised clinical trials. *BMJ Open* **2014**, *4*, e006140. [CrossRef] [PubMed]

36. Seo, B.K.; Lee, J.H.; Sung, W.S.; Song, E.M.; Jo, D.J. Bee venom acupuncture for the treatment of chronic low back pain: Study protocol for a randomized, double-blinded, sham-controlled trial. *Trials* **2013**, *14*, 16. [CrossRef] [PubMed]

37. Seo, B.K.; Lee, J.H.; Kim, P.K.; Baek, Y.H.; Jo, D.J.; Lee, S. Bee venom acupuncture, NSAIDs or combined treatment for chronic neck pain: Study protocol for a randomized, assessor-blind trial. *Trials* **2014**, *15*, 132. [CrossRef] [PubMed]

38. Bilo, B.M.; Rueff, F.; Mosbech, H.; Bonifazi, F.; Oude-Elberink, J.N.G.; EAACI Interest Group on Insect Venom Hypersensitivity. Diagnosis of Hymenoptera venom allergy. *Allergy* **2005**, *60*, 1339–1349. [CrossRef] [PubMed]

39. Raghuraman, H.; Chattopadhyay, A. Melittin: A membrane-active peptide with diverse functions. *Biosci. Rep.* **2007**, *27*, 189–223. [CrossRef] [PubMed]

40. Damianoglou, A.; Rodger, A.; Pridmore, C.; Dafforn, T.R.; Mosely, J.A.; Sanderson, J.M.; Hicks, M.R. The synergistic action of melittin and phospholipase A2 with lipid membranes: Development of linear dichroism for membrane-insertion kinetics. *Protein Pept. Lett.* **2010**, *17*, 1351–1362. [CrossRef] [PubMed]

41. Vila-Farres, X.; Giralt, E.; Vila, J. Update of peptides with antibacterial activity. *Curr. Med. Chem.* **2012**, *19*, 6188–6198. [CrossRef] [PubMed]

42. Blondelle, S.E.; Houghten, R.A. Hemolytic and antimicrobial activities of the twenty-four individual omission analogues of melittin. *Biochemistry* **1991**, *30*, 4671–4678. [CrossRef] [PubMed]

43. Dempsey, C.E. The actions of melittin on membranes. *Biochim. Biophys. Acta* **1990**, *1031*, 143–161. [CrossRef] [PubMed]

44. Boman, H.G.; Wade, D.; Boman, I.A.; Wahlin, B.; Merrifield, R.B. Antibacterial and antimalarial properties of peptides that are cecropin-melittin hybrids. *FEBS Lett.* **1989**, *259*, 103–106. [CrossRef] [PubMed]

45. Merrifield, R.B.; Juvvadi, P.; Andreu, D.; Ubach, J.; Boman, A.; Boman, H.G. Retro and retroenantio analogs of cecropin-melittin hybrids. *Proc. Natl. Acad. Sci. USA* **1995**, *92*, 3449–3453. [CrossRef] [PubMed]

46. Merrifield, R.B.; Wade, D.; Boman, H.G. Antibiotic Peptides Containing D-Amino Acids. US5585353 A, 17 December 1996.

47. Anju, G.; Reetu, G.; Sudarshan, K. Liposome-Encapsulated Antimicrobial Peptides: Potential Infectious Diseases Therapy. In *Hanbook of Research on Diverse Apllications of Nanotechnology in Biomedicine, Chemistry, and Engineering*; Soni, Shivani: Hershey, PA, USA, 2015; Chapter 14; pp. 301–332.

48. Stockwell, V.O.; Duffy, B. Use of antibiotics in plant agriculture. *Rev. Sci. Tech.* **2012**, *31*, 199–210. [PubMed]

49. Bardaji, E.; Montesinos, E.; Badosa, E.; Feliu, L.; Planas, M.; Ferre, R. Antimicrobial Linear Peptides. US8026219, 27 September 2011.

50. Badosa, E.; Ferre, R.; Planas, M.; Feliu, L.; Besalu, E.; Cabrefiga, J.; Bardaji, E.; Montesinos, E. A library of linear undecapeptides with bactericidal activity against phytopathogenic bacteria. *Peptides* **2007**, *28*, 2276–2285. [CrossRef] [PubMed]

51. Rubner, M.F.; Yang, S.Y.; Qiu, Y.; Lynn, C.; Lally, J.M. Method for making medical devices having antimicrobial coatings thereon. US20140112994, 24 April 2014.

52. Baghian, A.; Jaynes, J.; Enright, F.; Kousoulas, K.G. An amphipathic alpha-helical synthetic peptide analogue of melittin inhibits herpes simplex virus-1 (HSV-1)-induced cell fusion and virus spread. *Peptides* **1997**, *18*, 177–183. [CrossRef] [PubMed]

53. Wachinger, M.; Saermark, T.; Erfle, V. Influence of amphipathic peptides on the HIV-1 production in persistently infected T lymphoma cells. *FEBS Lett.* **1992**, *309*, 235–241. [CrossRef] [PubMed]

54. Saermark, T.; Erfle, V. Method and Composition for the Treatment of Mammalian HIV Infection. WO1991008753, 27 June 1991.

55. Hood, J.L.; Jallouk, A.P.; Campbell, N.; Ratner, L.; Wickline, S.A. Cytolytic nanoparticles attenuate HIV-1 infectivity. *Antivir. Ther.* **2013**, *18*, 95–103. [CrossRef] [PubMed]

56. Wickline, S.A.; Lanza, G.; Hood, J. Nanoparticulate-Based Contraceptive (Anti-HIV Composition and Methods. US20120100186 A1, 26 April 2012.

57. Zurbriggen, R. Compositions Comprising Melittin-Derived Peptides and Methods for the Potentiation of Immune Responses against Target. CA2543072, 2 June 2005.

58. Peterson, J.W.; Saini, S.S.; Wood, T.G.; Chopra, A.K. Anti-inflammatory and other therapeutic, prophylactic or diagnostic uses of synthetic melittin and new related peptides. WO/1998/056400, 17 December 1998.

59. Saini, S.S.; Peterson, J.W.; Chopra, A.K. Melittin binds to secretory phospholipase A$_2$ and inhibits its enzymatic activity. *Biochem. Biophys. Res. Commun.* **1997**, *238*, 436–442. [CrossRef] [PubMed]

60. Gabay, C.; Hasler, P.; Kyburz, D.; So, A.; Villiger, P.; von Kempis, J.; Walker, U. Biological agents in monotherapy for the treatment of rheumatoid arthritis. *Swiss Med. Wkly.* **2014**, *144*, w13950. [PubMed]

61. Ross, R. Atherosclerosis—An inflammatory disease. *N. Engl. J. Med.* **1999**, *340*, 115–126. [CrossRef] [PubMed]

62. Park, H.J.; Son, D.J.; Lee, C.W.; Choi, M.S.; Lee, U.S.; Song, H.S.; Lee, J.M.; Hong, J.T. Melittin inhibits inflammatory target gene expression and mediator generation via interaction with IkappaB kinase. *Biochem. Pharmacol.* **2007**, *73*, 237–247. [CrossRef] [PubMed]

63. Son, D.J.; Kang, J.; Kim, T.J.; Song, H.S.; Sung, K.J.; Yun do, Y.; Hong, J.T. Melittin, a major bioactive component of bee venom toxin, inhibits PDGF receptor beta-tyrosine phosphorylation and downstream intracellular signal transduction in rat aortic vascular smooth muscle cells. *J. Toxicol. Environ. Health A* **2007**, *70*, 1350–1355. [CrossRef] [PubMed]

64. Kim, S.J.; Park, J.H.; Kim, K.H.; Lee, W.R.; Kim, K.S.; Park, K.K. Melittin inhibits atherosclerosis in LPS/high-fat treated mice through atheroprotective actions. *J. Atheroscler. Thromb.* **2011**, *18*, 1117–1126. [CrossRef] [PubMed]

65. Gajski, G.; Garaj-Vrhovac, V. Melittin: A lytic peptide with anticancer properties. *Environ. Toxicol. Pharmacol.* **2013**, *36*, 697–705. [CrossRef] [PubMed]

66. Ladokhin, A.S.; White, S.H. Folding of amphipathic alpha-helices on membranes: Energetics of helix formation by melittin. *J. Mol. Biol.* **1999**, *285*, 1363–1369. [CrossRef] [PubMed]

67. Fukushima, N.; Kohno, M.; Kato, T.; Kawamoto, S.; Okuda, K.; Misu, Y.; Ueda, H. Melittin, a metabostatic peptide inhibiting Gs activity. *Peptides* **1998**, *19*, 811–819. [CrossRef] [PubMed]

68. Katoh, N. Inhibition by melittin of phosphorylation by protein kinase C of annexin I from cow mammary gland. *J. Vet. Med. Sci.* **2002**, *64*, 779–783. [CrossRef] [PubMed]

69. Gerst, J.E.; Salomon, Y. Inhibition by melittin and fluphenazine of melanotropin receptor function and adenylate cyclase in M2R melanoma cell membranes. *Endocrinology* **1987**, *121*, 1766–1772. [CrossRef] [PubMed]

70. Vitale, N.; Thierse, D.; Bader, M.F. Melittin promotes exocytosis in neuroendocrine cells through the activation of phospholipase A$_2$. *Regul. Pept.* **2010**, *165*, 111–116. [CrossRef] [PubMed]

71. Hait, W.N.; Grais, L.; Benz, C.; Cadman, E.C. Inhibition of growth of leukemic cells by inhibitors of calmodulin: Phenothiazines and melittin. *Cancer Chemother. Pharmacol.* **1985**, *14*, 202–205. [CrossRef] [PubMed]

72. Fadeel, B.; Gleiss, B.; Hogstrand, K.; Chandra, J.; Wiedmer, T.; Sims, P.J.; Henter, J.I.; Orrenius, S.; Samali, A. Phosphatidylserine exposure during apoptosis is a cell-type-specific event and does not correlate with plasma membrane phospholipid scramblase expression. *Biochem. Biophys. Res. Commun.* **1999**, *266*, 504–511. [CrossRef] [PubMed]

73. Song, C.C.; Lu, X.; Cheng, B.B.; Du, J.; Li, B.; Ling, C.Q. Effects of melittin on growth and angiogenesis of human hepatocellular carcinoma BEL-7402 cell xenografts in nude mice. *Ai Zheng* **2007**, *26*, 1315–1322. [PubMed]

74. Liu, S.; Yu, M.; He, Y.; Xiao, L.; Wang, F.; Song, C.; Sun, S.; Ling, C.; Xu, Z. Melittin prevents liver cancer cell metastasis through inhibition of the Rac1-dependent pathway. *Hepatology* **2008**, *47*, 1964–1973. [CrossRef] [PubMed]

75. Shin, J.M.; Jeong, Y.J.; Cho, H.J.; Park, K.K.; Chung, I.K.; Lee, I.K.; Kwak, J.Y.; Chang, H.W.; Kim, C.H.; Moon, S.K.; *et al.* Melittin suppresses HIF-1alpha/VEGF expression through inhibition of ERK and mTOR/p70S6K pathway in human cervical carcinoma cells. *PLoS One* **2013**, *8*, e69380. [CrossRef] [PubMed]

76. Huh, J.E.; Kang, J.W.; Nam, D.; Baek, Y.H.; Choi, D.Y.; Park, D.S.; Lee, J.D. Melittin suppresses VEGF-A-induced tumor growth by blocking VEGFR-2 and the COX-2-mediated MAPK signaling pathway. *J. Nat. Prod.* **2012**, *75*, 1922–1929. [CrossRef] [PubMed]

77. Yang, X.; Zhu, H.; Ge, Y.; Liu, J.; Cai, J.; Qin, Q.; Zhan, L.; Zhang, C.; Xu, L.; Liu, Z.; *et al.* Melittin enhances radiosensitivity of hypoxic head and neck squamous cell carcinoma by suppressing HIF-1alpha. *Tumour Biol.* **2014**, *35*, 10443–10448. [CrossRef] [PubMed]

78. Dunn, R.D.; Weston, K.M.; Longhurst, T.J.; Lilley, G.G.; Rivett, D.E.; Hudson, P.J.; Raison, R.L. Antigen binding and cytotoxic properties of a recombinant immunotoxin incorporating the lytic peptide, melittin. *Immunotechnology* **1996**, *2*, 229–240. [CrossRef] [PubMed]

79. Zhao, X.; Yu, Z.; Dai, W.; Yao, Z.; Zhou, W.; Zhou, W.; Zhou, J.; Yang, Y.; Zhu, Y.; Chen, S.; *et al.* Construction and characterization of an anti-asialoglycoprotein receptor single-chain variable-fragment-targeted melittin. *Biotechnol. Appl. Biochem.* **2011**, *58*, 405–411. [CrossRef] [PubMed]

80. Jin, H.; Li, C.; Li, D.; Cai, M.; Li, Z.; Wang, S.; Hong, X.; Shi, B. Construction and characterization of a CTLA-4-targeted scFv-melittin fusion protein as a potential immunosuppressive agent for organ transplant. *Cell Biochem. Biophys.* **2013**, *67*, 1067–1074. [CrossRef] [PubMed]

81. Holle, L.; Song, W.; Holle, E.; Wei, Y.; Wagner, T.; Yu, X. A matrix metalloproteinase 2 cleavable melittin/avidin conjugate specifically targets tumor cells *in vitro* and *in vivo*. *Int. J. Oncol.* **2003**, *22*, 93–98. [PubMed]

82. Holle, L.; Song, W.; Holle, E.; Wei, Y.; Li, J.; Wagner, T.E.; Yu, X. In vitro- and *in vivo*-targeted tumor lysis by an MMP2 cleavable melittin-LAP fusion protein. *Int. J. Oncol.* **2009**, *35*, 829–835. [PubMed]

83. Yang, L.; Cui, F.; Shi, K.; Cun, D.; Wang, R. Design of high payload PLGA nanoparticles containing melittin/sodium dodecyl sulfate complex by the hydrophobic ion-pairing technique. *Drug Dev. Ind. Pharm.* **2009**, *35*, 959–968. [CrossRef] [PubMed]

84. Hu, H.; Chen, D.; Liu, Y.; Deng, Y.; Yang, S.; Qiao, M.; Zhao, J.; Zhao, X. Target ability and therapy efficacy of immunoliposomes using a humanized antihepatoma disulfide-stabilized Fv fragment on tumor cells. *J. Pharm. Sci.* **2006**, *95*, 192–199. [CrossRef] [PubMed]

85. Barrajon-Catalan, E.; Menendez-Gutierrez, M.P.; Falco, A.; Carrato, A.; Saceda, M.; Micol, V. Selective death of human breast cancer cells by lytic immunoliposomes: Correlation with their HER2 expression level. *Cancer Lett.* **2010**, *290*, 192–203. [CrossRef] [PubMed]

86. Popplewell, J.F.; Swann, M.J.; Freeman, N.J.; McDonnell, C.; Ford, R.C. Quantifying the effects of melittin on liposomes. *Biochim. Biophys. Acta* **2007**, *1768*, 13–20. [CrossRef] [PubMed]

87. Soman, N.R.; Lanza, G.M.; Heuser, J.M.; Schlesinger, P.H.; Wickline, S.A. Synthesis and characterization of stable fluorocarbon nanostructures as drug delivery vehicles for cytolytic peptides. *Nano Lett.* **2008**, *8*, 1131–1136. [CrossRef] [PubMed]

88. Soman, N.R.; Baldwin, S.L.; Hu, G.; Marsh, J.N.; Lanza, G.M.; Heuser, J.E.; Arbeit, J.M.; Wickline, S.A.; Schlesinger, P.H. Molecularly targeted nanocarriers deliver the cytolytic peptide melittin specifically to tumor cells in mice, reducing tumor growth. *J. Clin. Investig.* **2009**, *119*, 2830–2842. [CrossRef] [PubMed]

89. Pan, H.; Soman, N.R.; Schlesinger, P.H.; Lanza, G.M.; Wickline, S.A. Cytolytic peptide nanoparticles ("NanoBees") for cancer therapy. *Wiley Interdiscip. Rev. Nanomed. Nanobiotechnol.* **2011**, *3*, 318–327. [CrossRef] [PubMed]

90. Huang, C.; Jin, H.; Qian, Y.; Qi, S.; Luo, H.; Luo, Q.; Zhang, Z. Hybrid melittin cytolytic peptide-driven ultrasmall lipid nanoparticles block melanoma growth *in vivo*. *ACS Nano* **2013**, *7*, 5791–5800. [CrossRef] [PubMed]

91. Luo, Q.; Zhang, Z.; Huang, C. Polypeptide carrying melittin, nanoparticle carrying melittin and use thereof. WO/2013/135103, 19 September 2013.

92. Zhang, Z.; Cao, W.; Jin, H.; Lovell, J.F.; Yang, M.; Ding, L.; Chen, J.; Corbin, I.; Luo, Q.; Zheng, G. Biomimetic nanocarrier for direct cytosolic drug delivery. *Angew. Chem. Int. Ed. Engl.* **2009**, *48*, 9171–9175. [CrossRef] [PubMed]

93. Varkouhi, A.K.; Scholte, M.; Storm, G.; Haisma, H.J. Endosomal escape pathways for delivery of biologicals. *J. Control. Release* **2011**, *151*, 220–228. [CrossRef] [PubMed]

94. Martin, M.E.; Rice, K.G. Peptide-guided gene delivery. *AAPS J.* **2007**, *9*, E18–E29. [CrossRef] [PubMed]

95. Sparrow, J.T.; Edwards, V.V.; Tung, C.; Logan, M.J.; Wadhwa, M.S.; Duguid, J.; Smith, L.C. Synthetic peptide-based DNA complexes for nonviral gene delivery. *Adv. Drug Deliv. Rev.* **1998**, *30*, 115–131. [CrossRef] [PubMed]

96. Rozema, D.B.; Ekena, K.; Lewis, D.L.; Loomis, A.G.; Wolff, J.A. Endosomolysis by masking of a membrane-active agent (EMMA) for cytoplasmic release of macromolecules. *Bioconjug. Chem.* **2003**, *14*, 51–57. [CrossRef] [PubMed]

97. Boeckle, S.; Fahrmeir, J.; Roedl, W.; Ogris, M.; Wagner, E. Melittin analogs with high lytic activity at endosomal pH enhance transfection with purified targeted PEI polyplexes. *J. Control. Release* **2006**, *112*, 240–248. [CrossRef] [PubMed]

98. Schlossbauer, A.; Dohmen, C.; Schaffert, D.; Wagner, E.; Bein, T. pH-responsive release of acetal-linked melittin from SBA-15 mesoporous silica. *Angew. Chem. Int. Ed. Engl.* **2011**, *50*, 6828–6830. [CrossRef] [PubMed]

99. Meyer, M.; Dohmen, C.; Philipp, A.; Kiener, D.; Maiwald, G.; Scheu, C.; Ogris, M.; Wagner, E. Synthesis and biological evaluation of a bioresponsive and endosomolytic siRNA-polymer conjugate. *Mol. Pharm.* **2009**, *6*, 752–762. [CrossRef] [PubMed]

100. Hou, K.K.; Pan, H.; Ratner, L.; Schlesinger, P.H.; Wickline, S.A. Mechanisms of nanoparticle-mediated siRNA transfection by melittin-derived peptides. *ACS Nano* **2013**, *7*, 8605–8615. [CrossRef] [PubMed]

101. Hou, K.K.; Pan, H.; Lanza, G.M.; Wickline, S.A. Melittin derived peptides for nanoparticle based siRNA transfection. *Biomaterials* **2013**, *34*, 3110–3119. [CrossRef] [PubMed]

102. Wooddell, C.I.; Rozema, D.B.; Hossbach, M.; John, M.; Hamilton, H.L.; Chu, Q.; Hegge, J.O.; Klein, J.J.; Wakefield, D.H.; Oropeza, C.E.; *et al.* Hepatocyte-targeted RNAi therapeutics for the treatment of chronic hepatitis B virus infection. *Mol. Ther.* **2013**, *21*, 973–985. [CrossRef] [PubMed]

103. Sebestyen, M.G.; Wong, S.C.; Trubetskoy, V.; Lewis, D.L.; Wooddell, C.I. Targeted *in vivo* delivery of siRNA and an endosome-releasing agent to hepatocytes. *Methods Mol. Biol.* **2015**, *1218*, 163–186. [PubMed]

104. Sponsor: Arrowhead Research Corporation. Study of ARC-520 in patients with chronic hepatitis B virus. NCT02065336, 26 January 2015.

105. Pease, J.H.; Wemmer, D.E. Solution structure of apamin determined by nuclear magnetic resonance and distance geometry. *Biochemistry* **1988**, *27*, 8491–8498. [CrossRef] [PubMed]

106. Habermann, E. Apamin. *Pharmacol. Ther.* **1984**, *25*, 255–270. [CrossRef] [PubMed]

107. Lazdunski, M. Apamin, a neurotoxin specific for one class of Ca^{2+}-dependent K^+ channels. *Cell Calcium* **1983**, *4*, 421–428. [CrossRef] [PubMed]

108. Lazdunski, M.; Fosset, M.; Hughes, M.; Mourre, C.; Romey, G.; Schmid-Antomarchi, H. The apamin-sensitive Ca^{2+}-dependent K^+ channel molecular properties, differentiation and endogenous ligands in mammalian brain. *Biochem. Soc. Symp.* **1985**, *50*, 31–42. [PubMed]

109. Habermann, E.; Fischer, K. Bee venom neurotoxin (apamin): Iodine labeling and characterization of binding sites. *Eur. J. Biochem.* **1979**, *94*, 355–364. [CrossRef] [PubMed]

110. Habermann, E.; Fischer, K. Apamin, a centrally acting neurotoxic peptide: Binding and actions. *Adv. Cytopharmacol.* **1979**, *3*, 387–394. [PubMed]

111. Lamy, C.; Goodchild, S.J.; Weatherall, K.L.; Jane, D.E.; Liegeois, J.F.; Seutin, V.; Marrion, N.V. Allosteric block of $K_{Ca}2$ channels by apamin. *J. Biol. Chem.* **2010**, *285*, 27067–27077. [CrossRef] [PubMed]

112. Adelman, J.P.; Maylie, J.; Sah, P. Small-conductance Ca^{2+}-activated K^+ channels: Form and function. *Annu. Rev. Physiol.* **2012**, *74*, 245–269. [CrossRef] [PubMed]

113. Wei, A.D.; Gutman, G.A.; Aldrich, R.; Chandy, K.G.; Grissmer, S.; Wulff, H. International Union of Pharmacology. LII. Nomenclature and molecular relationships of calcium-activated potassium channels. *Pharmacol. Rev.* **2005**, *57*, 463–472. [CrossRef] [PubMed]

114. Begenisich, T.; Nakamoto, T.; Ovitt, C.E.; Nehrke, K.; Brugnara, C.; Alper, S.L.; Melvin, J.E. Physiological roles of the intermediate conductance, Ca^{2+}-activated potassium channel Kcnn4. *J. Biol. Chem.* **2004**, *279*, 47681–47687. [CrossRef] [PubMed]

115. Schumacher, M.A.; Rivard, A.F.; Bachinger, H.P.; Adelman, J.P. Structure of the gating domain of a Ca^{2+}-activated K^+ channel complexed with Ca^{2+}/calmodulin. *Nature* **2001**, *410*, 1120–1124. [CrossRef] [PubMed]

116. Pedarzani, P.; McCutcheon, J.E.; Rogge, G.; Jensen, B.S.; Christophersen, P.; Hougaard, C.; Strobaek, D.; Stocker, M. Specific enhancement of SK channel activity selectively potentiates the afterhyperpolarizing current I_{AHP} and modulates the firing properties of hippocampal pyramidal neurons. *J. Biol. Chem.* **2005**, *280*, 41404–41411. [CrossRef] [PubMed]

117. Messier, C.; Mourre, C.; Bontempi, B.; Sif, J.; Lazdunski, M.; Destrade, C. Effect of apamin, a toxin that inhibits Ca^{2+}-dependent K^+ channels, on learning and memory processes. *Brain Res.* **1991**, *551*, 322–326. [CrossRef] [PubMed]

118. Deschaux, O.; Bizot, J.C.; Goyffon, M. Apamin improves learning in an object recognition task in rats. *Neurosci. Lett.* **1997**, *222*, 159–162. [CrossRef] [PubMed]

119. Ikonen, S.; Schmidt, B.; Riekkinen, P., Jr. Apamin improves spatial navigation in medial septal-lesioned mice. *Eur. J. Pharmacol.* **1998**, *347*, 13–21. [CrossRef] [PubMed]

120. Ikonen, S.; Riekkinen, P., Jr. Effects of apamin on memory processing of hippocampal-lesioned mice. *Eur. J. Pharmacol.* **1999**, *382*, 151–156. [CrossRef] [PubMed]

121. Inan, S.Y.; Aksu, F.; Baysal, F. The effects of some K^+ channel blockers on scopolamine- or electroconvulsive shock-induced amnesia in mice. *Eur. J. Pharmacol.* **2000**, *407*, 159–164. [CrossRef] [PubMed]

122. Stackman, R.W.; Hammond, R.S.; Linardatos, E.; Gerlach, A.; Maylie, J.; Adelman, J.P.; Tzounopoulos, T. Small conductance Ca^{2+}-activated K^+ channels modulate synaptic plasticity and memory encoding. *J. Neurosci.* **2002**, *22*, 10163–10171. [PubMed]

123. Deschaux, O.; Bizot, J.C. Apamin produces selective improvements of learning in rats. *Neurosci. Lett.* **2005**, *386*, 5–8. [CrossRef] [PubMed]

124. Brennan, A.R.; Dolinsky, B.; Vu, M.A.; Stanley, M.; Yeckel, M.F.; Arnsten, A.F. Blockade of IP3-mediated SK channel signaling in the rat medial prefrontal cortex improves spatial working memory. *Learn. Mem.* **2008**, *15*, 93–96. [CrossRef] [PubMed]

125. Vick, K.A., IV; Guidi, M.; Stackman, R.W., Jr. *In vivo* pharmacological manipulation of small conductance Ca^{2+}-activated K^+ channels influences motor behavior, object memory and fear conditioning. *Neuropharmacology* **2010**, *58*, 650–659.

126. Kallarackal, A.J.; Simard, J.M.; Bailey, A.M. The effect of apamin, a small conductance calcium activated potassium (SK) channel blocker, on a mouse model of neurofibromatosis 1. *Behav. Brain Res.* **2013**, *237*, 71–75. [CrossRef] [PubMed]

127. Dauer, W.; Przedborski, S. Parkinson's disease: Mechanisms and models. *Neuron* **2003**, *39*, 889–909. [CrossRef] [PubMed]

128. Cilia, R.; Cereda, E.; Klersy, C.; Canesi, M.; Zecchinelli, A.L.; Mariani, C.B.; Tesei, S.; Sacilotto, G.; Meucci, N.; Zini, M.; *et al.* Parkinson's disease beyond 20 years. *J. Neurol. Neurosurg. Psychiatry* **2014**. [CrossRef]

129. Doo, A.R.; Kim, S.T.; Kim, S.N.; Moon, W.; Yin, C.S.; Chae, Y.; Park, H.K.; Lee, H.; Park, H.J. Neuroprotective effects of bee venom pharmaceutical acupuncture in acute 1-methyl-4-phenyl-1,2,3,6-tetrahydropyridine-induced mouse model of Parkinson's disease. *Neurol. Res.* **2010**, *32*, 88–91. [CrossRef] [PubMed]

130. Kim, J.I.; Yang, E.J.; Lee, M.S.; Kim, Y.S.; Huh, Y.; Cho, I.H.; Kang, S.; Koh, H.K. Bee venom reduces neuroinflammation in the MPTP-induced model of Parkinson's disease. *Int. J. Neurosci.* **2011**, *121*, 209–217. [CrossRef] [PubMed]

131. Salthun-Lassalle, B.; Hirsch, E.C.; Wolfart, J.; Ruberg, M.; Michel, P.P. Rescue of mesencephalic dopaminergic neurons in culture by low-level stimulation of voltage-gated sodium channels. *J. Neurosci.* **2004**, *24*, 5922–5930. [CrossRef] [PubMed]

132. Toulorge, D.; Guerreiro, S.; Hild, A.; Maskos, U.; Hirsch, E.C.; Michel, P.P. Neuroprotection of midbrain dopamine neurons by nicotine is gated by cytoplasmic Ca^{2+}. *FASEB J.* **2011**, *25*, 2563–2573. [CrossRef] [PubMed]

133. Alvarez-Fischer, D.; Noelker, C.; Vulinovic, F.; Grunewald, A.; Chevarin, C.; Klein, C.; Oertel, W.H.; Hirsch, E.C.; Michel, P.P.; Hartmann, A. Bee venom and its component apamin as neuroprotective agents in a Parkinson disease mouse model. *PLoS One* **2013**, *8*, e61700. [CrossRef] [PubMed]

134. Hartmann, A.; Bonnet, A.M.; Schüpbach, M. Medicament for treating Parkinson's Disease. US8232249 B2, 31 January 2012.

135. Thomas, N.C.; Justin, D.O.L. Composition for Treatins Parkinsin's Disease. WO2013083574 A1, 13 June 2013.

136. Delgado, M.C.; Pitt, B. Composition and Methods for Preserving Red Blood Cells. WO2008089337 A1, 24 July 2008.

137. Ballabh, P.; Braun, A.; Nedergaard, M. The blood-brain barrier: An overview: Structure, regulation, and clinical implications. *Neurobiol. Dis.* **2004**, *16*, 1–13. [CrossRef] [PubMed]

138. McCarthy, D.J.; Malhotra, M.; O'Mahony, A.M.; Cryan, J.F.; O'Driscoll, C.M. Nanoparticles and the blood-brain barrier: Advancing from in-vitro models towards therapeutic significance. *Pharm. Res.* **2014**, *32*, 1161–1185. [CrossRef] [PubMed]

139. Cheng-Raude, D.; Treloar, M.; Habermann, E. Preparation and pharmacokinetics of labeled derivatives of apamin. *Toxicon* **1976**, *14*, 467–476. [CrossRef] [PubMed]

140. Wu, J.; Jiang, H.; Bi, Q.; Luo, Q.; Li, J.; Zhang, Y.; Chen, Z.; Li, C. Apamin-mediated actively targeted drug delivery for treatment of spinal cord injury: More than just a concept. *Mol. Pharm.* **2014**, *11*, 3210–3222. [CrossRef] [PubMed]

141. Cosand, W.L.; Merrifield, R.B. Concept of internal structural controls for evaluation of inactive synthetic peptide analogs: Synthesis of [Orn13,14]apamin and its guanidination to an apamin derivative with full neurotoxic activity. *Proc. Natl. Acad. Sci. USA* **1977**, *74*, 2771–2775. [CrossRef] [PubMed]

142. Oller-Salvia, B.; Teixido, M.; Giralt, E. From venoms to BBB shuttles: Synthesis and blood-brain barrier transport assessment of apamin and a nontoxic analog. *Biopolymers* **2013**, *100*, 675–686. [CrossRef] [PubMed]

143. Monteiro, M.C.; Romao, P.R.; Soares, A.M. Pharmacological perspectives of wasp venom. *Protein Pept. Lett.* **2009**, *16*, 944–952. [CrossRef] [PubMed]

144. Pfeiffer, D.R.; Gudz, T.I.; Novgorodov, S.A.; Erdahl, W.L. The peptide mastoparan is a potent facilitator of the mitochondrial permeability transition. *J. Biol. Chem.* **1995**, *270*, 4923–4932. [CrossRef] [PubMed]

145. Higashijima, T.; Uzu, S.; Nakajima, T.; Ross, E.M. Mastoparan, a peptide toxin from wasp venom, mimics receptors by activating GTP-binding regulatory proteins (G proteins). *J. Biol. Chem.* **1988**, *263*, 6491–6494. [PubMed]

146. Katsu, T.; Kuroko, M.; Morikawa, T.; Sanchika, K.; Yamanaka, H.; Shinoda, S.; Fujita, Y. Interaction of wasp venom mastoparan with biomembranes. *Biochim. Biophys. Acta* **1990**, *1027*, 185–190. [CrossRef] [PubMed]

147. Nakahata, N.; Abe, M.T.; Matsuoka, I.; Nakanishi, H. Mastoparan inhibits phosphoinositide hydrolysis via pertussis toxin-insensitive G-protein in human astrocytoma cells. *FEBS Lett.* **1990**, *260*, 91–94. [CrossRef] [PubMed]

148. Ozaki, Y.; Matsumoto, Y.; Yatomi, Y.; Higashihara, M.; Kariya, T.; Kume, S. Mastoparan, a wasp venom, activates platelets via pertussis toxin-sensitive GTP-binding proteins. *Biochem. Biophys. Res. Commun.* **1990**, *170*, 779–785. [CrossRef] [PubMed]

149. Weingarten, R.; Ransnas, L.; Mueller, H.; Sklar, L.A.; Bokoch, G.M. Mastoparan interacts with the carboxyl terminus of the alpha subunit of Gi. *J. Biol. Chem.* **1990**, *265*, 11044–11049. [PubMed]

150. Rocha, T.; de Souza, B.M.; Palma, M.S.; da Cruz-Hofling, M.A. Myotoxic effects of mastoparan from *Polybia paulista* (Hymenoptera, Epiponini) wasp venom in mice skeletal muscle. *Toxicon* **2007**, *50*, 589–599. [CrossRef] [PubMed]

151. Dongol, T.; Dhananjaya, B.L.; Shrestha, R.K.; Aryal, G. Pharmacological and immunological properties of wasp venom. In *Pharmacology and Therapeutics*; Joghi, S., Gowder, T., Eds.; INTECH: Rijeka, Croatia, 2014; pp. 47–81.

152. Cabrera, M.P.; Alvares, D.S.; Leite, N.B.; de Souza, B.M.; Palma, M.S.; Riske, K.A.; Neto, J.R. New insight into the mechanism of action of wasp mastoparan peptides: Lytic activity and clustering observed with giant vesicles. *Langmuir* **2011**, *27*, 10805–10813. [CrossRef] [PubMed]

153. Leite, N.B.; da Costa, L.C.; Dos Santos Alvares, D.; Dos Santos Cabrera, M.P.; de Souza, B.M.; Palma, M.S.; Ruggiero Neto, J. The effect of acidic residues and amphipathicity on the lytic activities of mastoparan peptides studied by fluorescence and CD spectroscopy. *Amino Acids* **2011**, *40*, 91–100. [CrossRef] [PubMed]

154. Yamamoto, T.; Ito, M.; Kageyama, K.; Kuwahara, K.; Yamashita, K.; Takiguchi, Y.; Kitamura, S.; Terada, H.; Shinohara, Y. Mastoparan peptide causes mitochondrial permeability transition not by interacting with specific membrane proteins but by interacting with the phospholipid phase. *FEBS J.* **2014**, *281*, 3933–3944. [CrossRef] [PubMed]

155. Lin, C.H.; Hou, R.F.; Shyu, C.L.; Shia, W.Y.; Lin, C.F.; Tu, W.C. In vitro activity of mastoparan-AF alone and in combination with clinically used antibiotics against multiple-antibiotic-resistant *Escherichia coli* isolates from animals. *Peptides* **2012**, *36*, 114–120. [CrossRef] [PubMed]

156. Mukai, H.; Suzuki, Y.; Kiso, Y.; Munekata, E. Elucidation of structural requirements of mastoparan for mast cell activation-toward the comprehensive prediction of cryptides acting on mast cells. *Protein Pept. Lett.* **2008**, *15*, 931–937. [CrossRef] [PubMed]

157. Jones, S.; Howl, J. Charge delocalisation and the design of novel mastoparan analogues: Enhanced cytotoxicity and secretory efficacy of [Lys[5], Lys[8], Aib[10]]MP. *Regul. Pept.* **2004**, *121*, 121–128. [CrossRef] [PubMed]

158. Jones, S.; Howl, J. Enantiomer-specific bioactivities of peptidomimetic analogues of mastoparan and mitoparan: Characterization of inverso mastoparan as a highly efficient cell penetrating peptide. *Bioconjug. Chem.* **2012**, *23*, 47–56. [CrossRef] [PubMed]

159. Avram, S.; Buiu, C.; Borcan, F.; Milac, A.L. More effective antimicrobial mastoparan derivatives, generated by 3D-QSAR-Almond and computational mutagenesis. *Mol. Biosyst.* **2012**, *8*, 587–594. [CrossRef] [PubMed]

160. Henriksen, J.R.; Etzerodt, T.; Gjetting, T.; Andresen, T.L. Side chain hydrophobicity modulates therapeutic activity and membrane selectivity of antimicrobial peptide mastoparan-X. *PLoS One* **2014**, *9*, e91007. [CrossRef] [PubMed]

161. Higashijima, T.; Burnier, J.; Ross, E.M. Regulation of Gi and Go by mastoparan, related amphiphilic peptides, and hydrophobic amines. Mechanism and structural determinants of activity. *J. Biol. Chem.* **1990**, *265*, 14176–14186. [PubMed]

162. Etzerodt, T.; Henriksen, J.R.; Rasmussen, P.; Clausen, M.H.; Andresen, T.L. Selective acylation enhances membrane charge sensitivity of the antimicrobial peptide mastoparan-x. *Biophys. J.* **2011**, *100*, 399–409. [CrossRef] [PubMed]

163. Guo, Y.B.; Zheng, J.; Lv, G.F.; Wei, G.; Wang, L.X.; Xiao, G.X. Experimental study on the antagonistic activity of cationic multi-peptide mastoparan-1 against lipopolysaccharide. *Zhonghua Shao Shang Za Zhi*, 2005; 21, 189–192. [PubMed]

164. Guo, Y.; Zheng, J.; Zhou, H.; Lv, G.; Wang, L.; Wei, G.; Lu, Y. A synthesized cationic tetradecapeptide from hornet venom kills bacteria and neutralizes lipopolysaccharide *in vivo* and *in vitro*. *Biochem. Pharmacol.* **2005**, *70*, 209–219. [CrossRef] [PubMed]

165. Guo, Y.B.; Zheng, Q.Y.; Chen, J.H.; Cai, S.F.; Cao, H.W.; Zheng, J.; Xiao, G.X. Effect of mastoparan-1 on lipopolysaccharide-induced acute hepatic injury in mice. *Zhonghua Shao Shang Za Zhi*, 2009; 25, 53–56. [PubMed]

166. Sample, C.J.; Hudak, K.E.; Barefoot, B.E.; Koci, M.D.; Wanyonyi, M.S.; Abraham, S.; Staats, H.F.; Ramsburg, E.A. A mastoparan-derived peptide has broad-spectrum antiviral activity against enveloped viruses. *Peptides* **2013**, *48*, 96–105. [CrossRef] [PubMed]

167. Wu, T.M.; Li, M.L. The cytolytic action of all-D mastoparan M on tumor cell lines. *Int. J. Tissue React.* **1999**, *21*, 35–42. [PubMed]

168. Jones, S.; Martel, C.; Belzacq-Casagrande, A.S.; Brenner, C.; Howl, J. Mitoparan and target-selective chimeric analogues: Membrane translocation and intracellular redistribution induces mitochondrial apoptosis. *Biochim. Biophys. Acta* **2008**, *1783*, 849–863. [CrossRef] [PubMed]

169. Danilenko, M.; Worland, P.; Carlson, B.; Sausville, E.A.; Sharoni, Y. Selective effects of mastoparan analogs: Separation of G-protein-directed and membrane-perturbing activities. *Biochem. Biophys. Res. Commun.* **1993**, *196*, 1296–1302. [CrossRef] [PubMed]

170. Yamada, Y.; Shinohara, Y.; Kakudo, T.; Chaki, S.; Futaki, S.; Kamiya, H.; Harashima, H. Mitochondrial delivery of mastoparan with transferrin liposomes equipped with a pH-sensitive fusogenic peptide for selective cancer therapy. *Int. J. Pharm.* **2005**, *303*, 1–7. [CrossRef] [PubMed]

171. De Azevedo, R.A.; Figueiredo, C.R.; Ferreira, A.K.; Matsuo, A.L.; Massaoka, M.H.; Girola, N.; Auada, A.V.; Farias, C.F.; Pasqualoto, K.F.; Rodrigues, C.P.; *et al.* Mastoparan induces apoptosis in B16F10-Nex2 melanoma cells via the intrinsic mitochondrial pathway and displays antitumor activity *in vivo*. *Peptides* **2014**. [CrossRef]

172. Ruoslahti, E.; Pierschbacher, M.D. New perspectives in cell adhesion: RGD and integrins. *Science* **1987**, *238*, 491–497. [CrossRef] [PubMed]

173. Wang, R.; Ni, J.; Yang, Z.; Song, J. Antimicrobial peptide AMitP with acid activation property and synthesis and application thereof in preparation of anti-tumor medicaments. CN102718844 B, 2 April 2014.

174. Moreno, M.; Zurita, E.; Giralt, E. Delivering wasp venom for cancer therapy. *J. Control. Release* **2014**, *182*, 13–21. [CrossRef] [PubMed]

175. Jones, A.T.; Sayers, E.J. Cell entry of cell penetrating peptides: Tales of tails wagging dogs. *J. Control. Release* **2012**, *161*, 582–591. [CrossRef] [PubMed]

176. Pujals, S.; Giralt, E. Proline-rich, amphipathic cell-penetrating peptides. *Adv. Drug Deliv. Rev.* **2008**, *60*, 473–484. [CrossRef] [PubMed]

177. Martin, I.; Teixido, M.; Giralt, E. Intracellular fate of peptide-mediated delivered cargoes. *Curr. Pharm. Des.* **2013**, *19*, 2924–2942. [CrossRef] [PubMed]

178. Sharma, G.; Modgil, A.; Zhong, T.; Sun, C.; Singh, J. Influence of short-chain cell-penetrating peptides on transport of Doxorubicin encapsulating receptor-targeted liposomes across brain endothelial barrier. *Pharm. Res.* **2014**, *31*, 1194–1209. [CrossRef] [PubMed]

179. Pooga, M.; Hallbrink, M.; Zorko, M.; Langel, U. Cell penetration by transportan. *FASEB J.* **1998**, *12*, 67–77. [PubMed]

180. Wierzbicki, P.M.; Kogut, M.; Ruczynski, J.; Siedlecka-Kroplewska, K.; Kaszubowska, L.; Rybarczyk, A.; Alenowicz, M.; Rekowski, P.; Kmiec, Z. Protein and siRNA delivery by transportan and transportan 10 into colorectal cancer cell lines. *Folia Histochem. Cytobiol.* **2014**, *52*, 270–280. [CrossRef] [PubMed]

181. Fanghanel, S.; Wadhwani, P.; Strandberg, E.; Verdurmen, W.P.; Burck, J.; Ehni, S.; Mykhailiuk, P.K.; Afonin, S.; Gerthsen, D.; Komarov, I.V.; *et al.* Structure analysis and conformational transitions of the cell penetrating peptide transportan 10 in the membrane-bound state. *PLoS One* **2014**, *9*, e99653. [CrossRef] [PubMed]

182. Pooga, M.; Kut, C.; Kihlmark, M.; Hallbrink, M.; Fernaeus, S.; Raid, R.; Land, T.; Hallberg, E.; Bartfai, T.; Langel, U. Cellular translocation of proteins by transportan. *FASEB J.* **2001**, *15*, 1451–1453. [PubMed]

© 2015 by the authors; licensee MDPI, Basel, Switzerland. This article is an open access article distributed under the terms and conditions of the Creative Commons Attribution (CC-BY) license (http://creativecommons.org/licenses/by/4.0/).

Editorial

An Introduction to the *Toxins* Special Issue on "Bee and Wasp Venoms: Biological Characteristics and Therapeutic Application"

Sok Cheon Pak

School of Biomedical Sciences, Charles Sturt University, Bathurst, NSW 2795, Australia; spak@csu.edu.au;
Tel.: +61-2-6338-4952; Fax: +61-2-6338-4993

Academic Editor: Bryan Grieg Fry
Received: 14 October 2016; Accepted: 19 October 2016; Published: 28 October 2016

Venoms, especially bee venom, have been used since ancient times as a healing treatment for various disorders. The therapeutic value of honey bee venom to improve the quality of life of patients has been acknowledged for over a hundred years. Modern approaches of venomics have allowed for the discovery of venom constituents that have proven to be of pharmacological significance and have opened the way to optimization of therapeutic strategies through the use of active components such as melittin and apamin. Subsequently, the application scope of honey bee venom has been expanding from conventional antinociceptive effect to degenerative diseases of the nervous system. This seems to be due to the properties of venom enzymes and peptides for their natural stability as injectable solutes, their effectiveness in reaching targeted tissues, and their ability to synergize their actions by enhancing cell–cell interactions. Expansion of the therapeutic application of bee and wasp venoms has advanced particularly far in recent years, so this is an opportune time to present this Special Issue on bee and wasp venoms, their biological characteristics and therapeutic application.

The venoms of bees and wasps are complex mixtures of biologically active proteins and peptides, such as phospholipases, hyaluronidase, phosphatase, α-glucosidase, serotonin, histamine, dopamine, noradrenaline, and adrenaline. However, melittin, apamin, and mast cell degranulating peptide are found only in bees, while mastoparan and bradykinin are exclusive to wasps. The recent review article on bee and wasp venoms for their potential therapeutic and biotechnological applications in biomedicine focuses on three major peptides, namely melittin, apamin, and mastoparan [1]. While mastoparan has been studied for its antimicrobial, anti-viral, and anti-tumor properties, melittin and apamin have a broad spectrum of therapeutic applications. To aid the reader in cross-referencing these applications, I present here a listing of the bee venom components used for different disease types and the latest references (Table 1). Interestingly, learning enhancement in animals was benefited exclusively from apamin, probably due to its highly specific mode of action in the brain.

Table 1. Application of honey bee venom components on different disease types.

Disease Type	Component	Reference
Parkinson's disease	Bee venom (BV)	[2]
	Apamin	[2]
	BV acupuncture	[3]
Amyotrophic lateral sclerosis (ALS)	BV	[4]
	Melittin	[5]
	BV acupuncture	[4,6]
Multiple sclerosis	BV	[7]
Cancer	BV	[8]
	Melittin	[9]
Liver fibrosis	BV	[10]
	Apamin	[11]
	Melittin	[12]
	PLA$_2$	[13]
Atherosclerosis	BV	[14]
	Apamin	[15]
	Melittin	[16]
Skin disease (acne vulgaris)	BV	[17]
	Melittin	[18]
Skin disease (atopic dermatitis)	BV	[19]
Learning deficit	Apamin	[20]
Pain	BV	[21]
	BV acupuncture	[22]
Lupus nephritis	BV	[23]

This Special Issue features eleven research papers, for which brief synopses follow. Jung et al. [24] provide details on the underlying mechanism of bee venom's effect against neuronal cell death. Mitochondrial dysfunction of cells was induced by rotenone which is an inhibitor of the mitochondrial respiratory chain complex and known to induce apoptosis via activation of the caspase-3 pathway. Pre-treatment with bee venom enhanced cell viability and ameliorated mitochondrial impairment in a rotenone-treated cellular model. Moreover, bee venom treatment inhibited the activation of JNK signaling, cleaved caspase-3 related cell death, and increased ERK phosphorylaton involved in cell survival in motor neuron cells.

Kim et al. [25] investigated whether melittin provides inhibition on cholangitis and biliary fibrosis in mice. Authors used a well-established animal model for cholangitis and biliary fibrosis, treated with a chronic feeding of 3,5-diethoxycarbonyl-1,4-dihydrocollidine (DDC). DDC feeding led to increased serum markers of hepatic injury, ductular reaction, induction of pro-inflammatory cytokines, and biliary fibrosis. However, melittin treatment attenuated hepatic function markers, ductular reaction, the reactive phenotype of cholangiocytes, cholangitis, and biliary fibrosis.

Cai et al. [6] present the latest insight into the effect of bee venom acupuncture on ALS. They used hSOD1^{G93A} transgenic mice as an ALS animal model and investigated the effect of bee venom in the central nervous system (CNS) and muscle. Bee venom acupuncture at ST36 enhanced motor function and decreased motor neuron death in the spinal cord compared with that observed in transgenic mice injected ip with bee venom. Furthermore, bee venom acupuncture eliminated signaling downstream of inflammatory proteins such as TLR4, CD14, and TNF-α in the spinal cord, confirming bee venom acupuncture as a chemical stimulant to engage the endogenous immune modulatory system in the CNS, at least in an animal model of ALS. The same group of researchers contributed another research paper [4] to this Special Issue. Given that immune dysfunction of organs from neuroinflammation is a consistent hallmark in ALS, bee venom acupuncture reduced pro-inflammatory proteins in the liver, spleen, and kidney and increased immune response. It was noted that bee venom treatment through an acupoint was more effective than an ip administration of bee venom.

Kang et al. [22] discuss advanced knowledge of the anti-nociceptive efficacy of bee venom. They designed the study to examine the potential anti-nociceptive effect of repetitive bee venom treatment in the development of below-level neuropathic pain in spinal cord injury rats which was induced by spinal hemisection. Hemisection of the rat spinal cord at thoracic level 13 produced prominent mechanical allodynia and thermal hyperalgesia. Repetitive bee venom acupuncture into the Zusanli acupoint on the same side as the spinal hemisection (ipsilateral side) twice a day from 15 to 20 days post-surgery suppressed pain behavior in the ipsilateral hind paw. A spinal hemisection-induced increase in spinal glia expression in terms of glial fibrillary acidic protein and glial ionized calcium-binding adaptor protein 1 was also hindered by repetitive bee venom acupuncture in the ipsilateral dorsal spinal cord. Moreover, bee venom acupuncture facilitated motor function recovery of affected hindlimb.

Li et al. [26] provide therapeutic options for the management of oxaliplatin-induced neuropathic pain. Oxaliplatin produces cold and mechanical hypersensitivity as side effects when used in the treatment of colorectal carcinoma. Their study is a follow-up of a previous evaluation of bee venom alleviating oxaliplatin-induced cold allodynia in rats via noradrenergic and serotonergic analgesic pathways. In the study, authors investigated whether phospholipase A_2 attenuates oxaliplatin-induced cold and mechanical allodynia in mice. While the significant allodynia signs were observed from one day after an oxaliplatin injection, daily ip administration of phospholipase A_2 for five consecutive days markedly attenuated cold and mechanical allodynia through the activation of noradrenergic system.

A detailed underlying mechanism of bee venom in ameliorating the renal fibrosis is presented by An et al. [27]. In an animal model of unilateral ureteral obstruction (UUO) for the development of progressive renal fibrosis, bee venom treatment markedly reduced the increased number of infiltrated inflammatory cells with UUO in the kidney tissues. The expression levels of TNF-α, IL-1β, TGF-β1, and fibronectin were significantly reduced in BV-treated mice compared with UUO mice. In addition, the expression of α-SMA was markedly withdrawn after treatment with bee venom. These data suggest that bee venom attenuates renal fibrosis and reduces inflammatory responses by suppression of multiple growth factor-mediated pro-fibrotic genes.

Lee et al. [28] introduce the application of bee venom in porcine reproductive and respiratory syndrome (PRRS) which is a chronic and immunosuppressive viral disease that is responsible for substantial economic losses in the swine industry. Utilizing the immunomodulatory property of bee venom, the study aimed at evaluating the effects of bee venom on the immune response and viral clearance during the early stage of infection with PRRS virus. Bee venom was administered subcutaneously via nasal, neck, and rectal routes, and the pigs were then inoculated with PRRS virus. In experimentally challenged pigs with virus, the viral genome load in the serum, lung, bronchial lymph nodes and tonsil was decreased as was the severity of interstitial pneumonia with bee venom administration. Furthermore, bee venom increased the levels of Th1 cytokines (IFN-γ and IL-12), along with the upregulation of pro-inflammatory cytokines (TNF-α and IL-1β).

Danneels et al. [29] confirm the anti-inflammatory action of *Nasonia vitripennis* venom and the potential anti-cancer role of wasp venom is displayed by the regulation of NF-κB pathway. The possibility of *Nasonia vitripennis* venom as therapeutic agent in some cancers is thus suggested.

Furthermore, Qian et al. [30] characterized two Kazal-type serine protease inhibitors (KSPIs) molecularly in *Nasonia vitripennis*. Most *NvKSPI-1* and *NvKSPI-2* mRNAs were expressed in the venom apparatus. The *NvKSPI-1* and *NvKSPI-2* genes were cloned into vector, and the recombinant products fused with glutathione S-transferase (GST) were purified. When tested against serine protease inhibitors, only GST-*NvKSPI-1* inhibited the activity of trypsin. In addition, both GST-*NvKSPI-1* and GST-*NvKSPI-2* inhibited prophenoloxidase activation in host hemolymph.

Matysiak et al. [31] confirm the influence of bee stings on the serum peptidome profile that broadens the understanding of the human organism's response to venom thanks to the mass spectrometry-based technique.

Additionally in this issue, there are five review articles. Lee et al. [32] provide current understanding of the mechanisms of the anti-inflammatory properties of bee venom and its components in the treatment of liver fibrosis, atherosclerosis, and skin disease.

Silva et al. [33] present the latest understanding of the mechanisms of action and future prospects regarding the use of new drugs derived from wasp and bee venom in the treatment of some neurodegenerative disorders such as Alzheimer's disease, Parkinson's disease, epilepsy, multiple sclerosis, and ALS. Given the severe adverse reactions of crude wasp venom, denatured wasp venom shows an antiepileptic effect by interacting with GABA receptors. Bradykinin isolated from wasp venom protects against ischemic brain injury and promotes neuronal survival.

Hwang et al. [34] discuss the scientific evidence of the therapeutic effects of bee venom and its components on allergic, autoimmune, inflammatory, and neurological diseases. Due to both therapeutic and allergy causing effects of bee venom, the optimal dose and treatment method without adverse reactions should be determined in each disease.

Moreau and Asgari [35] present the latest research on constituents from parasitoid wasp venom with an emphasis on their biological function, applications, and new approaches used in venom studies.

Finally, Moreno and Giralt [1] report on the practical applications of three valuable peptides from bee and wasp venoms. The antimicrobial properties of melittin from bee venom and mastoparan from wasp venom have been the most studied and developed among the components of venoms. Their mode of action on membrane has been developed into antimicrobial peptides, and they can be accepted as a unique solution to the growing incidence of antibiotic resistance.

It is my hope that this Special Issue will be a source for many years to those interested in the agricultural and pharmacological relevance of bee and wasp venoms. The currently available state-of-the-art approaches in proteomics and transcriptomics provide us with intricate mechanisms to further the application of venoms into diverse biomedicine.

Conflicts of Interest: The author declares no conflict of interest.

References

1. Moreno, M.; Giralt, E. Three valuable peptides from bee and wasp venoms for therapeutic and biotechnological use: Melittin, apamin and mastoparan. *Toxins* **2015**, *7*, 1126–1150. [CrossRef] [PubMed]
2. Alvarez-Fischer, D.; Noelker, C.; Vulinović, F.; Grünewald, A.; Chevarin, C.; Klein, C.; Oertel, W.H.; Hirsch, E.C.; Michel, P.P.; Hartmann, A. Bee venom and its component apamin as neuroprotective agents in a Parkinson disease mouse model. *PLoS ONE* **2013**, *8*, e61700. [CrossRef] [PubMed]
3. Khalil, W.K.; Assaf, N.; ElShebiney, S.A.; Salem, N.A. Neuroprotective effects of bee venom acupuncture therapy against rotenone-induced oxidative stress and apoptosis. *Neurochem. Int.* **2015**, *80*, 79–86. [CrossRef] [PubMed]
4. Lee, S.H.; Choi, S.M.; Yang, E.J. Bee venom acupuncture augments anti-inflammation in the peripheral organs of hSOD1G93A transgenic mice. *Toxins* **2015**, *7*, 2835–2844. [CrossRef] [PubMed]
5. Lee, S.H.; Choi, S.M.; Yang, E.J. Melittin ameliorates the inflammation of organs in an amyotrophic lateral sclerosis animal model. *Exp. Neurobiol.* **2014**, *23*, 86–92. [CrossRef] [PubMed]
6. Cai, M.; Choi, S.M.; Yang, E.J. The effects of bee venom acupuncture on the central nervous system and muscle in an animal hSOD1G93A mutant. *Toxins* **2015**, *7*, 846–858. [CrossRef] [PubMed]
7. Karimi, A.; Ahmadi, F.; Parivar, K.; Nabiuni, M.; Haghighi, S.; Imani, S.; Afrouzi, H. Effect of honey bee venom on lewis rats with experimental allergic encephalomyelitis, a model for multiple sclerosis. *Iran. J. Pharm. Res.* **2012**, *11*, 671–678. [PubMed]
8. Choi, K.E.; Hwang, C.J.; Gu, S.M.; Park, M.H.; Kim, J.H.; Park, J.H.; Ahn, Y.J.; Kim, J.Y.; Song, M.J.; Song, H.S.; et al. Cancer cell growth inhibitory effect of bee venom via increase of death receptor 3 expression and inactivation of NF-kappa B in NSCLC cells. *Toxins* **2014**, *6*, 2210–2228. [CrossRef] [PubMed]

9.   Jo, M.; Park, M.H.; Kollipara, P.S.; An, B.J.; Song, H.S.; Han, S.B.; Kim, J.H.; Song, M.J.; Hong, J.T. Anti-cancer effect of bee venom toxin and melittin in ovarian cancer cells through induction of death receptors and inhibition of JAK2/STAT3 pathway. *Toxicol. Appl. Pharmacol.* **2012**, *258*, 72–81. [CrossRef] [PubMed]

10.   Kim, K.H.; Kum, Y.S.; Park, Y.Y.; Park, J.H.; Kim, S.J.; Lee, W.R.; Lee, K.G.; Han, S.M.; Park, K.K. The protective effect of bee venom against ethanol-induced hepatic injury via regulation of the mitochondria-related apoptotic pathway. *Basic Clin. Pharmacol. Toxicol.* **2010**, *107*, 619–624. [CrossRef] [PubMed]

11.   Lee, W.R.; Kim, K.H.; An, H.J.; Kim, J.Y.; Lee, S.J.; Han, S.M.; Pak, S.C.; Park, K.K. Apamin inhibits hepatic fibrosis through suppression of transforming growth factor β1-induced hepatocyte epithelial-mesenchymal transition. *Biochem. Biophys. Res. Commun.* **2014**, *450*, 195–201. [CrossRef] [PubMed]

12.   Park, J.H.; Lee, W.R.; Kim, H.S.; Han, S.M.; Chang, Y.C.; Park, K.K. Protective effects of melittin on tumor necrosis factor-α induced hepatic damage through suppression of apoptotic pathway and nuclear factor-kappa B activation. *Exp. Biol. Med.* **2014**, *239*, 1705–1714. [CrossRef] [PubMed]

13.   Kim, H.; Keum, D.J.; won Kwak, J.; Chung, H.S.; Bae, H. Bee venom phospholipase A$_2$ protects against acetaminophen-induced acute liver injury by modulating regulatory T cells and IL-10 in mice. *PLoS ONE* **2014**, *9*, 27. [CrossRef] [PubMed]

14.   Son, D.J.; Ha, S.J.; Song, H.S.; Lim, Y.; Yun, Y.P.; Lee, J.W.; Moon, D.C.; Park, Y.H.; Park, B.S.; Song, M.J.; et al. Melittin inhibits vascular smooth muscle cell proliferation through induction of apoptosis via suppression of nuclear factor-kappaB and Akt activation and enhancement of apoptotic protein expression. *J. Pharmacol. Exp. Ther.* **2006**, *317*, 627–634. [CrossRef] [PubMed]

15.   Kim, J.Y.; Kim, K.H.; Lee, W.R.; An, H.J.; Lee, S.J.; Han, S.M.; Lee, K.G.; Park, Y.Y.; Kim, K.S.; Lee, Y.S.; et al. Apamin inhibits PDGF-BB-induced vascular smooth muscle cell proliferation and migration through suppressions of activated Akt and Erk signaling pathway. *Vascul. Pharmacol.* **2015**, *70*, 8–14. [CrossRef] [PubMed]

16.   Jeong, Y.J.; Cho, H.J.; Whang, K.; Lee, I.S.; Park, K.K.; Choe, J.Y.; Han, S.M.; Kim, C.H.; Chang, H.W.; Moon, S.K.; et al. Melittin has an inhibitory effect on TNF-α-induced migration of human aortic smooth muscle cells by blocking the MMP-9 expression. *Food Chem. Toxicol.* **2012**, *50*, 3996–4002. [CrossRef] [PubMed]

17.   Kim, J.Y.; Lee, W.R.; Kim, K.H.; An, H.J.; Chang, Y.C.; Han, S.M.; Park, Y.Y.; Pak, S.C.; Park, K.K. Effects of bee venom against *Propionibacterium acnes*-induced inflammation in human keratinocytes and monocytes. *Int. J. Mol. Med.* **2015**, *35*, 1651–1656. [PubMed]

18.   Lee, W.R.; Kim, K.H.; An, H.J.; Kim, J.Y.; Chang, Y.C.; Chung, H.; Park, Y.Y.; Lee, M.L.; Park, K.K. The protective effects of melittin on *Propionibacterium acnes*-induced inflammatory responses in vitro and in vivo. *J. Investig. Dermatol.* **2014**, *134*, 1922–1930. [CrossRef] [PubMed]

19.   Kim, K.H.; Lee, W.R.; An, H.J.; Kim, J.Y.; Chung, H.; Han, S.M.; Lee, M.L.; Lee, K.G.; Pak, S.C.; Park, K.K. Bee venom ameliorates compound 48/80-induced atopic dermatitis-related symptoms. *Int. J. Clin. Exp. Pathol.* **2013**, *6*, 2896–2903. [PubMed]

20.   Kallarackal, A.J.; Simard, J.M.; Bailey, A.M. The effect of apamin, a small conductance calcium activated potassium (SK) channel blocker, on a mouse model of neurofibromatosis 1. *Behav. Brain Res.* **2013**, *237*, 71–75. [CrossRef] [PubMed]

21.   Kang, S.Y.; Roh, D.H.; Yoon, S.Y.; Moon, J.Y.; Kim, H.W.; Lee, H.J.; Beitz, A.J.; Lee, J.H. Repetitive treatment with diluted bee venom reduces neuropathic pain via potentiation of locus coeruleus noradrenergic neuronal activity and modulation of spinal NR1 phosphorylation in rats. *J. Pain* **2012**, *13*, 155–166. [CrossRef] [PubMed]

22.   Kang, S.Y.; Roh, D.H.; Choi, J.W.; Ryu, Y.; Lee, J.H. Repetitive treatment with diluted bee venom attenuates the induction of below-level neuropathic pain behaviors in a rat spinal cord injury model. *Toxins* **2015**, *7*, 2571–2585. [CrossRef] [PubMed]

23.   Lee, H.; Lee, E.J.; Kim, H.; Lee, G.; Um, E.J.; Kim, Y.; Lee, B.Y.; Bae, H. Bee venom-associated Th1/Th2 immunoglobulin class switching results in immune tolerance of NZB/W F1 murine lupus nephritis. *Am. J. Nephrol.* **2011**, *34*, 163–172. [CrossRef] [PubMed]

24.   Jung, S.Y.; Lee, K.W.; Choi, S.M.; Yang, E.J. Bee venom protects against rotenone-induced cell death in NSC34 motor neuron cells. *Toxins* **2015**, *7*, 3715–3726. [CrossRef] [PubMed]

25. Kim, K.H.; Sung, H.J.; Lee, W.R.; An, H.J.; Kim, J.Y.; Pak, S.C.; Han, S.M.; Park, K.K. Effects of melittin treatment in cholangitis and biliary fibrosis in a model of xenobiotic-induced cholestasis in mice. *Toxins* **2015**, *7*, 3372–3387. [CrossRef] [PubMed]

26. Li, D.; Lee, Y.; Kim, W.; Lee, K.; Bae, H.; Kim, S.K. Analgesic effects of bee venom derived phospholipase A(2) in a mouse model of oxaliplatin-induced neuropathic pain. *Toxins* **2015**, *7*, 2422–2434. [CrossRef] [PubMed]

27. An, H.J.; Kim, K.H.; Lee, W.R.; Kim, J.Y.; Lee, S.J.; Pak, S.C.; Han, S.M.; Park, K.K. Anti-fibrotic effect of natural toxin bee venom on animal model of unilateral ureteral obstruction. *Toxins* **2015**, *7*, 1917–1928. [CrossRef] [PubMed]

28. Lee, J.A.; Kim, Y.M.; Hyun, P.M.; Jeon, J.W.; Park, J.K.; Suh, G.H.; Jung, B.G.; Lee, B.J. Honeybee (*Apis mellifera*) venom reinforces viral clearance during the early stage of infection with porcine reproductive and respiratory syndrome virus through the up-regulation of Th1-specific immune responses. *Toxins* **2015**, *7*, 1837–1853. [CrossRef] [PubMed]

29. Danneels, E.L.; Formesyn, E.M.; de Graaf, D.C. Exploring the potential of venom from *Nasonia vitripennis* as therapeutic agent with high-throughput screeing tools. *Toxins* **2015**, *7*, 2051–2070. [CrossRef] [PubMed]

30. Qian, C.; Fang, Q.; Wang, L.; Ye, G.Y. Molecular cloning and functional studies of two Kazal-type serine protease inhibitors specifically expressed by *Nasonia vitripennis* venom apparatus. *Toxins* **2015**, *7*, 2888–2905. [CrossRef] [PubMed]

31. Matysiak, J.; Świątly, A.; Hajduk, J.; Matysiak, J.; Kokot, Z.J. Influence of honeybee sting on peptidome profile in human serum. *Toxins* **2015**, *7*, 1808–1820. [CrossRef] [PubMed]

32. Lee, W.R.; Pak, S.C.; Park, K.K. The protective effect of bee venom on fibrosis causing inflammatory diseases. *Toxins* **2015**, *7*, 4758–4772. [CrossRef] [PubMed]

33. Silva, J.; Monge-Fuentes, V.; Gomes, F.; Lopes, K.; dos Anjos, L.; Campos, G.; Arenas, C.; Biolchi, A.; Gonçalves, J.; Galante, P.; et al. Pharmacological alternatives for the treatment of neurodegenerative disorders: Wasp and bee venoms and their components as new neuroactive tools. *Toxins* **2015**, *7*, 3179–3209. [CrossRef] [PubMed]

34. Hwang, D.S.; Kim, S.K.; Bae, H. Therapeutic effects of bee venom on immunological and neurological diseases. *Toxins* **2015**, *7*, 2413–2421. [CrossRef] [PubMed]

35. Moreau, S.J.; Asgari, S. Venom proteins from parasitoid wasps and their biological functions. *Toxins* **2015**, *7*, 2385–2412. [CrossRef] [PubMed]

© 2016 by the author; licensee MDPI, Basel, Switzerland. This article is an open access article distributed under the terms and conditions of the Creative Commons Attribution (CC-BY) license (http://creativecommons.org/licenses/by/4.0/).

MDPI AG

St. Alban-Anlage 66

4052 Basel, Switzerland

Tel. +41 61 683 77 34

Fax +41 61 302 89 18

http://www.mdpi.com

Toxins Editorial Office

E-mail: toxins@mdpi.com

http://www.mdpi.com/journal/toxins

www.ingramcontent.com/pod-product-compliance
Lightning Source LLC
Chambersburg PA
CBHW051837210326
41597CB00033B/5691